D0093174

A PEACE DIVIDED

The finest in Fantasy and Science Fiction by TANYA HUFF from DAW Books:

The Peacekeeper Novels:
AN ANCIENT PEACE (#1)
A PEACE DIVIDED (#2)

The Confederation Novels:
A CONFEDERATION OF VALOR
Valor's Choice/The Better Part of Valor
THE HEART OF VALOR (#3)
VALOR'S TRIAL (#4)
THE TRUTH OF VALOR (#5)

* * *

THE SILVERED

* * *

THE ENCHANTMENT EMPORIUM (#1)
THE WILD WAYS (#2)
THE FUTURE FALLS (#3)

* * *

SMOKE AND SHADOWS (#1)
SMOKE AND MIRRORS (#2)
SMOKE AND ASHES (#3)

* * *

BLOOD PRICE (#1)
BLOOD TRAIL (#2)
BLOOD LINES (#3)
BLOOD PACT (#4)
BLOOD DEBT (#5)
BLOOD BANK (#6)

* * *

THE COMPLETE KEEPER CHRONICLES
Summon the Keeper/The Second Summoning/Long Hot Summoning

* * *

THE QUARTERS NOVELS, Volume 1:
Sing the Four Quarters/Fifth Quarter
THE QUARTERS NOVELS, Volume 2:
No Quarter/The Quartered Sea

* * *

WIZARD OF THE GROVE
Child of the Grove/The Last Wizard

* * *

OF DARKNESS, LIGHT, AND FIRE
Gate of Darkness, Circle of Light/The Fire's Stone

TANYA HUFF

A PEACE DIVIDED

Peacekeeper: Book Two

DAW BOOKS, INC.
DONALD A. WOLLHEIM, FOUNDER
375 Hudson Street, New York, NY 10014
ELIZABETH R. WOLLHEIM
SHEILA E. GILBERT
PUBLISHERS

www.dawbooks.com

Jacket art by Paul Youll.

Jacket designed by G-Force Design.

Book designed by Elizabeth Glover.

DAW Book Collectors No. 1759.

Published by DAW Books, Inc.
375 Hudson Street, New York, NY 10014.

First Printing, June 2017
1 2 3 4 5 6 7 8 9

PRINTED IN THE U.S.A.

In memory of Larry Smith.

ONE

"GUNRUNNERS," Werst snarled, sliding over the almost buried shell of the APC as rounds impacted against the metal. "*Gunrunners*, they told us, not *users*."

"Logical progression." Ressk fired a quick burst through one of the second-floor windows on the ruined anchor, interrupting the gunrunners' fire long enough for Werst to get to cover. "Especially if they knew we were coming."

"How could they know we were coming?" Werst demanded.

"The Justice Department has a leak."

"A leak?" Werst leaned around the back end of a destroyed APC. "You think that's possible, Gunny?"

"They were a little too prepared," Torin admitted, helmet scanner registering heat signatures at the windows where they'd already identified shooters through the less technical method of being shot at. Unfortunately, if a scanner existed that could see through walls built to withstand both the rigors of space and an atmospheric entry, she hadn't been issued one. The building at the center of every new colony, the anchor, was a cross between a Marine Corps Susumi packet and a large vacuum-to-atmosphere transport. Thirty meters by twenty meters by six meters, it held everything the colony needed to get started and once emptied became a community center, a hospital, and—if necessary—a fortress. Designed to be nearly indestructible, it was part of the Confederation's promise to the Younger Races that they'd be supported as they spread out through known space. Nearly indestructible hadn't been enough for this particular anchor to entirely survive a Primacy landing force during the last year of the war.

Although, to be fair to the anchor's designers and engineers, it also had to survive the Confederation Marine Corps retaking the colony and no one had yet come up with anything—buildings, transportation, tech—that was Marine proof. Marine resistant, yes. Proof, no.

Again, to be fair, the anchor was in better shape than the rest of the colony.

Sh'quo Company, Torin's old unit, hadn't been part of the attack that had driven the Primacy out of Three Points, but she could read the story of the battle on the ruins and debris and she knew the weight of the senior NCO's vest, heavy with the number of bodies they'd carried out. Bodies reduced to their basic components for ease of transport and stored in small metal cylinders. No Marine left behind.

Her hands were steady on her KC-7, the familiar weight of the Corps primary weapon canceling the twitch toward the places on her own vest where her dead had rested. The combat vest was a recent addition to the Warden's uniform, as was the KC. Change came slowly to the Wardens, to the entire Confederation, but change came whether the Elder Races welcomed it or not.

Not that Torin expected anyone to welcome the need for armed response teams.

"*Gunny, I've got hostiles on the roof. Two, no three . . . moving a large rectangular crate up through the trap.*"

Boots on the ground, the angle kept Torin from picking up any of the action two stories up. In place on one of the remaining rock formations that had given Three Points its name, Binti Mashona had a clear line of sight. "Do you have a shot?"

"*No. They've got a good idea of where I have to be, and they're using the crate to . . . Fuk me sideways, it's a mortar.*"

Specs flashed along the lower edge of Torin's visor as the mortar came on line.

"Well, that answers a question we didn't give a shit about," Werst muttered. "One of the dirtbags was artillery."

"Not likely," Ressk argued as Torin squeezed off two quick shots—one to herd, one to hit. A di'Taykan screamed. "We're almost in the building with them and their structural integrity was breached before we got here."

"The glass was broken," Werst interjected.

"That's what I said. If one of this lot was artillery, they'd have known to open with the mortar."

Torin's team had almost reached the building, using the cover of darkness and the surrounding ruins, when the gunrunners had opened fire. They hadn't tripped a perimeter alert, and there'd been no sentries set to give the alarm. They might have been spotted through a second-floor window, but Torin doubted it. The response had been too fast, too accurate. For variable definitions of the word *accurate* given they had zero casualties to two gunrunners bleeding. Selling illegal weapons had taken precedence over practicing with them.

"I have a clear shot on the mortar, Gunny, targeting and ignition."

"Can you take it out?"

"Please, this close I could hit it with a rock."

"Take the shot."

Profanity followed close behind the impact of high speed metal on metal.

Ressk fired at the flicker of a shadow in one of the windows. "I was hoping for an explosion."

"Weren't we all."

Mashona fired again. *"Careless. One down. The other two hauled her back inside."*

Three gunrunners bleeding.

"All right, enough. Quick and quiet is a bust. Craig."

"Torin."

"Land it. Alamber, distraction on contact."

"You got it, Boss."

"Ressk, Werst, heat imaging off and get ready to move. Plan B." Her own scanner back to neutral, her eyes readjusting to the night, Torin adjusted both her weight and her grip on her weapon, ready to run. Shifting in place, she leaned away from the spray of dirt thrown up by a missed shot. It had missed by a smaller margin than previous shots—odds were good any ex-Marines in the anchor had begun to remember their training. On the one hand, it was about time; up until now, their aim had been embarrassing. On the other hand, as she was one of the targets they were aiming at . . .

She felt the shuttle's approach as much as heard it, a deep hum in her bones that announced Craig was fighting gravity with everything the VTA had. The Navy surplus vacuum-to-atmosphere shuttle provided by the Justice Department had been straight up and down, sturdy enough to save their lives when it crashed, but with the flight capability of a brick. The Taykan-designed VTA they'd acquired next was faster, significantly less sturdy, and had been built with the added feature of horizontal travel at the bottom of a gravity well. It wasn't an attractive feature, she noted, as the VTA came into sight, but it got the job done.

"Blocking team implants in three, two, now," Alamber announced as the VTA descended toward the roof, his voice in her PCU barely audible over the roar. *"Distraction in three, two . . ."*

The raised metal edge crumpled under the weight, but the roof held as Craig set her down.

". . . boned the bad guy, Boss."

Sergeants and above came out of the military with jaw implants, full comm units set into the bone. The Justice Department had provided implants for their Strike Teams, but the expense of installation and upkeep prevented most civilians from using the tech. Including those civilians who used to be enlisted Marines. Odds were high that the pulse Alamber had sent over the most common military frequencies had knocked the fight out of the people making the decisions inside the anchor.

"Move!" Torin broke into a run, head down to protect her face from the airborne debris. Craig had brought the shuttle up on their one eighty using the anchor to block the exhaust, but it had still thrown an impressive amount of heated grit into the air. The grit would nullify the gunrunners' heat imaging, had any of them managed to keep their attention on the job at hand while a few metric tons of VTA landed on the roof and their leaders writhed on the floor.

She was close enough now to hear the screaming.

Human, very probably male, and a Krai, no idea of gender. Eleven years on various battlefields had allowed her to add *can identify species by sounds of pain* to her skill set. Three years out of the Corps and it remained useful.

The air lock on the narrow end of the anchor had been blown apart either by the Primacy or the Confederation or a combination of both. The reality of war meant the winner often held real estate that had been destroyed in the taking or in the retaking. The first-floor common room had long, narrow windows, an obvious entry point given the lack of glass, but the gunrunners had reinstalled the exterior shutters that essentially made the wall a spaceship hull. Impenetrable to anything Torin's team had with them.

Except . . .

During destruction of the air lock, the end wall had buckled enough to twist the nearest window a centimeter off square, the shutter not entirely secure, a triangle of light visible at the upper right and lower left corners.

Torin pulled the coil of wire from her vest as she ran, whipped it out to its two-meter length as she reached the anchor, dropped to one knee to slide it through the lower gap, and thumbed the release on the capacitor before shoving it through hard enough to clear the interior sill. Then she stood and braced her forearms against the wall.

"Distraction's shut down, Boss."

They'd spent part of the trip out here arguing the fine line between pain as distraction and pain for the sake of causing pain. None of them had much sympathy for the gunrunners; they spent too much time dealing with their customers.

Using fingers and prehensile toes, Werst reached the second-floor window as the wire ignited.

"Hope they weren't stupid enough to store their ordnance in the unstable corner," Ressk muttered as his foot gripped her shoulder.

Torin hoped so too. The Justice Department insisted that property damage be kept to a minimum, and Torin didn't want to spend another afternoon justifying an accidental explosion. When Ressk pushed off, she caught the line Werst sent down and went up hand over hand until she could grab the windowsill and haul herself over.

"Almost Krai-like," Ressk told her as her boots hit the floor.

"I can fake anything for two meters." Torin resettled the weight of her vest on her shoulders, swung her KC back around, and waved the two Krai toward the door.

The room was still configured as a barracks, Three Points having barely moved beyond the entire colony living in the anchor when they were attacked. Given that space was large enough to keep any one system in the OutSector from having much of a strategic significance in an interstellar war, the Confederation had assumed the attack had been over real estate with a proportionate nitrogen/oxygen atmosphere, a gravity within specific tolerances, and readily available water. Turned out, the assumption had been incorrect. There'd been no logical reason for the attack as the war had been run as a social experiment by sentient, polynumerous molecular polyhydroxide alcoholydes—a discovery no one would have believed had Torin not got the shape-shifting, organic plastic hive mind to admit it on camera moments before they departed known space to analyze the accumulated data. She'd been cleaning up the mess they'd left behind ever since.

The second-floor hall was empty. Scanners showed two thermal signs behind the closed door of the anchor's infirmary—one Human, one di'Taykan—and the blood that had drawn a dotted line between the stairs leading to the roof and the infirmary suggested they weren't doing what a Human and the most enthusiastically indiscriminate species in known space were usually doing behind a closed door. Torin pointed at the lock. Ressk moved forward, touched his slate to it, and rewrote the code. The coiled spring latches rang out as they slammed into place, metal against metal—not a lot of what went into space could be called delicate, and that included most of the people.

At the clang Torin switched her attention to the main stairs, but it seemed no one on the lower level had heard the clang over the shouting. For the most part, they were shouting about the explosion as well as someone named Ferin's inability to keep watch, summed up at high volume. ". . . lazy, blind, *serley chrika*! Get your head out of your own ass!"

Two locked in the infirmary, four downstairs standing, three on the ground. All nine gunrunners accounted for.

Except . . .

The infirmary windows faced away from Mashona's position.

"Craig, keep an eye on the north side of the building. We've got two hostiles locked in the infirmary and the odds are good the more mobile will make a run for it."

"No honor among thieves?"

She could hear the smile in his voice and answered it with one of her own. "Not that I've ever noticed."

"Only four dirtbags left to take out." Werst drew his lips back off his teeth. "Hardly worth a team effort. Want us to wait up here, Gunny?"

In answer, she started down the stairs, and they fell into position behind her.

Their orders were to apprehend the gunrunners. Where apprehend meant bring them in alive or face the staggering amount of paperwork required to document every corpse. Their task made more difficult given that the people they were trying to *apprehend* shot to kill.

"Ferin, Yizaun, check the weapons are secure. Mack, get that shutter dogged in. Shiraz, you're bleeding all over the fukking floor, do something about it."

"Who put you in charge, Harr?"

Harr paused at the foot of the stairs, facing back into the community hall. "That'd be when those fuktards took the chief out."

Torin could see a line of blood running from the corner of his mouth where he'd driven his teeth through his lower lip, but a Krai jawbone was one of the toughest organic substances in known space and the pulse Alamber had sent through his implant had done a lot less damage than it would have to a Human or di'Taykan. It had done enough damage, however, that Harr was on the bottom step before he noticed them pressed along the right wall.

His eyes widened, his nostril ridges began to close, and Torin grabbed him around the throat, yanking him forward into the butt of Werst's KC. She'd stepped out into the community hall before he hit the floor.

Shiraz, slumped against the wall, awkwardly trying to wrap a blood-soaked cloth tighter around her shoulder, Torin ignored. Mack, his broad back toward her, muscle straining the seams of his shirt, was going to be more of a problem.

She couldn't shoot a man in the back.

So she shot him in the back of the knee.

He screamed, hit the floor, rolled, and came up holding . . .

Torin had no idea what it was, but she'd looked down enough muzzles while in the Corps and after to recognize one now. It was small, dwarfed further by Mack's hand, and it was definitely a weapon. An easy to conceal and therefore illegal weapon. His first shot hit the wall behind her and ricocheted, drawing an impressive string of profanity from Werst. Pain had Mack's arm shaking like a recruit's knees, and Torin figured if he hit her at all, he'd hit her by accident. As it happened, it was an accident she didn't want to have.

"Rehab can rebuild your knee," she snapped, "not your head."

Might've been the threat, might've been the pain—the odds were about even as his arm dropped to the floor with an impressive thud. Ressk kicked the weapon away.

The Krai and the di'Taykan charging in through the door at the far end of the hall should have taken a shot from the doorway first. Should have. Didn't. The belief in their own invulnerability—and Harr's reference to the chief—said Navy.

"Werst, go high."

"Going high, Gunny."

There hadn't been a lot of ship-to-ship boarding parties during the war. The Artek had managed to latch on and get successfully through one hundred and fourteen hulls in five centuries, but the training required to fight giant bugs translated badly into fighting bipedal mammals. Other boarding parties had left no survivors, so the lessons learned were lost. On the rare occasions that ships had been deployed without Marines on board, Navy boarding parties would have been sent to defend or retake stations. As those boarding parties would be all the close combat most Navy personnel would ever see, Torin could only hope that these two had never been part of one.

The Krai—Ferin, given the Krai fondness for r's and the lack in Yizaun—launched toward Werst, grabbing impressive air. Yizaun, the di'Taykan, reached for Torin, ready to grapple, intent on pulling her close enough for pheromones to shift her attention. Torin kicked Ferin in the stomach. Werst climbed Yizaun like a tree, wrapped both feet around his neck, and dropped down his back, using weight and momentum to drop the di'Taykan to the floor, rolling free at the last moment.

The gasping Krai's wrists and ankles secured, Torin turned her—she was almost certain the gray on green mottling was a female pattern—onto her side so she wouldn't choke on her own vomit, and then reached over her to crank the di'Taykan's masker up from the lowest setting.

"Thanks." Werst slapped a zip-tie around the di'Taykan's slender wrists. "*Serley chrika* fights dirty."

"Smartest thing any of them did."

Werst adjusted the crotch of his trousers, lips drawn back off his teeth. "Yeah, well, your affected bits are less protruding."

The di'Taykan were the most sexually nondiscriminating species in the Confederation. The pheromones that let their own species know they were available, made every other species incredibly receptive—and, so far, it had been *every other species*. The maskers let the rest of the Confederation make a choice about accepting or declining the perpetual invitation. Using the pheromones as a weapon was considered to be, as Werst had said, fighting dirty.

Personally, however uncomfortable she might be for the next little while, Torin approved of fighting to win. She saw little point in doing it otherwise.

Ferin drew in a long shuddering breath and, when Werst looked down, opened her nostril ridges, licked her lips, and purred out what sounded like an extended series of consonants in the Krai's most common language. Torin could only assume Ferin intended seduction, given the words she recognized. Werst laughed. Ferin bared her teeth, nostril ridges slamming shut.

"Yeah, yeah, I'm threatened," he muttered searching the di'Taykan for weapons while Torin patted Ferin down, carefully staying away from her mouth. The Krai could bite through bone. Torin had seen one amputate the leg of a bondmate when no other option had been available. "Looks like you've got things under control, Boss."

Pulling a knife that was more an all-purpose tool than a weapon from a pocket on Ferin's jumpsuit, Torin twisted far enough to see Alamber standing by the stairs, the muzzle of his KC pointed at the floor, his finger nowhere near the trigger. She hadn't wanted him taught—a large part of her wanted no one taught to use a weapon ever again, but the practical part, the part that acknowledged the amount of old damage needing to be dealt with before that could happen, the part that wouldn't kick the young and damaged di'Taykan from her team as long as he wanted to stay, well, that part wanted to give Alamber his best chance of staying alive. And sometimes, regardless of her personal preference, he couldn't stay in the ship.

Sometimes Craig couldn't either, but that was an entirely different problem.

"Help Ressk with the wounded." She straightened and stepped back as Ferin lunged for her ankle. "I can't handle another three days writing reports."

Alamber stared down at the bloody cloth wrapped around Shiraz's shoulder, his pale blue hair flicking back and forth in disdain. "They don't have sealant?"

"Didn't expect to get shot," Ressk grunted emptying another tube around Mack's knee, immobilizing it.

"But they fired first!"

"So they aren't too bright." Ressk's upper lip pulled back off his teeth as he secured Mack's massive wrists with a double zip-tie. "And it sounds like they were Navy."

"Good call about the window, Torin. The di'Taykan was out on the knocker. I convinced him he'd be healthier if he stayed inside."

With two gunrunners still unaccounted for, Torin headed for the single door in the wall opposite the windows. "You think he'll stay?"

"I dropped a shuttle on the roof—he's dummied out that I'll drop it on him if he tries."

"That'll work." Torin beckoned Werst over and, when he was ready, opened the door.

The anchors were designed to be similar, not identical, and she had no idea what this internal room had been intended to hold. Currently, it held twelve crates of weapons as well as an unconscious Human and a wounded Krai sitting, back braced against the closest case, holding a KC pointed at the door.

Torin preferred to believe that an ex-Marine in the same circumstances would have pulled the trigger the moment the door opened. She locked eyes with the Krai, who was young and terrified and in pain, and she raised a single brow. The Krai swallowed, audibly, then carefully set the weapon down and raised both hands, fingers and nostril ridges trembling.

"Werst, see to . . ."

"His," Werst interjected.

". . . his wound." The fabric over his left hip had been cut away, and a sloppy bandage seeped blood. "I'll see to the chief."

Process of elimination—of the others, only Harr had an implant and they'd heard Harr say the chief had been taken out.

Even in profile, the chief looked like a lifer, forced to retire after thirty with another hundred years in front of him and no idea of what to do next. Lifers shared a common expression built out of three decades of shared experiences, an expression Torin had intended to wear had the plastic aliens not cut her career short. Although she'd have found a fukking hobby before running guns.

The chief's pale hair had been buzzed close, a glimmer over his scalp, and an impressive amount of blood had pooled under slack features. A quick check pulled his cheek from the floor with a sucking sound, and determined the blood had come from his nose. Past tense.

Nothing to worry about, then. His pulse was thready, but his position had kept his breathing clear.

"We'll corral them all in the common room." Gripping the chief's wrists, she lifted his upper body far enough off the floor that his head was in no danger of further damage, and dragged him toward the door. Growing up on Paradise, the first of the Human colony worlds, had given her a minor strength advantage over the original Human design. Unfortunately, gravity at the Three Points Colony was almost exactly the same as it was at home, canceling her advantage. "I want them where we can keep an easy eye on them."

"And away from the guns," Werst added, lifting the injured Krai until he could balance on his good leg. The numbing agent in the sealant now covering his wounded hip had gone a long way toward bringing some green back into his face.

"Are you going to kill us?"

Torin dropped his age down a few years. "Kid, if we wanted you dead, you'd already be dead."

"You know, the chief said we couldn't trust the Berin gang. He said, given a chance, you'd most likely turn on us."

"The chief's probably right. But we're not the Berin gang." She met Werst's gaze and grinned. "We're the Wardens."

"The Wardens don't . . ."

"They do now," Werst grunted.

Out in the common room, they propped five of the seven gunrunners against the end wall. Harr, eyes unfocused, nostril ridges bleeding, snarled inarticulate profanity under his breath, only pausing long enough to add another mouthful of blood and saliva to the pattern on the floor. Yizaun slumped, defeated, ends of his fuchsia hair twitching. Shiraz, loosely secured because of her shoulder, had nearly cut Ferin's wrists free before anyone noticed their position had nothing to do with comfort.

"Alamber!"

"I searched her, Boss, I swear." Pale blue hair hugging his head, he dragged Ferin out of reach, then dug his thumb into the muscle of her calf to get her to release his ankle. While not as articulated as their hands, Krai feet could grip with more force. "I don't know where she was hiding it."

Ferin yelled out something in Krai. Werst laughed.

"Doesn't translate," he said, when everyone turned to look.

"And it's anatomically impossible," Ressk added.

"Yeah, like that's ever . . ."

Raising a hand, Torin cut Alamber off. "We don't need to know."

Shiraz smirked.

Torin rolled her eyes, cut the old ties—the Mictok webbing that had replaced the old plastic ties needed a specialized blade—and secured Shiraz's wrists to the sides of her belt, then her belt to her clothes. Minimum strain on the injured shoulder, but she'd have to strip to get free.

The kid watched it all with wide eyes, curled into the angle between the floor and the wall, trying to make himself look smaller.

The chief and Mack had been laid out on the floor, the chief's head turned to one side in case his nose started to bleed again. Eyes closed, Mack hummed a song popular when Torin was in school and looked happily stoned.

"Heavy-duty pain blockers," Ressk explained. "The good drugs. Should last three hours, maybe two given his size. Seemed safer."

Torin took another look at the two Humans. Lying side by side, the chief, who wasn't a small man, barely came up to Mack's shoulder. "Good call."

"You want us to get the two upstairs?"

"No, they're not going anywhere." She took off her helmet, leaving the PCU in her ear, and picked up the tiny weapon Mack had been using. It looked larger in her hand than it had in his, but not by much.

"You ever see one of those before, Gunny?"

"No." Anyone familiar with weaponry would recognize it as a gun. Barrel, trigger, firing chamber, magazine . . . Torin thumbed the magazine release and popped the first round, frowning at it cupped in the palm of her hand. Although smaller than the rounds used in the KC-7, it was, once again, easily recognizable. A distance weapon, small enough to be concealed, broke any number of Confederation laws. Torin thought one law would have been enough—*don't make distance weapons small enough to be concealed*—but she wasn't a politician. "Mashona, hostiles are contained. Come in."

"Coming in, Gunny."

"If you don't kill us, we'll come after you. We'll get our merchandise back!"

Ferin elbowed Shiraz in her bad arm. "Shut up!"

"Why should I?" Her eyebrows, dyed an iridescent purple, folded into a deep vee as Shiraz glared at Torin. "They can't just waltz in here and take what's ours and expect us to like it!"

"Seriously, shut the fuk up!"

"No. This was our big score, and they can't . . ."

"Do you want them to kill us, you dumb shit!"

That shut her up.

Alamber cocked his head, pale blue eyes darkening, lid to lid, as more light receptors opened. "Why would you think we're going to kill you?"

"Why wouldn't you?" Ferin muttered. "No one believes that honor among thieves shit."

"Boss?"

"They think they've been betrayed by the people, and I use the term loosely, that were supposed to buy the guns. We're not the Berin gang." Torin swept a gaze over the conscious. It was a gaze she'd used on Marines who'd thought they could buck the Corps and the shit had risen over their heads. Navy or not, all five of them recognized it. Harr tried to square his shoulders. Now she had their attention, she tapped the insignia on her chest. "We're Wardens."

Ferin snapped her teeth. "*Gren sa talamec!* Peacekeepers aren't armed."

"Surprise. Craig, let the C&C know we're ready for them."

"Will do. Any dead?"

"Not yet."

"Aces. Good for you." He kept his tone light, but she knew he meant it.

Torin turned the small weapon over in her hands and checked the chamber was clear. The rounds were shorter, blunter, and given the length of the muzzle, accuracy would be unlikely over thirty meters. This wasn't a weapon for war. At least not the kind of war she knew. Tucked in her waistband, under a jacket, no one would ever know she was carrying it until she used it. Memory replayed countless lessons from children's programming right up to basic training, lessons about fair play and respect and acknowledgment of intent. Her grip tightened, metal digging into her palms. "I want a look in those cases. Ressk, with me. Werst, Alamber, if you have to shoot them, don't destroy body parts that can't be fixed."

Werst grinned. "Wasn't Major Svensson pretty much a brain in a jar before the Corps tanked him?"

"Good point. Don't destroy body parts the Justice Department can't afford to fix."

The cases were unlocked, ready, she assumed, for the Berin gang to inspect the merchandise. Eight held KC-7s, the ninth KC-12s, the larger weapons used by the heavy gunners, exoskeletons augmenting their strength. There were no exoskeletons, and she assumed another group dealt with that half of the equation. The tenth, eleventh, and twelfth cases held ammunition. Standard rounds for the sevens, and for the twelves . . .

"Impact boomers. Grenades." Ressk straightened. "There's burners in there, too, Gunny."

Rules of combat disallowed flamethrowers against living targets. Both Torin and Ressk had seen the rule broken. As neither the gunrunners nor the Berin gang were operating within any rules at all, Torin carefully didn't consider what the burner rounds could be used for. It would do nothing to improve her mood, and Justice also had paperwork for when prisoners arrived unnecessarily broken.

There were no more of the small weapons.

She set a KC carefully back into the crate and rubbed her fingers together. "Factory sealant."

"Brand-new," Ressk agreed. "Not stolen from a Corps depot, then." The Corps had the KCs out of the box, cleaned and sighted before the supply sergeant finished filing the delivery documents. "On the way to a Corps depot, maybe?"

"No. The crates are too clean." It had been a long time since Torin had seen a weapon that wasn't ready to fire, but she remembered a stack of crates in supply, delivery information stenciled on all six surfaces, a low- tech solution to prevent potential high-tech interference from the enemy.

They'd been sent out after gunrunners. Armed and violent and ready to shoot first, the gunrunners were exactly the sort of dangerous criminal the Strike Teams had been formed to deal with.

Their brief had said only, *stolen weapons*—where they'd been stolen from, unimportant.

Wiping her fingers on her thigh, Torin had a feeling it was about to become important. "Remember when Big Bill getting his grubby hands on a Marine armory was the worst possibility we'd ever encountered? Good times."

"At least they only had the one mortar. They don't seem to be selling the big stuff." When Torin turned toward him, Ressk shrugged, a human motion the Krai had adopted but had never really mastered. "No harm in looking on the bright side. These assholes get their hands on a sammy and there goes our arrest record. Well, not these assholes," he amended after a moment. "But there's plenty of assholes out there."

"Not arguing." There seemed to be more assholes, and more ambitious assholes appearing every day. "We have any information on the Berin gang?"

"Not a word, Gunny. Could be new."

"And expecting to be very well armed."

"Unless these guys are wholesale and the Berin are the distributors."

Nine cases of weapons. Three cases of ammo. Enough to fight a small war. Torin drummed her fingers against a case. "They're organizing."

"If they are . . ." He shrugged again when she turned to look at him. "Not enough data. You can't gut feeling a broad social analysis. What am I saying," he added before she could speak, "you can, but that's only a single data point, Gunny. I can tell you there's probably a Human at the root of this."

"Can you?"

"Twelve cases. You lot have a hindbrain attachment to your duh-zen. No other species defaults to it."

She stared at him for a long moment. "Don't even . . ."

"Humans First was pretty pissed when Richard Varga died in rehabilitation."

"We took that organization down."

"And the remnants reorganized pretty damned quickly when Varga died," Ressk reminded her. "Put out a new mission statement, removed the apostrophe, started recruiting."

"Son of a fukking . . ."

"Torin, C&C's on the way in. Three hours twenty-seven to orbit."

"Why weren't they in orbit?" The Clean-up Crew, C&C, didn't hit dirt until the shooting ended, but they usually matched orbits with the *Promise* to cut the wait time.

"Supply drop to a mining platform. Company called in a favor. Justice agreed since we were in-system anyway."

"Roger that." Torin pressed her palm against the case of guns and shook her head, although she wasn't sure what she was denying. "Come on." Ressk fell into step beside her. "We need to get those two in the infirmary down," she called to Werst as they reentered the common room, "before they come up with a way to use damaged, abandoned medical supplies as a defense."

Werst snorted, nostril ridges flapping. "Not likely. They're Navy."

"Fuk you," Shiraz muttered, chin on her chest.

"Oh—and, Torin, Analyzes Minutiae to Discover Truth wants me to remind you that contact destroys evidence."

"The evidence is nine cases of stolen weapons, three of ammunition, and a firefight."

"I'll tell him you said so, then."

"You didn't identify yourselves as Wardens," Ferin snarled, responding to Torin's side of the conversation.

"You opened fire before we had the chance."

"We thought you were the Berins!"

"You're not allowed to shoot them either."

"They'd have shot us!" She clenched her feet. "Chief got word they were coming in early. We wouldn't have engaged if we knew you were Wardens. And you never read us our rights!"

"I don't have to read you your rights. I have to not shoot you in the head."

"But if you're arresting us . . ."

"Subduing you."

Ferin's nostril ridges flared open. "So we're not under arrest? We can . . ."

"Shut up!" It seemed Harr had pulled himself together while Torin and Ressk were in the storeroom. "You don't tell them shit!"

"That's right," Alamber said cheerfully. "You don't. You tell it to Asks a Million Questions Until You Find Yourself Spilling Your Guts."

Harr frowned. "That's not a real name."

"No," Alamber admitted, "but I've been trying to convince her to change it."

"Dornagain?"

"Duh."

"Coming here?"

"We take you down, they write you up." He grinned and his eyes lightened. "I'm thinking of having buttons made."

When Ferin continued to look confused, Torin leaned in. "After you're written up, you'll be under arrest."

Every clean-up team included at least two Dornagain Wardens to deal with the extended bureaucracy involved in sending Confederation citizens into rehabilitation.

"Dornagain," Harr repeated. He swept an angry gaze over Torin's team. "We'd rather deal with you fukkers."

Torin smiled down at him. "You'd be surprised how often we hear that."

Last of her team out of the anchor, Torin let the deep rumble of Finds Truth Through Inquiry's instructions to the captives carry her up the stairs and onto the roof.

"No, as a felon injured while in a position of temporary authority due to the chain of command being disrupted by the Strike Team, you were intended to fill out form FFA334. It appears you filled out FFA333. You'll have to begin again. We need three copies."

She'd bet her pension Harr was thinking, *Shoot me now.*

Both hands resting on top of her weapon, Binti Mashona waited by the ramp into the VTA. "I don't want to be the sniper anymore, Gunny. I never get to have any fun."

"You shot out the mortar."

"Okay, hardly any fun."

"Good shot, by the way."

"Only kind I make." She grinned and followed Torin inside, one of five people Torin gave no second thought to having armed at her back.

"At least no one was shooting at you. If they'd had TI, we'd have been holed." Unarguably, her people were better shots, but thermal imaging made it a lot easier to find enemies in the dark. Fortunately for them, the ex-military, violent offenders the Strike Teams dealt with had a strange distaste for wearing helmets.

"Hey, no one was shooting at me, but I took damage. There are bugs on this planet. Bugs that bite." She thrust out an arm, uniform sleeve tugged up.

Dark skin and the pale dawn light combined to hide any evidence. "There are bugs that bite on every planet. Suck it up."

"Harsh."

Torin secured both their weapons while Binti dogged down the hatch. "We have a seal," she called when the lights turned orange—a Taykan-built VTA used orange and blue rather than green and red. Humans adapted. It was one of the things they did best. "Run decontamination."

"Running decon." Alamber twisted around and peered over the top of the second command seat. "How long, Boss?"

"All the way back to the *Promise*." The ship was in a low orbit, only fifteen hundred kilometers up. Hopefully, the trip would last long enough for decon to do its job. "There are bugs on this planet that bite," she added as his hair flattened and he made an unhappy noise. "And I don't like the thought of planetary microbes mutating in space, joining together, gaining sentience, and starting another war."

His hair flipped. "Yeah, like that'll happen again."

"Long odds occasionally pay off. 'S why chumps play the lottery." Craig tipped his head back. Torin bent forward and kissed him. "Strap in," he said against her mouth. "Let's rattle our dags and get out of here before cleanup needs another form filled out. Plastic?"

"Some." She dropped back into her seat and dragged the crash harness down over her shoulders. "All inert."

They both knew she hadn't touched every piece of plastic in the anchor and in the ruined APCs around it. They both allowed the comfort of close enough. They had to, given that the marker left in their brains when they'd passed through a section of sentient, polynumerous molecular polyhydroxide alcoholydes arranged into the shape of a spaceship didn't always evoke a reaction. Torin wanted a reaction. She wanted a reaction, and then she wanted enough of them in one place to hold down and . . . And what? And demand an apology from the remnants of the plastic aliens lingering in known space who'd thrown two galactic civilizations into war as a social experiment? Demand restitution? Demand that they restore the lives lost on both sides while they accumulated data? No. Not only was there no way for them to make amends for the centuries of death and destruction they'd caused, for the continuing violence from a population trained to war, but previous interactions proved they felt they had nothing to make amends for.

Torin kept that *hold down and . . .* open ended, certain she'd resolve it in time. Sooner or later, they'd have an ass she could kick.

Meanwhile, she touched—and Craig touched—as much plastic as they could. Just in case.

Not one of the xenoneurologists who'd tested them, singly and collectively, had been able to determine how the protein marker worked or why it only worked part-time. They didn't know why the alien in General Morris' office had reacted to Craig or why the aliens in the prison hadn't reacted to Torin until she'd forced the issue. The general consensus, based on sweet fuk all as far as Torin could see, was that the aliens were running smaller experiments within the larger. She believed that as well as being patronizing, probably telepathic, voyeuristic hive minds, they were also contrary, psychopathic shitheads.

The Confederation had collectively decided to get rid of as much plastic as possible.

Impossible to get rid of it all, all at once.

Research into plastic substitutes tossed out something new every couple of tendays.

Natural materials had made a huge comeback. So had environmental protections.

"I almost forgot," she added as the VTA shuddered and lifted off, heading back to the *Promise*, "there's a chance Humans First is involved in the gunrunning."

"Fukking Humans First," Craig sighed. "They make the rest of us look bad."

"Or good in comparison," Binti offered. "And at least they got rid of the apostrophe."

Thirteen hours later, Torin pushed wet hair back off her face—she'd been in the shower when the alert sounded—and frowned at the information about the incoming ships scrolling across *Promise*'s main board. "Their public files are too perfect; they're shouting nothing to see here."

"You have a suspicious mind, luv." Craig dragged a shirt down over damp skin.

"It makes me good at my job."

"Should we tag them?"

They had no way of knowing if the pair of ships heading in-system held the Berin gang, but even if the bullshit in their public files turned out to be true, their trajectory, riding the momentum from their Susumi wave, meant they hadn't used the buoy and that meant an unregistered jump. As this was a MidSector system and both gas giants had mining platforms, the ships were, at the very least, guilty of a traffic violation.

She was starting to think like a Warden.

Torin wasn't one hundred percent sure how she felt about that. The Confederation Marine Corps had been the core of her identity for a long time.

"They just snagged our PFs." The Strike Teams had slightly less than perfect public files, designed to be seen and forgotten.

"So they're scanning." Wardens scanned automatically, filing information on every ship in range. Most ships didn't. Civilian salvage operators might, or might not, have developed a program Justice Department scans slid past. The CSOs were prickly about government interference, and Torin had heard about the possibility through unofficial channels, so she was ignoring it for the moment.

"We could engage." Craig dropped into the pilot's seat. Metal joints creaked a protest, the sound a background noise rather than a warning as the chair had held them both during some athletic maneuvering.

Rather than sit in the second chair, Torin rested her arms on the duct-taped upper curve behind Craig's head and, in turn, rested her chin on her arms. "And end up in a high-speed chase across the galaxy?"

"Why not? We'll use Presit's equations to follow them through Susumi."

"We'll jump out, smack into a solid object, and die young." Pilots who missed a decimal point in a Susumi equation were given memorial services.

"Then I herd them in too close to a gravitational field to make their jump."

"You'll match speeds, hook them with a grappling cable, and reel them in?"

"Don't be daft, I'd have to bugger their engines first."

"With?" The Wardens' ships weren't armed.

"Force of personality."

She rose up and leaned down to kiss the top of his head. "Well, we don't deal with traffic violations and until they start shooting at us, they're not our problem so . . ."

His hands stilled on the board. "If they *are* the Berins, they'll want those weapons."

C&C had minimal defenses, a single member of the Younger Races there to protect the nonviolent species that had, until a year ago, been the only Wardens. They were still working the kinks out of the new system, but not even an ex-Marine could stand between two shiploads of criminals and a dozen cases of arms and ammunition.

"*Ablin gon savit!*"

"I get all aroused when you swear in Taykan, Boss."

"You get all aroused when you read training manuals," Torin said absently, staring down at what their scanner continued to pull up about the pair of distant ships.

"Depends on who's being trained in what." Alamber dropped into the second chair and yawned, eyes paling as light receptors closed. "Who'd we scan?"

"Possibly the Berin gang."

"We taking off in hot pursuit?"

"You two need to cut back on your *Star Wardens.*" The vids had gotten very popular since the addition of the Younger Races to the Justice Department, although given that the dialogue tended toward *"He's the law, I'm the enforcement."* Torin wasn't sure why. Still, they couldn't leave the clean-up team undefended. "Trajectory?"

"Theirs?" Craig pulled a line of streaming data up to the front of the screen. "Close enough to Three Points for government work."

"ETA?" Riding their Susumi wave, the potential Berins were moving a lot faster than the *Promise*.

"Five, give or take."

"Can we intercept?"

"If I micro jump us from here to here . . ." He pulled up a hard light projection. ". . . we can cut them off."

A wise man back on Terra had once declared space was big. The corollary to *big* was *empty,* at least outside the Core. Even in the MidSectors the distance between ships—unless those ships were very unlucky—made anything but fictional pursuit impractical. Although micro jumps into Susumi space could cover millions of kilometers in a few moments, they fell somewhere between suicidal and *are you fukking kidding* on the dangerous ideas scale and while all six of the Strike Team pilots had done simulations, an actual micro jump remained theory. So far, circumstances had never fallen on the jump side of *is this worth dying for.*

Still hadn't, Torin noted silently. "What are the odds that C&C will be dirtside when they arrive?"

"There were nine bad guys down there, Boss. Odds are Finds Truth will still have them filling out their WGB78s."

Craig was ship's captain; his was the final word within *Promise*'s bulkheads, but she commanded the team, so this was her call. She straightened, squaring her shoulders. "Drop our public files. Let them see who we are."

"You think the Berins will run?"

"From Gunnery Sergeant Torin Kerr and Strike Team Alpha?" Alamber replied before Torin could. "If they're smart enough to dress themselves, they're smart enough to run. You good with me trying to ride our scan into their OS, Boss?"

"Knock yourself out." He hadn't managed it yet, but if . . . *when* he did, they'd be able to take control of any ship close enough to scan and enjoy a brief window where their quarry couldn't escape into the vast emptiness of space. And then a coder on the other side would develop a block. Torin didn't subscribe to the belief that criminal meant stupid; Alamber was proof of that. He'd been part of Big Bill's pirate support organization, pulling his own weight from the moment his *vantu* had died, leaving him alone, young, and unethically brilliant. He'd helped Torin's team escape, engineering his own escape at the same time. Technically, Torin should have handed him over to the Wardens and rehabilitation, but she'd promised him a place for as long as he wanted it and she'd done everything she could to keep that promise.

If the Justice Department suspected Alamber's background—and Torin wasn't arrogant enough to assume they didn't—they either refused to act without proof or they considered his inclusion in Strike Team Alpha to be a part of his rehabilitation. She was good with either possibility.

"They're bringing their engines on. Canceling momentum." Craig leaned back, right hand splayed across the bottom of the screen, fingers bracketing the data from the long-range scanners. "And they're running."

"Back to their entry point?"

"It's the easiest math. Ninety-eight percent probability they jump out where they jumped in."

"You made that number up."

"I did. Still true."

"Tag them."

Craig had been a civilian salvage operator before Torin, before the Primacy, before the gray plastic aliens and the end of the war. He'd worked alone, tagging and collecting battle debris and then selling it back to the military. He claimed he could hit a moving piece of live tech the size of his fist from ten kilometers out, and Torin had no reason to doubt him. Unless the Berins had a CSO in their crew, they had no idea of what he could do.

"Tags away."

The tags were small and fast, small enough that when their vector intercepted the vector of the ships in—Torin leaned over the back of the chair and scanned the screen to find the relevant data—in three and a half hours, they'd read like space dust impacting against the outer hull. In spite of that, they were large enough to send out a tight beacon only the Navy and the Wardens could read. Alamber called them HI PAMS—essentially short for Hello, I'm a Bad Person, Arrest Me. The name had been slow to catch on. Werst insisted it was because he'd left out the B. *"Bad person; that's the relevant part right there. Makes no fukking sense without the B, kid."*

"And that's another section of space preserved for truth, justice, and good government." Craig stretched, spine cracking. "Go us."

TWO

"**SO, MIDDLE OF THE NIGHT** when we could logically expect even gunrunners to be tucked into bed dreaming of mayhem, they were all up, waiting for you. Lovely." Lanh Ng looked down at his desk where Strike Team Alpha's reports took up most of the surface, and frowned. As the first Human Warden, he'd been tapped to command the six newly formed Strike Teams although, with the teams having been structured for independent operations, his job seemed heavily weighted toward translating the actions of the Younger Races under his nominal leadership to the Elder Races who made up the bulk of the Justice Department. Torin wondered if they'd given him a choice. "You're certain you didn't trip a sensor?"

"They left no security in orbit, the ship was inert, and Alamber swears it was clear dirtside."

"So either they were more alert than you believe, or someone warned them you were coming. I don't like that word, *someone.* A few gunrunners trying to turn a profit on death and destruction; means and motive are clear. Throw in too many *someones* and the conspiracy theorists come out of the duct work."

"It might not be that complicated," Torin offered.

"Humans First?" He rearranged the reports, files flashing across the desk. "Ressk makes a good point about the dozen crates—Human involvement, however, doesn't necessarily mean Humans First. You captured most of the core group, and we've had no indication . . ." His lip curled; *no indication* was Warden-speak for *nothing to bring them in on yet*. ". . . no indication that the new membership is doing anything other than recruiting and issuing the occasional manifesto reminding us all that Humans died in the Elder Races' war . . ."

Ignoring that Taykan and Krai died as well, Torin added silently.

". . . and we deserve more than we're getting."

"They removed the apostrophe." Torin had been willing to take them down for calling themselves Human's First, but the Justice Department had waited until rhetoric had turned to action, waited until Human's First had swarmed stations, killed those who stood in their way, and stolen ships with Susumi capabilities before sending her team in.

A dark brow rose. "Your point?"

"They've gotten smarter, and they weren't happy about Richard Varga dying in rehab."

Back when Richard Varga had taken control of Human's First, he'd fed the self-entitlement of a few discontented speciesists and built them into a movement, the bulk of his recruiting aimed at veterans damaged in service. Human's First under Varga had been too busy whining about how *it wasn't fair* to organize more than random violence. Humans First, after Varga, had found a focus.

"No one was happy about Varga dying in rehab," Ng pointed out, as though he'd followed her thoughts to the same conclusion. "His death pulled the scattered remnants of the cause together and gave them a piece of grit to recruit around—if he'd lived, they'd have died out before he finished his program."

"Justice ruled his death natural causes."

Ng met her gaze and held it. "The investigation didn't find enough evidence to rule otherwise."

Which meant only that they hadn't *found* enough evidence to rule.

Nothing she could do, nothing any of them could do, but wait until Humans First crossed the line again.

"So just for fun . . ." He leaned back in his chair, fingers linked on the edge of the desk. ". . . let's say Humans First tipped your gunrunners off. How did they know your team was on the way in?"

"Scooped our signal." Ressk and Alamber could make code sit up and beg; they were good, better than good, but they weren't unique. "If they're rebuilding, they'll want to keep as much of an eye on us as we do on them."

"Possible."

"And . . ." Torin had to consciously relax her jaw to get the words out. ". . . there's always the chance that one or more of the Humans in the Justice Department is sympathetic to the cause."

He stared at her for a long moment. "Not possible. The department runs extensive background checks on every employee."

"Yes, sir."

Commander Ng hadn't been Corps, his rank might as well have been supervisor or project manager, but he'd learned to interpret the variations of that response faster than many officers ever managed. "I won't be able to convince my superiors to have the behavior of every Human on this station examined for suspicious activity. The violations of privacy alone . . ." He rubbed his temples. "Your gunrunners included Krai and di'Taykan."

"They could be hiring out the dirty work to less . . . uniform groups. Rebuilding in the background, setting up others to take the heat."

The noise he made wasn't so much agreement as acknowledgment that he'd heard her. "While it would make our jobs easier, it's unlikely Humans First is behind every group of armed lunatics we apprehend. And remember, the burden of proof remains with us."

Humans First. The Berin gang. Gunrunners. Centuries of war. Groups of mercenaries without names or secret handshakes. Millions of people from under-evolved species trained to violence and not all of them willing to give it up. The largest difference between those people and the Strike Teams created to bring them in was their targets, Torin acknowledged. The level of violence was much the same. "The Elder Races had a good thing going until sentient plastic got involved and they had to let the badly socialized kids join the club."

Ng opened his mouth. Closed it again. Shook his head—although not, Torin realized, in denial. He took a deep breath and squared his shoulders, physical cues to a change of topic. One hand flat on the desk, he shuffled reports with the other. "The after actions look complete for a change. Pass on my thanks to Warden Mashona for adding details beyond the number of shots she fired, the mapped trajectories, and a complaint about her perch. If I have any questions after C&C comes in, I know where to find you."

"I'll leave you to your filing, then, sir." Ng didn't consider the *sir* necessary. After working under him for the last two years, Torin did.

She touched her slate to his desk, signed out of the meeting, and headed for the door.

"I see Ryder offered to micro jump."

And stopped. "It wasn't necessary. He had no trouble tagging."

Ng made the noncommittal hum again. "Every Strike Team pilot we have declares the need for a micro every time they go out. Gossip has the Navy putting a research vessel together, and they want Justice to make the first jump."

"That's good."

Frowning, he looked up from his desk—although his hands continued moving over the surface. "What is?"

"That means even the three ex-Navy pilots on the Strike Teams consider themselves to be Wardens now." Old loyalties could twist teams out of alignment and the lines between *us and them* had to be clearly drawn when live ammunition was involved.

Brows at his hairline, Ng locked his gaze on her face. "So you're saying we should beat them to the jump?"

"Hell, no! It's micro suicide. Let the Navy do it."

"This place is so shiny new, I'm afraid to touch anything," Craig grumbled as Torin closed the hatch to their quarters behind her.

Berbar, Justice's station in MidSector Seven, had expanded when the three contract teams became six Strike Teams, resulting in forty-two new Wardens—Humans, di'Taykan, and Krai—added to the roster. Justice stations in other MidSectors were implementing teams using Berbar's protocols.

The quarters were generous, if unimaginative, painted in muted shades of gray or blue or green for the Humans and the Krai, and considerably brighter shades of purples and yellows for the Taykan. Justice—and Torin assumed Justice in this case meant specifically the Elder Races on station—wanted its new Wardens to feel calm and serene when not actively utilizing the violent skills they needed to do the job.

Strike Team Alpha had moved into their new new quarters immediately before leaving for Three Points, and they still looked like barely finished construction. Shiny new, as Craig said.

"I keep thinking," he added, "that I can smell wet paint."

"Paint doesn't smell . . ."

He turned to face her. "Depends on where you've salvaged it from. Pedro pulled six containers from a shipyard once. We had to evacuate half the station until it dried. And just so we're clear on this," he continued before Torin could respond, "I'm calling plausible deniability, so you're hanging Bony."

"That's fair." She shrugged out of her uniform jacket, crossed the room, and lifted the Silsviss skull off the entertainment center onto the installed supports. She'd had to get a special dispensation to display the remains of a sentient species, but Justice, after more meetings than anyone should have to attend, had finally been convinced that insulting the Silsviss—who had, after all, presented her with the trophy—would be a diplomatic faux pas likely to delay an already complicated integration. Back in the Corps, the brass had briefly considered the implications of pissing off an entire planet of warlike giant lizards and signed off on the trophy without discussion. "Given what the Ciptrans display . . ."

"Expect it's different when you're displaying your own species. How'd the final debrief go?"

The full team debrief occurred as soon after arrival as possible, before memories had begun to shift, but Ng liked to sit down with his Strike Team Leads, one on one. Torin suspected he'd do personal debriefs with every team member had there not been only twenty-seven hours in a day.

"No sudden insights." She stepped back until they were standing shoulder to shoulder, staring up at the skull. "There's too many outstanding variables to accuse Humans First of being involved. Hopefully, Finds Truth will bring back a detail we can hang an investigation on." The Dornagain's interrogation style reminded Torin of water against rock. In the end, water always won.

Craig leaned into her space, his arm a line of heat around her waist. "So, fifty-four hours of obligatory, post-mission down time."

His mouth traced a warm, wet line along the curve of her throat. The rest of the team had scattered and they had space in their new quarters for a full-size bed. "I can think of a way to fill a few hours," Torin allowed.

He hummed an agreement, the buzz against her skin lifting the hair on the back of her neck. "There's four new episodes of *Star Wardens* on the EC."

Sliding her hand under his waistband, Torin tugged him closer. "And I should hit the range."

"Need to supervise the QSM restocking the *Promise*." They separated long enough for him to drag her shirt over her head. "Last time the fukkers shorted the beer."

"I've got three new Wardens who have to retake How the Justice Department Defines Excess Force and I'll need to knock some heads together before they retain anything." She tightened her grip and he shuddered.

"Later?"

"Works for . . . me."

• —◆— •

"*Harveer*, I've found something!"

Harveer Arniz roused herself out of a half nap and cautiously stretched her aching back. Maybe her department head had been right. Maybe she was too old to go tramping across the galaxy. Maybe she should have given her place on the team to a younger scholar.

"*Harveer*!"

She snorted. Maybe her department head was a politically motivated asshole. "What is it, Dzar?"

Across the shelter that soil sciences had claimed as their lab, Dzar bounced on her stool, tail lashing. "High levels of urea, but that's . . ."

"That's not unexpected." Arniz blinked her inner eyelid, counted to three, and wondered, not for the first time, when they'd started accepting ancillary students right out of the egg. She was sure ancillaries used to be . . . older. "The soil you're analyzing came from a sealed latrine." The defined edges made for an easy introduction to fieldwork. They'd mapped it, recorded all flora and fauna within its perimeter, and drawn a core for soil profiling. "I'd be more surprised if you weren't reading high levels of urea."

"But these levels are *extremely* high! Carnivore high!"

Keeping her own tail still—she'd been out of the egg for more years than she wanted to recall—Arniz carefully moved around the piles of trays and half assembled equipment and waved Dzar off the stool. "Then there's a possibility we've learned something new about the pre-destruction inhabitants of the city." Adjusting the focus and silently cursing the inconvenience of elderly eyes, she ignored the quiet protest of how there hadn't been any post-destruction inhabitants, assuming she hadn't been intended to hear it. "The analysis of latrines can uncover a great deal of unexpected information. People dump everything, including other people, into the pri . . ." She flicked her inner eyelid closed, once, twice, and squinted at the data on the screen. "Plastic?"

"That's what I was trying to tell you!"

"Were you? Well, then, I apologize for interrupting."

"This is the only trace of plastic that's ever been discovered in the pre-destruction layer! I've read the Mictok reports; after three years on site, they're certain the pre-destruction civilization hadn't reached a tech level necessary for the production of plastic."

True enough, although that hadn't been how they'd phrased it.

"And as all evidence suggests our site is the same age . . ."

The concurrent ages having been the reason the university had finally been given permission to begin exploration, Arniz added silently.

". . . and as there's been no evidence to date that civilizations in this hemisphere were any further advanced . . ."

They'd been on the plateau at the edge of the jungle for less than four tendays, much of that time spent either grumbling about the paths laid out by the botanists or arguing over how the early results by the geophysicists mapped onto observable surface features. They had a preliminary layout for the area of the city not covered by jungle, but evidence of anything else was speculative at best.

". . . we need to ask ourselves, why is there plastic in this latrine? And not only that, it's surrounded by components, organic polymers of high molecular mass, as though there was a higher concentration present at one point."

"The components are hardly rare." Arniz took advantage of Dzar's pause for breath. "Let's concentrate on the actual trace of plastic, shall we?"

"But, *Harveer* . . ." Dzar slumped, her tail tip touching the ground, when Arniz turned to glare. It was a good glare; she'd perfected it over three decades of teaching. "Yes, *Harveer*. But why . . ."

"First, we're going to make absolutely certain the site wasn't contaminated. And if it wasn't . . ."

"It wasn't."

"Of course, it wasn't. It was fifteen point two seven centimeters down in an undisturbed latrine, and I trained you. That said, you need to put together the data to prove it. Go all the way back to the initial Ministry for the Preservation of Pre-Confederation Civilizations survey. While you're working on that, I'll go talk to Dr. Ganes." Ganes, an ex-Navy engineer, kept the ridiculous amount of equipment they'd dragged along with them up and running. He was also responsible for uploading communications to the university's orbiting Susumi drones. Most of the scientists on the dig kept a wary distance from Ganes, the only member of the Younger Races on the team. Arniz appreciated his efficiency, tried to stay upwind of the damp musk the Human exuded in the heat, and thought he was too damned tall. She could see no evolutionary reason that any species needed to be much more than a meter high and no reason for the Ciptran to exist at all. As they did, and as she wasn't an evolutionary specialist, she let it go. But not happily.

"Why talk to Dr. Ganes, *Harveer*?"

She blinked and shook herself free of reflections. "He needs to shift data allocation from one of the other teams. It's not like architecture is using much." Centuries of exposure had worn the arc of the city out on the plateau down to surface and subsurface, and, so far, subsurface referred only to foundation stones. Orbital scans had found actual buildings in the jungle, ranging from ruins of walls to intact interiors, but their permissions held them to the plateau. *Harveer* Salitwisi had become even more annoying than usual, knowing there were visible ruins under the canopy and yet being unable to explore them.

"We'll upload the data about the plastic?"

"No, I thought I'd upload the urea stats, seven or eight thousand times." When Dzar looked confused, right thumb rubbing at the smooth scales on her left inner wrist, Arniz sighed. She was a good kid, from the same southern province as Arniz, her name a sensible length rather than the random spew of letters they used in the north. Sheltered, and shy around other species, she had the makings of a halfway decent scientist; easier to work with, though, if she'd develop a sense of humor and a few less obvious ticks. "Yes, Dzar, we'll be uploading all information about the plastic, about your discovery of the plastic, as well as the proof that the site hasn't been contaminated. I want a clear and detailed procedural trail."

"To prove that the plastic aliens were here pre-destruction . . ."

"Dzar." Arniz cut her off. "Let's just concentrate on the plastic you found. On the science, not on crazy theories of how the plastic aliens might be responsible for the sudden planet-wide destruction that wiped out hundreds of separate civilizations, thousands of kilometers apart."

Dzar's mouth dropped open, her tongue flicking out to taste the air. "I hadn't thought of that."

Turning her head so as not to fog the primitive, static display, Arniz sighed again. Young. So very young.

•—◆—•

Commander Ng hadn't been happy when Torin had informed him that if Justice wanted the Strike Teams to continue tidying up the post-war peace, their skills in controlled violence needed to be maintained. They needed more than the gym, provided for *fitness* and funded by denial, they needed a range. She didn't know how Ng had convinced the Justice Department, but given how the Elder Races felt about weapons, she suspected simple logic hadn't been enough. She also suspected Ng, a former lawyer, had enjoyed doing the browbeating even if he refused to admit it. Eventually, the Justice Committee to Examine a Proposed Expansion of Strike Team Training Facilities had agreed to add a range to the new section of the station and, in a surprising triumph of need over budget, had paid a premium to bring in the construction company used by the Corps.

Torin had no complaints about either setup or safety—the finished range was a copy of the Level 26 range on Ventris Station—but she continued to agitate for a planet-based training facility. Firing live ammo in a high-tech container surrounded by vacuum had always seemed like tempting fate; no matter that the bulkheads had been lined with an organic gel interlaced with Mictok webbing, the semisolid mass designed to absorb impact.

Given that Mictok webbing was extruded from the ass of giant, sentient spiders, Torin had never asked who or what provided the gel. As long as it worked, she was happier not knowing.

The surrounding vacuum had been why she'd handed the testing of the small gun they'd confiscated at Three Points to Binti. Just as Torin knew she could hit almost anything she aimed at with a KC-7, she knew Binti could hit anything she aimed at. Period. Corps sniper training after their "diplomatic mission" to Silsviss had honed a natural talent.

Binti's first shot with the small gun hit the outer ring on the left side of the target at the fifty-meter flag. Her second moved right two centimeters. Her last three made a tight group a centimeter left of dead center.

"Honestly, Gunny . . ." She straightened, removed the magazine, checked the chamber, and laid the weapon on the table behind the shield. ". . . if you're not me, I'd say maximum effective range is no more than seventy meters. I could maybe make the shot at a hundred, hundred and twenty, but the short sight radius on this thing is stupidly limiting."

"And with practice?" They'd found nine rounds in the small gun's magazine, a tenth in the chamber. They'd been lucky it hadn't gone off again when Mack had thrown it to the floor.

"Meh. It's just not very accurate."

Torin reached past Binti's hip and picked it up. Even considering it was missing the magazine, it was lighter than it looked, made of the same alloys as the familiar KC-7. Same alloys. Same firing mechanisms. So much smaller.

"It kicks," Binti continued as Torin closed her fingers around it. "There's no room for dampeners, and all the movement fuks with accuracy."

"It doesn't need to be accurate at distance. You can hide this. Get close."

"Get close? While being shot at?"

"It's not a soldier's weapon." It wasn't a weapon for war, or at least not the kind of war both sides knew they were fighting. The implications were as ugly as the small gun.

"I think it's a prototype."

"Based on?"

"The three ways to improve it I'd worked out by the second shot." Binti recorded the hits, then cleared the target. "I get why some people are into miniaturization, but I don't get why anyone would make a working model of something like this."

Torin did. "Violence has been spreading now the war is over. The Wardens have brought in specialists to deal with it." That was the opening line of the vid Justice had put together to introduce the Strike Teams to the public. Torin had argued they should stay covert. Commander Ng had decided to use their existence as a deterrent. With no way to measure uncommitted crime, there was no way to discover if it had worked.

Arms folded, Binti stared at her for a long moment, both brows up, then huffed out a sigh. "You think this is for protection against a perceived rise in violence? At an effective range of seventy meters? Against what?"

"Other people with these."

"Yeah, given the shit we have to go through before and after we use our weapons—and we're the law—not going to happen."

"Legally." Torin set it back down on the table. The last thing the Wardens needed were easy to conceal weapons loose among a civilian population. Actually, fuk the Wardens; it was the last thing anyone needed.

"You have a suspicious nature, Gunny."

"So I've been told." It wasn't paranoia if it kept people alive.

They turned together as the hatch opened and Dr. Deyell entered, high stepping over the lip, feathers ruffled, head twitching back and forth as he tried to see into all corners of the range at once. "This one isn't mistaken about being allowed to enter?"

"Light was off. You're good, Doc." The young Rakva was the only member of Strike Team R&D who enjoyed interacting with the Younger Races. All other interactions were tainted with apprehension, as though R&D expected the Strike Teams to explode into unprovoked violence at any moment. The Rakva, like the Katrien and the Niln, weren't among the original members of the Confederation, and, considered less Elder by some, were often referred to unofficially as the Mid Races. It had been a Rakva doctor who'd worked his feathers off to keep her desperately outnumbered platoon alive during the Silsviss attack and a Rakva doctor who'd saved her after severe radiation poisoning. Torin had a soft spot for the species.

Deyell's rudimentary beak curved into a smile. "Ah, Strike Team Alpha Lead, this one finds it very convenient to find you already on the range."

Torin suspected the station sysop had fingered her. Grabbing a handful of Binti's shirt as the sniper began to move away, she muttered, "Where are you going?"

Binti grinned. "That one finds it convenient to find the Strike Team Alpha Lead on the range. As I'm not leading anything, I'm out of here."

Teal-and-gray feathers smoothed out as Deyell crossed to set the case he carried on the table. "If we're going to test fire the new prototype, this one will need you as well, Warden Mashona. You'll put the dart exactly where this one tells you, removing a potential variable."

"Damn."

"Ah." His crest rose and he poked at the small gun. "This one sees you have found a pistol."

"A what?"

"A pistol. A weapon your species used in the past. Well, all the Younger Races had versions of it, but this particular example seems to be very Human in design, so this one uses the Human word. This one combed countless historical records in order to develop a functional delivery system for a tranquilizer and found a few remaining references to the pistol." Feathers fluffed on his brow. "The implication was that all three species seemed absurdly fond of the design. But . . ." He gave himself a shake and the feathers settled. ". . . this one isn't here for a history lesson."

Pistol. It had a name. Torin thought the "name it, know it" belief was bullshit, but it would make it easier to write the report. In her single year as a Warden, she'd written more reports and filled out more forms than in fifteen years as a Marine and was in favor of anything that got her through them faster.

Deyell opened the case. "Now, to the new design for the delivery system."

None of the other designs had worked, but Justice remained unhappy about their Strike Teams using lethal force, so R&D kept trying. *"They're using lethal force against us"* had been considered, at best, an irrelevant argument.

"Unlike the previous attempt which fired compressed air from an HPA . . ."

"And had an effective range of fifteen meters," Binti said, picking up a dart and flicking the green fluff on the blunt end.

Deyell plucked the dart from her hand. ". . . we used a rifle blank and a 1CC dart loaded with a new compound that guarantees an almost instantaneous result on Humans."

"Almost?" Torin's brows rose. "And what about non-Humans?"

"Or Humans who mass less than we do," Binti added. "Collins, the pilot on *T'Jaam*, is barely a meter and a quarter."

"We've calibrated for mass and metabolism and worked up a variety of darts. Not only for Humans, but for Krai and for di'Taykan as well."

For the violent Younger Races brought in to fight a war the Elder Races were too socially evolved to fight themselves. The only races the Strike Teams were sent out to stop.

Torin shook her head. "A variety of darts that we'd need to load into a single-shot weapon . . ."

"The labeling is very clear."

". . . while under fire."

"This one admits it continues to be a work in progress. If you could . . ." He held out the tranquilizer gun.

Binti took it, fingers wrapped possessively around the barrel and the grip. "Your turn at the pointy end, Gunny."

Torin sighed and signed the KC that had been modified for the tests out of lockup. It shot dissolving rounds at a considerably lower velocity than the usual 953 mps. The rounds stung, but beyond a minimum of fifty meters would leave only bruising even if they hit bare skin. In the interest of not losing an eye, Binti pulled on a helmet and activated the visor.

So did Torin. Given that she was on the pointy end of the exercise.

"Effective distance?" Binti asked as Torin began to walk up range.

Deyell tapped his fingers against his beak. "Numbers suggest sufficient accuracy for testing at 80 meters."

"Numbers suggest?"

"The lack of empirical data is why this one is here."

"Go out to seventy, Gunny."

"We can regrow arms and legs and organs, and yet this one wonders why we can't equip our Wardens with nonlethal weapons."

Binti snorted. "False comparison."

"Humor this one. It's been a long day."

"Ready, Gunny?"

Torin turned at the seventy, feet braced, weapon hanging at her side, muzzle pointed toward the floor, her finger away from the trigger. Fighting both training and experience that demanded to know what the hell she thought she was doing becoming a target, she squared her shoulders. "Ready."

The dart hit her in the meat of her stomach, to the right of her navel. She huffed out a breath at the pain of impact, and, as the cold spread, snapped up her weapon and got off three shots before Binti could duck back behind the shield.

Vision blurring, she heard Binti snarl, "Fukking ow, that stings!"

Then her knees buckled, and the last thing she heard as she made contact with the floor was a morose, "Still too slow taking effect. This one has to admit he expected that."

"You're drooling."

Torin blinked, focused first on Craig's face and then on the rest of the team gathered behind him. Either she'd rolled when she fell or someone—probably Craig, on his knees beside her—had flipped her onto her back. She attempted to swipe at a line of warm moisture running toward her ear, assumed that her unresponsive arm meant R&D had increased the paralytic, then she blinked again. Her eyelids felt like they weighed five kilos each. "Why are you all here?"

"You were out for a while this time." Binti shrugged. "We worry."

Craig took a deep breath and let it out slowly, not bothering to hide either his anger or his concern. "Allowing R&D to drug you is not part of your fukking job."

"Actually," Ressk broke in before she could respond, "it's covered under cooperating with the continuing development of law enforcement techniques."

"Seriously?" Werst snarled.

"Did you not read your contract?"

Alamber pushed between them, slate in his hand. "I talked to Dr. Deyell . . ."

"We all talked to Dr. Deyell," Ressk interrupted.

Craig smiled. "Werst threatened him."

Torin's face seemed to be working, so she returned the smile. "You didn't?"

"We thought it would be more effective coming from Werst."

"*I* talked to Dr. Deyell," Alamber repeated and Torin managed to raise a finger to keep anyone from cutting him off again, "and while you were napping did some research of my own on the pre-Confederation pistol. Their destruction was one of the conditions of the Younger Races being allowed onto the playground. It took almost a century before the last remnants were destroyed—little less for the di'Taykan, little more for the Krai—and another before the pistol was forgotten."

"Concealed weapons, concealed anything that affects another person, is just wrong," Werst growled.

"Yeah." Alamber waved his slate. "That's what they wanted you to think."

"All things considered," Torin sighed, flexing her toes as more feeling in her extremities returned, "they were awfully fussy about how we kill people."

Craig lifted her hand to his lips and kissed her knuckles. "You're still drooling," he said.

Finds Truth Through Inquiry fluffed the long red-gold fur on her arms and sighed at Torin's question, even though Torin had every right as Strike Team Lead to participate in C&C's debriefing. Ng refused to make attendance mandatory, but Torin felt the whole point behind making informed decisions involved being informed. Craig had pointed out her control freak tendencies as an alternative reason, and Torin had pointed out in turn that anyone who could do the job better could do it with her blessing.

Most of the other leads had followed her example. The holdouts had fallen before her logic.

Finds Truth Through Inquiry sighed again when Ng indicated she should answer, showing the closest thing to impatience Torin had ever seen from a Dornagain. "No, the *pistol* . . ." She popped the 'p' of the Human word, "wasn't stolen from the Marteau Industries warehouse in Sector Eight with the other weapons. According to the gunrunner, ex-Corporal Mthunz Mackenzie, he received it anonymously."

"Anonymously?" Torin asked when the final pause continued long enough the odds were high Finds Truth had finished speaking. Although the Dornagain in the Justice Department continued using the unhurried cadence that matched their slow and deliberate movement, they'd grown more concise. Torin assumed it was at least partially in self-defense as all three Younger Races interrupted the moment they felt the point had been made.

Shifting the top third of her massive body, highlights rippling through her fur, Finds Truth Through Inquiry fixed Torin with an irritated gaze. "It means, Strike Team Alpha Lead, Warden Torin Kerr, that the agency through which he received it remains unknown."

Torin had tried to get Finds Truth to drop everything but the Warden and the Kerr but the Dornagain, like a number of the Elder Races, asserted their individuality within the Confederation through verbal ticks. Like many of the Younger Races, who all spoke tick-free Federate, Torin found it an annoying affectation. It had taken a good ten tendays of argument before the Dornagains attached to the teams had agreed to the shortening of their own names in informal conversation.

"As he indicated on form AR77B . . ." A long arm stretched out, fringe of fur swaying, and the C&C lead lightly touched the edge of Ng's desk with a blunted claw. Light flickered as the forms shifted. ". . . he found the *pistol* three tendays ago, in among his possessions along with a container of ammunition. He remained adamant during further questioning and while completing form AR78A that he has no idea of who might have placed either the pistol or the ammunition with his belongings. Although he was unable to correctly spell container, he refused to substitute the word *box* on AR77B and within the specific description of the item on AR77BX."

"He was wounded."

The corner of Ng's mouth twitched. "Are you suggesting the injury may have affected his ability to spell, Warden Kerr?"

"No, sir, but it may have affected his willingness to cooperate."

Ng allowed her a single nod of acknowledgment. "Granted."

"We've confiscated the remaining ammunition from him," Finds Truth Through Inquiry continued pointedly, her attention back on Torin, "and we'd appreciate the *pistol* returned to evidence at your earliest possible convenience, Strike Team Alpha Lead, Warden Torin Kerr. The chain of evidence has been irreparably broken. You know better than to have taken it from an active crime site without following procedure."

True. In all honesty, Torin wasn't sure why she had. She liked to think it was because she'd been appalled by the existence of the weapon, but she suspected curiosity had as much to do with it.

"And," the Dornagain Warden continued, "Analyzes Minutiae to Discover Truth requires copies of all data gathered during the testing of said weapon in order to complete his documentation and has requested I mention that includes security recordings made at the range."

"Commander Ng has the weapon . . ." Torin had never intended to keep it. ". . . and the data is available to anyone with Justice Department access."

"He requires *copies*," Finds Truth Through Inquiry repeated, rising far enough off her haunches to tower over Torin, which wasn't far given their relative sizes. "As well as new DNA samples from everyone who came in contact with the evidence. To be sent to his desk at your earliest convenience."

"Of course." If that's all it took to smooth ruffled fur, Torin wouldn't complain.

"Copies have already been sent to our analysts." Eyes on his desk, Ng flicked the case files from one side to the other. It took a while. Dornagain, as a species, had an obsession with details; it made them excellent Wardens, as long as the crimes investigated were neither violent nor time sensitive, and they were neither once the Strike Teams had finished with them. "The analysts will apply the specifics to other, unsolved transgressions where weapons have been involved."

"To determine if ex-Corporal Mthunz Mackenzie has participated in previous criminal activity, Strike Team Commander Lanh Ng?"

"That, and we need to be sure there aren't more of these pistols in circulation."

Finds Truth Through Inquiry's small, rounded ears flattened against her head, and she rattled her heavy front claws against each other. "That would be . . . bad."

The following pause was almost Dornagain in length. "Bad would be one way of describing it," Ng said at last, voice dry enough Torin wouldn't have been surprised to see sparks rise off shifting fur. "New DNA samples first, Warden Kerr. Discovers Truth Through Minutia . . ."

"Analyzes Minutiae to Discover Truth, sir."

"Of course. Analyzes Minutiae to Discover Truth will run the weapon in question through a cleansing as soon as possible. I want the bio evidence narrowed down quickly."

Finds Truth Through Inquiry rolled her eyes. "In two, maybe three tendays." When Torin's brow rose, she met her gaze. "We aren't the only team using the lab, Strike Team Alpha Lead, Warden Torin Kerr."

"Before you point out that one lab for six teams is five labs too few, under the mistaken belief I can't count," Ng said, before either Warden could respond, "I've spoken to the Strike Team Oversight Committee about it and have a guarantee that expansion will be discussed for the next budget. However, even working within our current restrictions, we'll have the information at the end of a tenday."

Behind the neutral mask she wore for superiors, Torin laughed at the C&C Lead's blatant disbelief.

"If we get lucky, we'll find bio evidence unrelated to that of the gunrunners and will be able to search military records for a match. Given how many of the Younger Races have served over the centuries, that should throw up at least a partial."

Fathers, mothers, sisters, brothers, cousins, uncles, aunts, nieces, nephews . . . Torin's hand twitched toward the combat vest she wasn't wearing, toward the pockets where she'd carried out the cylinders of ash, all that remained of lost Marines. She gripped her thigh, digging her fingers into the muscle, to keep her hand from rising.

Finds Truth Through Inquiry made a dismissive sound, unexpectedly small given her size. "Considering the high percentage of ex-military responsible for the current violence the Strike Teams are called to end, I fully expect a search of military records for a match will easily identify the anonymous provider of the *pistol*."

From the way the fur rose along Finds Truth Through Inquiry's spine, the growl surprised her.

It surprised Torin, too, and she'd made it.

"Kerr . . ."

"No, Strike Team Commander Lanh Ng, the fault was mine." It wasn't easy to read facial expression under fur, but Torin had plenty of practice over the years and Finds Truth Through Inquiry looked embarrassed. "I referred to numbers only, millions who've served, hundreds who broke under the weight of war. I forgot that to Strike Team Alpha Lead, Warden Torin Kerr those numbers referred to comrades, many of whom didn't survive to see peace." She laid a large hand over Torin's, the heat from the hairless palm relaxing Torin's hold on her thigh. "My people's history teaches that we voted against bringing the Younger Races into Confederation. In due time, because they would be worthy members, yes. Merely to fight our war, no. Better we die, my ancestors thought, than have the young die for us.

It's why so many of us . . ." Her other hand covered the badge hanging from the lanyard around her neck. ". . . go into public service."

Torin met her gaze. "Not a species obsession with bureaucratic detail?"

The Dornagain had evolved on proteins from shellfish and insect larva, nothing they had to chase, but their smiles still held a lot of teeth. "Yes. That, too."

"Are you done?" They turned together to find Ng wearing a bland expression of polite disinterest. "While I approve of you two building a better working relationship, I'd approve more if you'd do it in a bar like everyone else and not during a debriefing we need to finish before my ten o'clock with yet another political subcommittee worried we're promoting aggression."

"We're done." Torin gave Finds Truth Through Inquiry's hand a brief squeeze as she lifted it.

"And only an idiot would believe the Strike Team Program promotes aggression," Finds Truth Through Inquiry added. "We harvest the aggression the war germinated."

"Do you hear me arguing?" Ng shook his head—at the analogy, Torin assumed—and took up the thread of the debriefing. "If we find nothing in the military databases, we'll have to do it the hard way and hope Mackenzie gives us something during further interrogation."

"I look forward to speaking with ex-Corporal Mthunz Mackenzie again."

Lots of teeth.

Torin ran back over what they knew, tuning out a detailed list of the forms Mackenzie would have to fill out during a second interrogation. The poor bastard. "Sir?"

"Kerr."

"We know the crates of weapons were stolen from Marteau Industries' warehouse."

"Your point?"

"If MI can manufacture large guns, why not small? The principle's the same, the materials are the same; only the size is different and that thing had to have come from somewhere."

"Granted. But all weapon designs have to be approved by a Parliamentary committee . . ."

"Sir, I think we can assume that whoever created this pistol did it without seeking Parliamentary approval."

"Better to ask forgiveness than delay production of a prototype?"

"Or he didn't give a shit."

"Also possible, but—if by *he* you mean Justin Marteau—Marteau Industries isn't the only company continuing to manufacture weapons post war."

"Yes, sir, but Sector Eight already has Wardens on site dealing with the unsolved theft. As long as they're there, why not have them check the manufacturing records and the subatomic signature of their materials? If there's no match, we can eliminate that particular production line as the source of the pistol without needing to commit more personnel."

"Sucking up to budget, Kerr?"

"Will it get us new labs faster?"

"Unlikely."

"Then, no." She took a deep breath. Thought of the unmistakable and highly visible bulk of the KCs and the people trained to use them. Trained when to use them and, more importantly, when not to. Thought of the pistol and how it fit into a waistband or a pocket. "We need to find who made it so we can find out why."

"To see if they could?"

"I hope so." It was, after all, curiosity that had gotten every species in the Confederation up out of their gravity well. "But then why give it to Mackenzie?"

"Field testing."

"Then who's retrieving the data? And how?"

"Excellent questions." Ng folded his hands together and rested his steepled forefingers against his upper lip. After a moment, he nodded. "I'll send a message to my counterpart at Sector Eight Station, she can pass it on to the investigative team. Although I can't see Marteau Industries or any of the big armament companies creating an illegal product; they've got too much to lose."

Justice wouldn't stop at rehabilitating the manufacturer of the *pistol*. Not when the Elder Races had tried to wipe out all knowledge of it the first time around.

"I wonder . . ." Finds Truth Through Inquiry slouched and fluffed the fur on her arms again, ". . . why Marteau Industries continues to make weapons if the war is now over. Surely there's weaponry already available to deal with any final, lingering violence."

"You'd think so." Ng pulled Marteau's logo up into the air above his desk, the MI centered within the shifting stars of an exploding galaxy.

Torin watched the galaxy explode and felt two puzzle pieces shift closer together. "I wonder if there's a connection between the successful theft of these weapons and the attempted theft of the ancient H'san weapons." They were no closer to discovering who'd financed the latter attempt. Not even Alamber had been able to follow the money back to a source although, before Torin called him off, he'd bent a few laws and had dug one layer deeper than the forensic accountant Justice had put on the trail. "It could be they've had to downgrade from the legendary to the modern."

"It's possible there's a connection, but finding it isn't our problem. You . . ." Ng pointed at Torin. ". . . neutralize violence. You . . ." His finger moved to indicate Finds Truth Through Inquiry. ". . . make sure it stays neutralized. And, if we're done, I have half a dozen bureaucrats to appease while wondering why I ever agreed to take this job."

"Boredom, Strike Team Commander Lanh Ng?"

"I wasn't aware they offered you a choice, sir."

"Get out of my office. Both of you."

"No one died on your last mission." Dr. Allan tapped a fingernail against the edge of her slate, the faint plink, plink not quite retreating to background noise. "How do you feel about that?"

Torin maintained her neutral expression. "I haven't lost a team member since we started."

"Ah, yes, but I mean that none of the opposing forces died this time."

"Opposing forces? You mean the gunrunners?"

"Yes."

"This isn't a war."

"Isn't it?"

The Justice Department had concerns about the mental health of personnel involved in an institutionalized armed response. Since it was an institutionalized armed response that had created the need for the Strike Teams in the first place, Torin acknowledged their concerns had merit. She found less merit in their insistence on mandatory therapy for all Strike Teams and Cleanup Crews, but—to their credit—Justice had staffed the new support department broadly enough everyone could see a therapist of their own species. Torin was of two minds about that now her court-appointed therapy with a Marine psychologist had ended, mostly because she found Dr. Erica Allan, who'd never served, annoyingly superficial. And occasionally perky.

Torin felt her endurance of that intermittent perkiness was a good indicator of how mentally stable she actually was.

Still, she'd followed worse orders than "go to therapy." If things changed and she came to believe she needed help to keep functioning, she'd switch to Dr. Verrir. The Krai therapist had fought at Fortune Station.

"Torin?"

"Doctor."

Finally realizing she wasn't going to get an answer to her question, Dr. Allan tried another. "Do you think guilt brought you to the Wardens as much as a sense of responsibility?"

"For the people I couldn't save?"

"Yes and no. For not realizing sooner that both the Confederation and the Primacy were being manipulated by the plastic aliens."

"How was I supposed to realize that any sooner?"

The doctor smiled, radiating approval. "And that's my point."

Only decades of facing officers, staring at the wall over their shoulders while they ignored reports from the front and laid out plans to win the war by *sectus*, kept Torin from responding.

While Ressk's circlet of leaves hung rakishly over one ear, Werst's head looked as though it were being devoured by a bush. Torin hadn't written off sentient plant life—as circumstances continued to remind them, space was big—and there was nothing to say it wouldn't be carnivorous if and when they found it. That said, Feerar, the station's chief agricultural scientist, must have all but stripped hydroponics to provide wreaths for the number of Krai gathered in the bar.

Torin knew the eight Krai from the other Strike Teams on station—Delta, out tracking the Berins, had sent regrets—as well as a couple of the Krai lawyers—property damage caused by a Warden was significantly more complicated to resolve than property damage caused by a Marine—and she knew one of the aides working at entry-level politics because Sarrk, the aide in question, had turned out to be part of Ressk's extended *jernine*. She knew them, but she'd have a hard time identifying them with the wreaths covering the scalp mottling other species used as identifiers.

For all that it was a Krai ceremony, for all that at least half the people in the bar were Krai, only Ressk, Werst, and Sarrk stood apart. The crowd itself remained mixed, undivided along species lines. Torin would have liked to have seen more of the Elder Races, given their dominant numbers on the station, but the Niln and the Dornagain and the Rakva from C&C had come out and eight Dornagain was about two more Dornagain than the bar could comfortably hold.

"Time has passed!" Sarrk raised her glass as the crowd echoed her words. Werst appeared to be checking for escape routes, but his nostril ridges were open and his fingers interlaced with Ressk's. "We celebrate the continuance of . . ."

Krai names were matrimonial and long. From a top-of-the-tree family, Ressk's should have taken the rest of the evening to recite. Sarrk, showing an affinity for the politics she studied, shortened it until it was barely longer than Werst's.

"May they climb side by side toward the sun!"

"*Harr tur*!" The Krai shouted it first. The non-Krai echoed it. The Dornagain politely modulated their volume so as not to deafen everyone in the room.

Sarrk drained her glass and set it down. "Now, we eat!"

"Short and sweet. Aces." Craig touched his glass to Torin's as the Krai began eating their wreaths. Which, to be fair, Torin should have expected. The Krai had the most efficient gut in known space and could digest anything organic—taking the "only omnivores have achieved sentience" belief to the extreme. While escaping Big Yellow, Werst had eaten what they'd later discovered to be a piece of the plastic aliens to no ill effect.

"You're frowning."

"I just realized we should have checked Werst's shit to see if the plastic reformed after digestion."

Craig stared at her for a long moment, then shook his head. "Your thought processes scare me sometimes. Come on, let's abuse your rank, get a table, and bog in."

Musselman's, the bar on Berbar Station claimed by the Strike

Teams, was no Sutton's. It was smaller, darker, and had only two small vid screens at either end of the long, carbon fiber bar. But the beer was good, the food cheap and filling, and Paul Musselman, the owner, had adapted to the sudden influx of ex-military by hiring more staff and extending hours to cover variable schedules. Barely making ends meet before Werst wandered in while exploring the station, he managed to combine appreciation for the new business and a polite refusal to put up with shit into the kind of welcome that bureaucracy-weary veterans could appreciate.

And he was likely to make a small fortune off the *vertrasir,* Torin noted. The food provided for the ceremony was predominately Krai, much of it imported, and could only be improved by enough alcohol to dull the taste buds—although every non-Krai in the bar knew to stay away from the *hujin* chips. Even the Dornagain had been warned—Torin having canceled the intended trial by fire. A gassy, five-hundred-kilo forensic pathologist in a sealed environment wasn't funny.

She was defending her last piece of *trupin* from Craig and arguing with Captain Ranjit Kaur of Strike Team Ch'ore about how it seemed likely the gunrunners had been given a heads up when Werst dropped into one of the empty seats at the table.

"Gunny, Ryder, Cap." He snagged her *trupin,* and Craig's beer, and side-eyed the captain's remaining trio of giant shrimp. "When I catch the fukker who added pine to my wreath," he growled around the *trupin*, "I'm going to add my foot to their ass."

"You can't eat pine?"

"I don't like pine. It's like chewy turpentine."

"When did you taste . . ." Captain Kaur shook her head. "Never mind." She pushed her shrimp over. "Your need is greater."

He stuffed one into his mouth. "Sun in your face, Cap."

"Bend without breaking, Werst."

He waved the second shrimp at Torin. "Notice that, Gunny? Officers get taught manners."

"You weren't, so what difference does it make?"

"It's my *vertrasir.* I'm faking it." He snagged a small, round loaf of bread out of the air, and waved his thanks to the server who'd thrown it. "Ah, *gheran*. What are we talking about?"

"Gunrunners who knew we were coming."

"And who spilled," Craig said. "That's the question."

"Spoiled for choice." Werst set the *gheran* down and spread his hands when they turned to look at him. "Justice is a government department. Government funded means open book. Privacy redaction within the department is minimal. Information's there if anyone wants to look, so anyone with enough time to sift out the crap could've told them. Or blabbed in a bar to some asshole who passed it on in turn."

"Keeps slipping my mind you're smarter than you look."

"Fuk you, Ryder." He finished the beer and added, "Or it's a Humans First conspiracy, and we're headed for shit up to our chins. That one's for you, Gunny."

"Thanks." Torin caught a server's attention and waved her empty glass. "Conspiracy theories call for another drink. Cap?"

She shook her head. "No thanks, Gunny. I've reached the line between beverage and intoxicant."

"Craig?"

"If everything we do is public record . . ." He emptied his glass and slid it across the table. ". . . and some poor fukker has to comb through the resulting shitload of files, we should make it interesting for them."

Cap's brows went up. "With alcohol?"

Craig winked at her. "It's a start."

It didn't take long after Werst and Ressk left—to an unsurprising number of explicit suggestions—for the crowd to thin down to more bearable numbers. The lawyers and politicians were gone, scientists and researchers condensed around one table, and Captain Kaur had called it a night when it became clear she was in a sober minority of one.

"Commander Ng needs to find a few more Sikhs for this gig," she'd said as she stood.

Post-Confederation, some religions had disappeared and some had evolved after deciding faith and a wider universe were not mutually exclusive.

Enough room having opened up in the bar to minimize potential accidents, Binti and three other snipers had begun a game of darts, all four of them drunk enough to find knocking each other's darts out of the air the funniest thing they'd done all evening. Torin didn't worry about Binti, because she understood Binti. She'd been with the platoon pinned down by a thousand adolescent Silsviss. One of the three survivors when Sh'quo Company had been destroyed. Though Torin had given the order, she'd taken the shot that had destroyed the *Heart of Stone*. She had her demons; no one came out of a war without a few, but she had them in hand.

"Is that safe?" Craig asked as a dart ricocheted into a pitcher of beer.

"Depends on how you define safe."

"Fair go you'd have more than one definition."

Under the table, she pressed her thigh against his. "I try to stay flexible."

A low growl cut Craig off before he could respond. "Alamber di'Crikeys is not yours."

Torin pressed harder, thigh to thigh, this time to keep Craig in his seat. Ex-Petty Officer Gamar di'Tagawa was a mean drunk, heavier built than most Taykan, and had the close combat training to win most of the fights he started. If it ever came to it, Torin wasn't certain she could beat him. "Your point?"

He leaned forward, knuckles braced on the table, still smart enough not to get right into Torin's face. "Alamber di'Crikeys is not yours."

"So you said." Torin glanced over to where Alamber stood in front of the vid screen, wrapped around three other young di'Taykan, cheering a . . . a Taykan sport of some kind, she assumed. "He's his own."

"Release him!"

She took a long swallow of beer before answering. "From what?"

Gamar's deep green hair flicked out around his head. "You don't even have sex with him!"

"He likes you best," Craig translated, leaning back, muscular arms spread wide, narrowed eyes suggesting he wasn't as relaxed as he appeared. "Buddy's jealous."

"Yeah, I'm getting that." Torin took a drink and kept her hand loosely wrapped around the glass, a potential weapon against the di'Taykan's longer reach. "Alamber makes his own choices, Gamar." For the most part, he slept communally on station, but still occasionally climbed into team beds. If the Confederation had a lock Alamber couldn't get through, he hadn't found it yet.

Gamar slammed both fists down, hard enough the empty glasses rattled together. "You're why he won't choose me!"

Torin smiled. "You're why he won't choose you."

Gamar's eyes darkened, the green disappearing as light receptors opened, his shoulders flexed against straining fabric.

"Seriously?"

As he lunged forward, the two di'Taykan who'd been lingering behind him grabbed his upper arms.

"Gam! Stand down!" Lemon-yellow hair flattened against her head, Sirin di'Hajak, Ch'ore's pilot, used her weight to swing Gamar away from the table. "You said you were going to talk to her!"

"I talked," he snarled. "Let me go!"

"No. She'll kick your ass, and I don't want to deal with the resulting embarrassment."

"I fought deck to deck on the *Straightaway*!"

"Good for you. She'll still kick your ass." Sirin nodded at the pink-haired di'Taykan on Gamar's other arm, and adjusted her grip. He sucked in a startled breath. "Come on. Sorry about that, Gunny. I honestly thought he was just going to talk."

Torin raised a single brow.

Sirin grinned. "Yeah, well, booze makes me optimistic." She tugged Gamar into step beside her and then across the bar to the exit. The pink-haired di'Taykan—Bartua? Bartun?—muttered an additional apology and followed behind.

"I don't remember the Corps achieving these levels of soap opera," Torin murmured into her glass.

Craig's arm slipped around her waist. "Selective memory."

"Probably."

THREE

IT HAD TAKEN HIM YEARS to appreciate why the government restricted the robotics industry, why research into Artificial Intelligence was so obsessively overseen and invariably buried. It wasn't that the Elder Races feared a machine uprising, ancient civilizations falling to the rule of mechanical overlords; it was that wetware was easier to manipulate than hardware. A conclusion so obvious, he was embarrassed it had taken him so long to realize it.

Aren't we wonderful, said the Elder Races to the Younger. *Look what we've done for you. Here's what you can do for us in return.*

The list of what the Younger Races couldn't do was significantly longer than what they could.

Pausing in front of his favorite piece of ancient plastic, he gently touched a pale pink curve. Now only slightly darker than his fingertip, analysis indicated the small, misshapen horse had once been much brighter, the missing mane and tail made up of thin, plastic strands. Although this piece had retained all four stubby limbs and its head through the centuries—replacing a headless purple horse in pride of place—he had people watching for one in even better condition.

No one person had enough information to lead anyone else to a hidden treasure. But a great many people gathering piece after piece—a rumor, a drunken confession, a conversation overheard, a communication diverted. Small plastic horse collecting was cutthroat in its competitiveness, and he would not allow any more pieces of Human history to be lost to alien hands.

When they'd accepted the Confederation's terms and flung

themselves from the close confines of Terra into a wider universe, Humans had lost culture, they'd lost languages, they'd even lost technology, throwing aside their own creations for the bright and shiny and new. Myths and stories began to change, their historical narrative edited to fit the new reality. And if that wasn't enough, more and more of what made Humans unique wore off in close contact with the Krai and the di'Taykan—their pride, their ambition, their awareness of their intended place in the universe. Too many Humans were happy to see themselves as equal to degenerates of lust and gluttony. Happy to be cannon fodder for the Elder Races, second-class citizens to the fur- and scale-bearing. To the multilimbed. To insects.

Insects, for fuksake.

He had thousands of people seeded through the Confederation; each using unique search parameters to sift through the masses of public information that resulted from an official policy of transparency.

Thousands of people. Among the pieces of information they sent him, a familiar discontent.

Men and women who'd bled for specialized training they could no longer use.

Men and women who needed something to believe in now they knew they'd fought and died for nothing.

An out-of-work actor who felt he'd never received the recognition he deserved.

He'd had Richard Varga nudged toward Human's First, and watched discontent become action.

Nothing more would be lost.

A shove against the small horse's flank, rocked it on its disproportionate, flat feet. Too old to be one of the enemy, to be part of the organic plastic hive mind that had manipulated the Elder Races with such ease and been responsible for how Humans had entered the Confederation.

The plastic aliens had precipitated the loss of Human identity and would be dealt with.

As the Elder Races would be dealt with.

Humans would reclaim their place.

Neither the plastic nor the Elder Races had any idea of what they'd unleashed.

He'd had to martyr Varga sooner than he'd wanted to, but the rise to leadership of a true believer from out of the ranks of the renamed Humans First had happened gratifyingly quickly.

He'd barely had to tweak the algorithms to include searches in obscure libraries and articles in fringe publications, to find a discredited scholar of ancient H'san.

To find a despairing Marine officer who'd been filling the data stream with increasingly desperate pleas to preserve her family name.

To send the scholar and the Marine to gather weapons unconstrained by false moralities and gelded technology.

"Gelded." He poked the small plastic horse on its genderless back end. "You'd know about that."

Unfortunately, retrieval of the H'san weapons hadn't gone according to plan. There hadn't been enough pieces of data retrieved to tell him exactly what had happened, and the names of citizens entering rehabilitation were shielded from the public. The visit of a high-ranking H'san to Berbar Station within the interval projected for the expedition's return, however, had resulted in an eighty percent probability of the Wardens being involved. Given that the Wardens' new Strike Teams had originated at Berbar Station, and given the personnel involved in the first of those Strike Teams—their serious faces above their brand-new uniforms having made at least a brief appearance on every news feed he followed—there was a ninety-two-percent probability that Gunnery Sergeant Torin Kerr had been involved.

As she'd been involved in identifying the sentient, polynumerous molecular polyhydroxide alcoholydes.

And ending the war.

And capturing Richard Varga.

What accounted for the missing eight percent, he had no idea.

New algorithms were in place should Joesph Dion, Major Sujuno di'Kail, or any of her team resurface, but he'd directed minimal resources toward it, cutting his losses. Willing to share only the technology they considered suitable, the H'san wouldn't permit a second attempt.

Hands curled into fists, he walked to the only piece of ceramic in the room. Older even than the plastic, pre-Confederation rather than pre-diaspora, it looked harmless, yet it could generate enough directed heat to melt through stone. He knew weapons, there was no one better, but he had no idea of how it worked and the possibility he still wouldn't understand—even if it were in pieces—kept him from demolishing it. The H'san had once had weapons powerful enough to dominate or destroy all of known space, to shift the borders into the configuration they desired. But they'd given them up. Buried them. Depended instead on the good nature and cooperation of the community they'd established.

Of the species they'd subjugated.

Of the behaviors they'd manipulated.

Cheese-loving sociopaths.

He raced against time now. Raced to be ready when the Elder Races realized what they'd unleashed and tried to restrain it.

The Wardens, the Strike Teams, were better than he'd anticipated. He wouldn't make that mistake again. Should his new operation be discovered, the Wardens' attention would be directed where he wanted it and Strike Team Lead Torin Kerr would dance to the tune he played.

He had no patience with those who failed to plan for every contingency, including failure.

Checking that the ceramic bowl remained cool—his major domo continued to be annoyed about the loss of the pre-diaspora oak table and a square meter of marble flooring—he walked away.

He had places to be.

Potential allies to recruit.

He'd recently received confirmation that the future was in plastics.

Vimtan Is In the Details huffed out a sigh deep enough to ruffle the deep brown fur on his chest, exposing the paler undercoat. "Yes," he said, "you are performing essentially the same job as you did in the military, answering force with force. But in upholding the law, force has become the last response, no longer the first."

"Not for us," Werst pointed out, lifting his cup of sah in salute. "We've always been the last response."

The Dornagain shook his head, and Torin felt a little sorry for him. Her people had embraced the new skills required by the job, but Justice had trouble believing they'd internalize new ideology as easily. Once every tenday they were required to attend guided discussions to reinforce the philosophical differences between being a Warden and being in the military. Vimtan Is In the Details was the latest discussion leader. Named after an ancient Dornagain god known for a sense of humor, there was a chance Vimtan might last long enough to beat the two-discussion average.

"Ah, but remember, Werst, you're the last response of the full unit, of the Wardens Who Guard the Peace and Integrity of the Confederation. The Strike Teams aren't deployed until it's been determined by the Justice Department that force against force is the *only* remaining answer."

Binti waved a hand to catch his attention—unnecessary, but it made the Dornagain jumpy and that amused her. "So, you walk softly; we carry a big stick?"

"We are the big stick," Cap pointed out.

"Well, some of us are."

Torin wasn't sure which di'Taykan had beat the rush to make the smug announcement since every one of them then expanded on the response. The noise level in the room rose, and Vimtan began to shift from side to side on his haunches, claws rattling against the floor.

Torin could have stopped it. She didn't.

One of the differences between being a Warden and being in the military was that they could take the piss during the weekly "discussion" if they wanted to.

"Sirin thinks he's insane."

Torin shrugged, watching Craig in the center of a group of pilots, sketching flight paths through a holographic display. "He flew for years alone, formulating his own Susumi equations."

"I suspect that's what she's basing it on," Captain Kaur said thoughtfully.

All pilots who flew boats with a Susumi engine theoretically knew how to formulate their own equations, but there were parsecs between theory and actually trusting a pilot's knowledge of both high-level mathematics and his ship. Military pilots relied on Susumi engineers.

"He survived a Susumi miscalculation," Captain Kaur continued.

"He's also lucky."

"He has the best spatial relationship to the immediate area around his ship I've ever seen."

"Years of flying alone through debris fields. Developed his peripheral vision." Torin could feel the weight of the captain's gaze on the side of her face, but she kept her eyes locked on Craig, enjoying the way he held the attention of his audience.

"Seriously?"

"That's what he told me. As far as I know, he's never lied to me."

"The qualifier kills the romance, Gunny."

"It'll take more than a qualifier, Cap."

They stood in silence for a moment as Porrtir, the Beta Team pilot, traced a line slightly to the right of Craig's. As the simulation ran, Porrtir's line turned red and then exploded, scattering debris. It was a good-looking bit of programming.

"Three ex-Navy, two ex-Marines . . ." Captain Kaur took a long swallow of coffee. "Not one of them thought they could learn anything from a CSO." Civilian salvage operators cleared debris fields and sold the scrap back to the military. Within the military, they were considered carrion feeders.

Torin lifted her cup in salute; across the mess, Craig grinned and winked. "Surprise."

"He's never seen battle."

"He wasn't in the war," Torin allowed, turning to face the captain. "He was captured by pirates and tortured. He jumped blind after a Primacy ship. He got us safely dirtside through H'san security satellites and back up again in an unfamiliar shuttle. He gave up safety and familiarity to help clean up the shitstorm we're in."

"Your point?" When Torin's brow rose, Captain Kaur shook her head. "Don't even try to tell me you don't have one, Gunny. You always have one."

"Our definition of valor needs to change."

Craig peered through the screen into the evidence locker, then over at Many Pieces Make a Whole—who was ignoring them—then up at the list of regulations printed out in a large, definitive script and hung on the wall, and then, finally, he turned his attention back to Torin. "Why would Marteau want both of us to rock up here? I didn't see his guns dirtside, I never left the VTA."

"The commander didn't say." Torin finished her coffee, dropped the cup into the recycler, and rolled her shoulders to work the stiffness out. The Corps, and the family farm before that, had made her a morning person, but morning was relative over the vast distances of the Confederation and the CEO of Marteau Industries was arriving at 0400 station time on his way to a meeting at the Ministry of Defense. Justice had been asked to move him in and out without delay.

Without delay; currently defined as *too damned early.*

"I checked." Craig leaned back against the wall, arms folded. "MI makes an empty bucket-load of shit I can fly."

The Confederation as a whole—Elder and Younger Races both—disapproved of profiting from war and Parliament, propelled by that disapproval, prevented any one company from controlling too much of the support manufacturing. Guns, but not ships. Ships, not tactical gear. Tactical gear, not field rations. Because they were the first of the Younger Races and had sent three generations of their young to fight and die before the Krai came in, a disproportionate number of military suppliers were Human. Torin personally disliked the concept of profit from death, realized that if government wouldn't build what the military needed then private industry had to, and believed far too many people who'd never worn the uniform could be considered sanctimonious assholes.

"Strike Team Commander Lanh Ng and Per Anthony Justin Marteau are in the link and descending." Many Pieces Make a Whole rubbed a claw over the graying hair of her muzzle, sounding less than thrilled. Torin doubted the Evidence Lockup Administrator would normally work this shift, so she'd also been hauled out of bed. Or whatever Dornagain slept on. In.

Craig pushed off the wall. "Tah, Pieces. You still on for the Threesday game?"

"Unless my employer, the Justice Department, in its bureaucratic *sutain*-licking wisdom, decides once again they can't trust my subordinates with a simple inspection, yes."

Torin adjusted *less than thrilled* to pissed. She leaned in toward Craig. "Pieces plays in your poker game?"

"Why not? Ressk discards slower than a geriatric *woma* and we let him play. Pieces is damned near speedy in comparison. Besides . . ." He grinned, corners of his eyes crinkling. ". . . she can't bluff for shit."

"I heard that, Strike Team Alpha Pilot, Warden Craig Ryder."

The distinctive sigh of the link arriving cut off Craig's response.

While Many Pieces Make a Whole pointedly locked her gaze on her slate, ignoring the new arrivals, Torin and Craig turned together to face the entrance hatch.

Marteau was short, almost as short as Ng, but he exuded an impression of tallness. Torin hadn't cared about height since her second day in basic when her platoon of raw recruits had been verbally eviscerated by a Krai DI, so she noted the effort and then ignored it. She put Marteau's age mid first century, but he was wealthy enough that he could've been considerably older. He had dark hair, pale gray eyes, and skin the pale brown of the Human default overlaid with the opalescent shimmer of those who seldom saw natural sunlight. She hoped his decorative facial hair hadn't given Craig any ideas.

As the commander introduced them, Marteau extended his hand.

It took Torin a moment, and had she not spent deployment after deployment listening to Sergeant Hollice expound on Terran customs and popular culture, she'd have had no idea of what to do. Before Marteau could withdraw his hand, she clasped it in hers.

He had interesting calluses. Familiar calluses. He knew how to use his own weapons: an insight as reassuring as it wasn't. On the one hand, the Corps' weapons were vetted at source. On the other, Anthony Marteau had never been Corps, or even Navy, and Torin disliked the idea of armed civilians—given her current occupation, they'd likely end up shooting at her.

"I'm pleased to meet you Warden Kerr, Warden Ryder." Marteau's smile was a toothy challenge.

She wondered if he smiled like that at the Krai, but she had no way to tell if he'd regrown fingers. When she released his hand, he extended it again toward Craig who shrugged and copied Torin's response—although the movement of their knuckles suggested Craig had given in to the temptation to posture. Craig was a big man who still did most of the maintenance on his ship. Marteau had to be stronger than he looked.

"Your weapons are beyond the screen." Ng walked to the desk, turned, and indicated a blue square illuminated on the deck. "We need a body scan to proceed, Per Marteau."

"I would rather you released the screen."

"And I would rather not compromise security. Your bio signature won't be entered into the system—one use only."

"I'll have to take your word for that, won't I?"

"If you want to inspect your stolen property before it's released, then yes, you will."

Marteau looked thoughtful as he aligned his feet parallel to the sides of the square. "I'm curious; inorganics?"

Ng frowned. "What about them?"

"I assume they don't need to be scanned?"

His frown deepened. "No."

"I see. Then I can only conclude that a thief could defeat this system with a suction cup arrow and a string."

"That what went down at your place, mate?"

Marteau twisted to face Craig, his expression patronizing enough it set Torin's teeth on edge, but not so much she could call him on it. "I have an MI security system installed on that warehouse, Warden Ryder. A much smarter system than the one in use here, and the thieves who defeated it are clearly smarter than the Wardens on site as those Wardens haven't yet worked out how the theft occurred."

"Maybe Warden Kerr and I should put eyes on it, then." Craig's voice had dropped into his chest, a low rumble.

"I'd appreciate that." The toothy challenge returned. "I believe the lack of familiarity with weapons I manufacture may be hindering the investigation by Wardens Vesernitic and Nubaneras. The Niln were not among those willing to fight for the Confederation."

"Wardens Vesernitic and Nubaneras are investigating the theft, not the weapons." Ng jerked his head toward the screen. "Kerr. Ryder."

Their bio scans already on record, Torin stepped into the lockup, Craig a meter behind. During the battle on V667, Torin's company had been swarmed by tiny insects every dusk and dawn, insects that buzzed as they collectively walked their thousands of feet across any bare skin. Back home, it had been easier for the Human Marines to depilatory their heads than try to get all the bugs out of their hair. Passing through the screen felt like a return to V667.

On the inside of the security screen, the lockup smelled of new weapons and Dornagain.

Ng waited for Marteau to pass through, then followed. Torin led the way to the dozen crates.

"I appreciate you recovering these, Warden Kerr, Warden Ryder." Fingertips splayed around the company logo on the closest crate,

Marteau leaned in. The locks released. “I have a personal override on everything that goes out of my factories,” he explained. “I see they were considerate thieves,” he added, stepping back, “and did as little damage as possible. I have to appreciate that.”

At Ng's signal, Torin opened the crate. Orders were to keep Marteau's involvement with the evidence to a minimum, reminding him that these weapons weren't currently his.

Marteau raised his slate. “I assume there's nothing missing, but I'm sure you understand that I need to know you've recovered everything that was taken.”

Commander Ng folded his arms. “Or you could assume I had the contents verified.”

“I could.”

He didn't. Torin and Craig shifted weapons and ammunition around in the shielded crates while Marteau scanned serial numbers. To his credit, he worked quickly and efficiently.

“I believe it's all here.” Eyes locked on his slate, he watched the numbers scroll by. “I'd have had a much larger problem if they'd broken up the crates and sold the KCs off one by one.”

Behind Marteau's back, Torin shut Craig down with a look, before he voiced his *no fukking shit* expression.

As Craig closed the last crate, Marteau drew in a deep, obvious breath. “I enjoy the family perfume,” he explained when he noticed he'd attracted attention. “It's the solid lubricant introduced into the metallic matrix during the microstructure phase. My father and my grandmother always smelled of it. New processes make it less prevalent on the manufacturing floor, but I've always found the smell reassuring.”

So did Torin. Only without the jargon.

“Speaking of new processes.” Ng indicated the pistol, resting in its own containment field. “Do you recognize this?”

Marteau moved closer, stopping only millimeters short of the security perimeter. “I believe this is a pistol, isn't it?”

“You know what it is?”

“I, like my father and grandmother before me, have studied the weapons of Human history.”

“Except that this weapon was wiped from Human history,” Torin pointed out.

“I'm a third-generation weapons manufacturer, Warden Kerr.”

Marteau smiled. "I have resources the general public does not. Personal resources."

"We'd like to see those resources." Ng had a reputation for getting straight to the point; just one of the reasons he'd been given command of the Strike Teams. Torin wondered if the early hour had added the *I don't have time for this shit* emphasis.

His attention back on the commander, Marteau's smile grew distinctly patronizing. "I said *personal*, Commander Ng."

"I heard you." Ng nodded at the containment field. "Was this pistol manufactured by Marteau Industries?"

Marteau's expression was a mix of annoyance and amusement. "I'm not personally familiar with every individual weapon Marteau Industries manufactures, Commander. MI supplies a great many weapons to both branches of the Confederation military and has for centuries. I am, however, aware that the pistol, speaking generally and not specifically, is illegal under the articles the Terran government signed in order to be welcomed into the Confederation. Which would make me very stupid should I admit this specific pistol came from MI." He slid his hands into his jacket pockets. They were deep, Torin noted. Deep enough to conceal an illegal weapon? "I am not stupid," he continued, "however, as you have no reason to take my word for it, I assume Wardens Vesernitic and Nubaneras will take time from their investigation of the theft to check accumulated manufacturing data and the molecular signature of my raw materials."

"You assume correctly." Ng didn't bother matching Marteau's conversational tone.

"I might be better able to ascertain its maker if I could take it . . ."

"It doesn't leave this facility."

"I see. Then I'd like to speak to the gunrunners you have in rehabilitation."

"You'd have to apply to both the Justice Department and the gunrunners for permission. Unfortunately, your assistant and the Parliamentary Secretary were adamant about the time constraints you're currently under."

The microshifts in Marteau's expression suggested an unfamiliarity with being denied. "I won't need more than a moment."

"Processing will take a minimum of eighty-one hours and there's no guarantee permissions will be given," Ng said blandly.

After a long moment, Marteau nodded, once. "I bow to my assistant and the Parliamentary Secretary and their knowledge of my schedule. I'm sure you'll eventually encourage your prisoners to talk."

As Torin understood it, they couldn't get Haar to shut up. Although he hadn't said anything useful. Or without profanity. "I expect you've already waved dinner and a dildo at the Krai and di'Taykan."

Torin had years of practice ignoring stupidity. Craig had never bothered. "If that's a joke, mate," he growled, "you have one hell of a poker face."

"I do, don't I." Sliding his slate into an inner pocket, Marteau headed back toward the security screen. "I'd rather not wait until the case is closed in order to have my property returned."

Ng fell into step beside him. "The Justice Department will do its best to expedite the process. Many Pieces Make a Whole has the forms you need to fill out."

"I'm sure Many Pieces Make a Whole does."

The disdain was almost obvious, masked barely enough for denial. To be fair to Marteau, not everyone enjoyed bureaucracy.

Tiny insects buzzed against Torin's skin. She remembered Glicksohn scratching insects out of the two-centimeter circle of stubble on his jaw where the depilatory wore off first and Chigma clutching his mug of sah with both feet, scowling at the floating scum of the drowned.

As Marteau stopped at the podium, Many Pieces Make a Whole drew herself up to her full height, leaving him with the choice of stepping back, staring at her paler stomach fur, or tipping his head up at a painful angle. He stepped back. "This will only take a moment, Per Anthony Justin Marteau, CEO of Marteau Industries. There are forty-seven forms in total."

That was about half of what Torin had anticipated. Someone wanted to go back to bed.

"Please place your slate on the desk, Per Anthony Justin Marteau, CEO of Marteau Industries. The contact will facilitate the upload."

"I think you should be warned, Commander, I have unbreakable encryption."

Ears flattened—reaction either to Marteau speaking to Ng instead of her or to the assumption she'd need to break an encryption to do her job—Many Pieces Make a Whole purred, "My mistake, it seems there are sixty-three forms." She paused pointedly. "Slate on the desk, Per Anthony Justin Marteau, CEO of Marteau Industries."

"I can't be the only one who finds the recitation of name and titles annoying," he declared, setting his slate on the podium. When no one answered, he added a quiet, "Tough room."

The sound of claws emphatically transferring documents, nearly drowned him out.

One hand maintaining contact with his slate, Marteau turned and swept a measuring gaze over first Torin and then Craig. "I'm sure you're both wondering why I asked to meet you." His tone suggested the curiosity should be killing them. A lot of things had tried to kill Torin; curiosity about Anthony Justin Marteau was not one of them. "I collect Human history, plastics for the most part." He smiled as though he were sharing something important with them. "I'd like you to come and touch my collection."

Torin was impressed Craig limited his reaction to a derisive snort.

"The Wardens' time is spoken for," Ng replied, reading Torin's reaction off her face and shooting her a silent warning. "Our Strike Teams need to move out at a moment's notice. Having them away from the station complicates that."

"I'm sure they have vacation time, unless the Justice Department doesn't follow government regulations."

"The Justice Department . . ."

"We touch your plastic, we have to touch everyone's." Craig interrupted. He frowned, folded his arms, and thrust out his chin. Torin hid a smile and suspected his reaction was as much a result of how that had sounded spoken aloud as of his dislike of Marteau.

Marteau swept a pale gray gaze over Craig from head to foot and back again. "I can pay significantly more than *everyone* for your time."

"Not interested, mate."

"I have a number of the oldest Human artifacts currently in private hands. I think *everyone* should be interested in discovering how far back the infestation goes."

The plastic aliens had started the war. Maintained the war. Had they also been responsible for Humans agreeing to fight the war?

Marteau smiled. Not a challenging smile. Or a salesman's smile. A smile that spoke of fellowship and actually reached his eyes. "I see you understand, Gunnery Sergeant Kerr."

"*Warden* Kerr," Ng snapped, "and Warden Ryder have interacted with artifacts held by the Human Historical Resource Council."

Some of what had survived the diaspora had been very strange. A few members of the Human Historical Resource Council had been stranger.

Still smiling at Torin, Marteau said quietly, "I'd like to know what you discovered, if you can tell me."

They'd touched one hundred and twenty thousand and thirty-two items made predominantly or entirely of bioplastic. Calling the aliens organic plastic was redundant, the Council had informed them. Given that all plastics were composed of a high percentage of carbon, they were all, by composition, organic. "I discovered that calling the aliens bioplastic is never going to catch on."

"I meant . . ."

"I know what you meant. It was all inert."

"I see."

"Besides, if the plastic had cocked up our past . . ." Craig folded his arms, sleeves stretched tight over his biceps. ". . . they'd have tucked into Human brains. You won't find them in you-gart containers."

"I . . ." Marteau frowned. "I hadn't considered that."

"Most don't," Craig agreed. "Most sleep at night. A little knowledge and all that shite. Now that's settled, question for you." He jerked his head back toward the crates. "That kit's new. Why make more weapons if the war's over?"

"I'm in the business of making weapons, Warden Ryder. I make fewer sales to the Confederation military, true, and there's too few Strike Teams as yet to make a dent in my inventory, so its fortunate that peace has opened up new markets. Many of the Primacy member species are less civilized than the Elder Races consider *us* to be." His reassumed mask of rich industrialist did nothing to hide his disdain.

As someone who'd fought and killed and seen her people killed and worked to keep her people from being killed because the Elder Races were too civilized to fight their own battles, Torin understood where Marteau's disdain came from.

Many Pieces Make a Whole rumbled deep in her chest.

On the other hand . . . "You're selling weapons to the enemy."

"I'd think that you of all people would know the war is over, Warden Kerr."

She knew. She also knew better than most it wasn't that easy. "Selling to the Primacy will make you no friends."

"I disagree. My employees—and many of them are veterans—are happy to have work. The Confederation's guaranteed income doesn't fill your day or give you a reason to get up in the morning. Also . . ." He waggled a finger in a *you know better* kind of way. ". . . I'm sure I saw an interview where you announced that we—the Confederation and the Primacy—need to learn to work together. I have weapons to sell; they're willing to buy." He sketched a circle in the air. "I'm making connections. For as long as I can do that legally . . ." His eyes narrowed. "I see you haven't heard. Parliament is considering a bill to make all weapons manufacture illegal. They want the Younger Races disarmed."

"Anyone with a functioning brain wants us disarmed," Craig muttered.

"I agree, Warden Ryder. And if you can tell me how to disarm every one of the Younger Races simultaneously, I'd love to hear your plan because I somehow doubt the socially misaligned persons your Strike Teams deal with would hand over their weapons on command. I don't believe you can stuff this particular genie back in the bottle . . . which is what I'll be telling the ministers when I meet with them."

Craig drew in a breath. Torin pressed her elbow against his ribs. Marteau wasn't finished, and she wanted to hear what he had to say.

"I think it would be naive to assume the weapons the government has allowed the Younger Races are the only weapons there are."

The H'san had a built a storage facility for their pre-Confederation weapons within a planet of their dead. Torin knew for a fact the weapons still worked.

Given his expression, the commander had known of Parliament's plans.

"Sir?"

He tipped his head slightly to one side. "You hate politics, Warden Kerr."

He wasn't wrong.

His back to the Dornagain, Marteau swept a surprisingly sincere gaze over the three Humans. "I know what I'm talking about when I say many weapons aren't particularly difficult to make. I don't have to tell the three of you that legal weapons are necessary to keep illegal weapons in check." His volume dropped; for emphasis, Torin assumed. Or he was unaware of how well the Dornagain could hear. "I also think it's naive to assume that the Younger Races will never need to defend their position in the Confederation now our services are no longer required."

Ng folded his arms, mirroring Craig. Torin remained standing easy. "The structure will hold, Per Marteau."

"I hope you're right, Commander."

"Your forms have been uploaded, Per Anthony Justin Marteau, CEO of Marteau Industries. We will require all eighty-one filled out, three copies made, and each copy signed, where indicated. All signatures must be original." The high-end slate looked tiny within the cage of claws. "Have a pleasant day."

He opened his mouth, closed it again, and silently took back his slate; smart enough, Torin noted, to keep his required number of forms from rising to triple digits.

"If you've seen what you needed to, Per Marteau . . ." *And you have,* added Ng's subtext. ". . . I'll escort you to your ship."

"I only need another moment, Commander." He turned his attention back to Torin. "I'd like to give you my contact information, in case you or Warden Ryder change your minds."

"About?"

"Touching my plastic." A dimple flashed outside the edge of the facial hair. "Not a euphemism."

"We can find you."

"I don't doubt you can. I am, however, surrounded by people who filter my contacts, and I'd like to keep you from a hundred conversations where no one believes you're who you say you are."

"I'd only have one." She let the implication hang.

"Per Marteau, if I could remind you of the word adamant."

"I'm not finished here, Commander."

Torin stared at his offered slate, stared past him at the commander's *I want him off station* expression, then snapped her own slate off her belt. "Fine." He couldn't get further than a temporary file outside the firewalls.

A certain gleam in his eye suggested he knew exactly what she was thinking. He had the arrogance to assume he—or his people—could break her encryption, but he'd never had the tutelage under Big Bill Ponner that made Alamber's countermeasures so vicious.

Information exchanged, he slid his slate into an inner pocket, crossed to where Ng waited by the hatch, and turned. "I'd like to assure you the offer to touch my plastic remains open, to both of you, and I remain hopeful we'll meet again. If we don't, be careful out there."

The soles of his very expensive shoes were soft enough that Ng's heels drowned out the sound of his footsteps as they walked away.

Craig's shoulder pressed warm against hers. Many Pieces Make a Whole tapped her claws against the desk. The faint scent of ozone from the screen drifted past, caught in the currents from the air recyclers. Finally, they heard the lift leave.

"Offer to touch his plastic remains open," Craig muttered. "And doesn't he have a personality like a hat full of assholes."

"He's . . ."

. . . now the Confederation no longer needs us.

. . . assume that the Younger Races will never need to defend their position in the Confederation.

The pause stretched a little too long.

"Torin?"

"He made some good points."

. . . a chance to find out how long the plastic aliens have been screwing with us. I'm sure that's information you'd like to have as well.

It was.

He transferred the eighty-one forms to his PA's slate as she fell into step beside him, traded for a quick synopsis of pertinent happenings while he'd been off ship. Of his three PAs—travel, business, and social—Orina Yukari had been with him the shortest amount of time and was already using the contacts she made through him to line up a better position. It was easy enough to train a new PA and much harder to deal with one who'd learned too much.

"We have a scheduled jump on the coreward buoy for 1800, and Minister Weta'na has sent regrets and will not be able to join you for a meal after you address the council."

No surprise. Why would a member of the Elder Races agree to eat at the children's table? He stepped through the hatch into the small outer office and crossed toward the inner. "And Per Honisch?"

"Her office has agreed to reschedule, but the administrator has personally approved your suggestion to, and I quote, put eyes on at

Justice. She adds that the Wardens, and particularly the Strike Teams, operate with a shocking lack of oversight."

"I'm thrilled she approves." He stepped through the inner hatch alone, considered continuing to his quarters, and crossed to his desk.

Gunnery Sergeant Kerr had an excellent poker face, but she wasn't entirely unsympathetic.

Had she—and Ryder—accepted his invitation, they could have discussed her misplaced need to serve and protect a corrupt government. But *not unsympathetic* wasn't enough, not on its own, and, in spite of the possibilities, she was too dangerous to leave in play. Still, it had been worth a shot. He could have ended up with Kerr and Ryder under his control, the original plan easily adapted to taking out a different Strike Team. The fewer Strike Teams the better, as far as he was concerned.

Ryder wasn't worth much, but he was Human.

"Sir." Yukari appeared in the open hatch. "The pilot reminds you that until you're belted in, we won't be permitted to leave."

The law was ridiculous. Years ago, a shipboard gravity generator had failed when disengaging from the station and a few people had been hurt and now he had to strap in until the all clear. Yet another shining example of the Elder Races' disregard for personal responsibility.

Waving Yukari back to her own seat, he dropped into his desk chair, secured the straps, and lifted a double plastic tube out of a tray on his desk. Fused together now, the smaller tube had once slid in and out of the larger. With dozens scattered through every collection of ancient Human items, he felt comfortable traveling with them despite their age.

He thumbed his desk on. "Yukari. I want a message sent to the Justice Department the moment the station has returned control of the ship."

"Sir?"

"I'm donating the stolen weapons to their armory. I want the Strike Teams to use the best equipment available rather than the best a government budget can afford. I've kept them alive for years, I'm proud to keep doing so."

"That's a large donation, Per Marteau."

"It's an investment, Yukari. Long-term."

Gunnery Sergeant Torin Kerr was not the only Human Warden and should she be taken out of the equation, some of the others might be more . . .

. . . amenable.

• ⸻◆⸻ •

"Are we expecting a supply shuttle, *Harveer*?"

Basking in the midafternoon sun, Arniz opened an eye and squinted up into the arc of brilliant violet. "Supply shuttle?" She'd been dreaming of her first *turl*. Of claw caressing scale. "What?"

"I'm sorry, *Harveer* Arniz, I didn't intend to interrupt your nap."

"Nap?" Both Arniz' eyes snapped open, inner eyelids closing to block the light. "I wasn't napping. I was considering this morning's finds and saw no reason to do it inside. Or with my eyes open. Nor do I need to explain myself to you." She waved away a small cloud of tiny blue insects—the most persistent of the insects on the plateau and off the menu on a Class 2 Designate—and held out an imperious hand. "Now, help me up. I'm not used to this gravity."

"But it's the same as . . ."

"Not at my age."

"Of course, *Harveer*." Dzar got a good grip, tail stretched out as a counterweight, and pulled.

This part of the planet had nothing so comfortable as a sand pile and, while being hauled to her feet, she wasted energy envying the Mictok team who were mapping ruins within the desert that covered the interior of the smallest continent. *They* had sand. Nine point two million square kilometers of it. Arniz scowled at the leading edge of jungle and gave thanks that their current license kept them out on the plateau. The humidity under the trees would be appalling. She hated feeling damp.

Eventually, she found herself on her feet staring up at a bright speck in the sky. The child had good eyes, she'd give her that, to have picked out the speck even higher up and not assumed it was a bird. As she watched, the shuttle dropped straight and fast and that meant a military pilot—they all flew like they still expected to be shot at on the way down with no regard whatsoever for the comfort of their passengers or the security of delicate equipment. Arniz had filed a complaint with the university about the young Krai who'd dropped them off at the beginning of the dig season, not that she expected anything to come of it.

"*Harveer*?"

"No, Dzar, we're not expecting a supply shuttle, and the university's budget most assuredly doesn't extend to additional visits."

"Do you think something's gone wrong?"

"Most of the time." The afternoon was far too warm for boots, so she stuffed her slate into a pocket on her overalls and stretched the stiffness out of her toes. "Have you got anything running you can't leave?"

"Me, *Harveer*?"

"No, I'm talking to the geologist, sanitarian, and silvoculturalists that Dr. Linteriminz said we couldn't afford this trip." The cheap *invic*. "Yes, of course you."

"This morning's samples are still running. I don't like to leave them, but . . ."

"But you can. Good." She'd gone half a dozen steps before she realized Dzar hadn't moved. "Well, come on."

"Where?"

"If strange shuttles are landing on a Class 2 Designate, I want to see who comes out of them. Scientific curiosity."

Dzar grinned. "If you say so, *Harveer*."

"Sass." Arniz tucked her hand in the crook of the younger Niln's elbow and touched her tongue to Dzar's cheek, making it clear she'd appreciated the comment.

By the time they reached the anchor, 493 meters from their site and an exact 100 meters where the Katrien geophysicists, Dr. Tyven a Tur durGanthan and Dr. Lows a Tar canHythin, had mapped the outer edge of the ruins on the plateau, the shuttle's descent had become a background rumble. Like thunder in the distance, Arniz noted as Dzar ran inside and brought out a stool. Like a storm coming in.

"They're moving fast." Gaze locked on the shuttle, *Harveer* Salitwisi's tail whipped back and forth.

"Water is wet."

He turned, inner eyelids flicking once across the black. "What are you talking about, Ganes?"

Standing apart from the group of gathered scientists, and half a meter taller than the Niln and the Katrien, Dr. Harris Ganes shrugged. "I thought we were stating the obvious."

Arniz snickered as she settled onto the stool. Credit where credit was due, Dr. Ganes accepted Salitwisi's rudeness better than she would have. Had Salitwisi snapped out her name without the honorific after only having known her for four tendays, she'd have snapped back. Had Dr. Ganes addressed Salitwisi in kind, dropping the *harveer*, they'd never hear the end of it. "Dr. Ganes."

He glanced over his shoulder at her.

"Do you recognize the ship?"

"It's a military design," he said, turning his attention back to the sky. "But that doesn't mean there's military in it."

"As technical support, oh, apologies, site engineer . . ." Salitwisi sketched the most sarcastic set of air quotes around the word engineer Arniz had ever seen, even given she'd spent eight long years teaching firsters. ". . . you didn't think you should contact them?"

The heads of Salitwisi's three ancillaries, two Niln and a Katrien, swiveled in unison from their mentor to the Human.

"I did. They didn't reply."

And back again.

"Are you certain you had the right channel?"

And to Ganes.

"I'm certain."

And back.

"And they didn't reply?"

"That are being illegal," Lows pointed out, interrupting the flow.

"It could be they're trying to surprise us." Arniz spread her hands when half the group turned toward her and smiled as sweetly as she could manage, her words dripping sarcasm. "A pleasant surprise."

Ganes grunted. There was really no other word for it, emphasizing the unfortunate fact that Dzar wasn't the only person on the dig without a sense of humor. Arniz watched how he stood, head up, gaze locked on the shuttle, the scholar and scientist disappearing behind the kind of training that allowed one sentient being to kill another. Or one hundred to kill one hundred. Or a thousand to kill a thousand. The Younger Races had been badly served by the Elder. While it had been necessary for the survival of the Confederation, that didn't make it any less sad.

Bow to stern, the shuttle was uninterrupted gray, scarred and ugly with years of hard use. It groaned as it touched down on the cleared and clearly designated landing site—which showed a decent spatial sense at the very least.

Ganes stepped back. Once. Twice. "We should take cover in the anchor."

"Take cover? Don't be ridiculous, Ganes."

As Team Lead, Salitwisi spoke for them all. Arniz was perfectly capable of speaking for herself, but she remained silent, tucking her arm into Dzar's as she stood.

Concrete slabs laid by the first crews in to protect the planet's surface pinged and popped as they cooled. A small flock of birds rose in the updraft created by the sudden, localized dry heat. The group of scientists ambled closer to the landing site, most of them chatting, enjoying the break, the difference in their day. *Harveer* Tilzonicazic had her head bent over her slate, a pair of ancillaries steering her around obstacles. Arniz suspected Tilzon had done some serious sucking up to her department head in order to bring two—although, to be fair, given that they were doing little more than mapping, the xenobotanist had been busy, plants being right out there in the open. They'd barely uncovered anything that could be considered architecture, however, making the more pertinent question why Salitwisi had been permitted to bring three. Most of the architecture on the plateau had been worn down to ground level, some of it only definable by soil analysis. If anyone should have gotten an extra ancillary, it should have been her.

"*Harveer*, you're pinching."

"Nonsense." But she loosened her grip on Dzar's arm.

Ganes had his slate in his hand, his thumb on the screen. Probably leaving a big, sweaty thumbprint as furless mammals did.

The first person out of the shuttle was Human. Much taller than Ganes, with broad shoulders and a narrow waist, the angle between the two significantly greater than the angle of Ganes' torso. The skin not covered by black fabric was a pale shade of nondescript beige, unpleasantly pasty in comparison to Ganes' lovely deep brown. A long time ago, Arniz had a TA from the eastern province who had scales that same warm shade of brown along his spine. Surrounded by the area's greens and grays and pale creams, he'd dressed to emphasize them. She'd been a lot younger then and a little jealous.

The new Human had close-cropped hair and large boots. Either

Ganes had very small feet or they were disproportionately large boots. Large enough to be kept outside the city limits.

She leaned into Dzar's space, her sense of smell having lost the ability to deal with the nuance of other species a decade or more ago. "Male?"

Dzar tasted the air. "I think so."

The Human moved quickly down to the insulated slabs, then more slowly toward them. He was followed by another three, one darker beige, one lighter, one a light brown—as many Humans as Arniz had ever seen in one place—then a pink-topped di'Taykan, two Krai, another two di'Taykan under deep blue and brilliant blue, respectively. She'd often wondered about how the colors fit with the Taykan ecosystem, but hadn't time to do research purely for curiosity's sake. There was never enough time. Less now in the later part of her life when time had begun to move faster.

"Those aren't scientists," Hyrinzatil muttered.

"Don't be a bigot," Salitwisi snapped at his senior ancillary. "Many of the Younger Races are very clever. No offense, Ganes."

"And yet, I'm offended. But he's right. They're not scientists."

Tilzonicazic looked up from her slate. "I won't speak with reporters."

"Reporters?" Arniz snorted. "What makes you think they're reporters?"

"We're on a Class 2 Designate. We're the first to study this hemisphere. That's news."

"News access is restricted," Arniz began.

Salitwisi cut her off. "And isn't maintaining that restriction Ganes' job? You handle communications, don't you?"

"What am I supposed to restrict them with, strong language?"

"Tell them to leave. It's my understanding most reporters are sentient beings."

Ganes shook his head, denying Salitwisi as much as the comment. "Those aren't reporters."

"Oh, please, you can't tell that from . . . what the *skisik* is that?"

That was a large quadruped covered in reddish brown fur with darker extremities that looked as if a mad scientist—insane, not angry, Arniz qualified silently—had removed the head of a large feline and stuffed the upper half of a biped into its place. It wore a vest on its upper body, muscular arms ending in broad hands with three thick fingers and the thumb required for fine manipulations. As it descended the ramp, she could hear claws against the metal. A thick mane no more than six centimeters long surrounded a face with a short, broad muzzle over a heavy jaw. From where she stood, Arniz couldn't see if they had tails, so she shuffled closer, dragging Dzar.

Another quadruped followed. And one more after that. The first, a deep, plush black and the second a mix of color—white and brown and gray and black and surprisingly attractive if a tad busy.

Dzar's arm stiffened under Arniz' hand. "Those three are definitely male."

"I hadn't asked." She hadn't needed to. The second quadruped to reach the ground had reared.

Her tongue tasted the air. "I don't like it, *Harveer*."

"Who would." They looked like the sort of species who enjoyed fighting. Whatever sort of species they were.

"Polint."

She frowned at Ganes—who'd apparently started reading minds.

"They're from the Primacy. One of their military species."

"I are thinking all the Primacy species are military," Lows murmured.

"I don't like any of this. I want to go back to the lab. The last sample should have finished running by now."

Arniz tightened her grip as Dzar tried to pull away. The child would never survive academia if she didn't overcome the urge to run when exposed to things outside her comfort zone. "I need you with me. Try to keep an open mind."

Last to emerge from the shuttle was a pale, hairless biped, dressed in many layers of red, its eyes as black as Arniz' own, but much larger. She noted the flicker of an inner lid with approval.

"Is that a Trun?" asked an ancillary.

"You are being an idiot. Trun are having fur," muttered a second.

"And tails," added a third.

"It's not a Trun." Something in Ganes' voice drew Arniz' attention off the newcomers. His eyes were locked on the shuttle, his thumb moving rapidly over his slate.

"Hello." It was the first Human. He was showing teeth, and Arniz felt her lips curl in response. She forced them down. She was far too old for posturing. "Sorry about interrupting science and discovery and all that shit, but we hear you've found a weapon that'll destroy the plastic aliens. We want it."

The silence lasted approximately ten seconds, then filled with everyone talking at once until Salitwisi's voice rose over the din. "What are you on about?"

"None of that." He waved a finger in a patronizing gesture. Arniz had waved enough similar fingers to recognize it. "Hand it over and we'll be off."

"This are being ridiculous." Tyven pushed her way forward. "We are having found no such thing. This planet are a Class 2 Designate, and you are not being allowed here. So you are going away. Now!"

He reached out, wrapped a hand around her head, fingers sinking deep into her fur, then he lifted her as though she weighed nothing at all and threw her aside. "Learn to speak Federate, you annoying furball."

The Katrien syntax could be annoying, Arniz admitted, but that was an out-of-proportion reaction. Apparently unhurt, Tyven scrambled to her feet, her bonded and their ancillaries surrounding her, the Katrien grooming any bits of fur they could reach, the Niln touching their tongues to her muzzle and hands.

"Eyes on me!" the Human barked. "We aren't leaving without that weapon!" He shifted his upper body and a long metal thing with a pipe on one end and handle on the lower edge of the other swung around on a strap. He tucked his finger through a half circle on the lower edge.

The others from the shuttle did the same.

"We can do this the easy way, or the hard way." He sounded entirely reasonable. "Your message to the university was intercepted. We know you have the weapon. You can't deny you have the weapon. Give us the weapon. That is the easy way."

"And the hard way?" Tilzonicazic muttered disdainfully. "Humans, so dramatic."

"*Harveer* . . ."

"Not now." Arniz tongued Dzar's cheek.

Ganes stepped forward, putting himself between the newcomers and his fellow scientists, standing very straight. Arniz relaxed. Ganes would deal with this and they could all get back to work. "I don't know what message you intercepted," he said calmly, "but you've misinterpreted it.

We've barely done more than begun to map the plateau. We've found no weapon of any kind."

The larger Human stared at Ganes for a long moment. "So, the hard way, then."

The pipe end rose.

Ganes straightened further. "That's enough. Now . . ."

The sound was more of a crack than a bang.

Ganes jerked.

Beside her, Dzar made a soft sound and crumpled to her knees. Gravity dragged her arm from Arniz's grip.

Blood dribbled from a red hole in the scales of her forehead and rolled down toward her left eye. She balanced on her knees and tail for a moment, then fell forward, exposing the bloody mess of bone and brain that had replaced the smooth curve of her skull. Arniz hissed, frozen in place, uncertain of what to touch, or if she should touch. She was old, she'd seen death, but never death like this. Never so deliberate. Never so violent. Never so . . . messy.

Arniz hissed again and spun around toward the Human, barely keeping her balance. "You murdered her!"

"I made a point. Give me the weapon." He enunciated each word clearly. As though they didn't understand Federate.

"There is no weapon! You murdered her for plastic remains found in a latrine!"

He rolled his eyes. "Give me the weapon responsible for the plastic remains in the latrine."

"Pay attention!" Arniz could feel warm hands on her arms, pulling her back. "If it was the sort of plastic that *can* be killed, and we have no way to prove that, we don't know what killed it!"

"Pay attention," he mocked. "It's not an it. It's a them. It's always a them—shape-shifting molecular plastic *hive mind.*" He seemed unaffected by having killed and pleased by Ganes on the ground at his feet gasping for breath, blood running from his nose. "Who found the plastic?"

Dzar's blood had begun to soak into the ground. "She did, you idiot!"

"One has been sacrificed for the safety of the many." A deep voice boomed out of the shuttle.

"Yeah." The Human smiled. "What he said."

"Reports from the Ministry for the Preservation of Pre-

Confederation Civilizations has shown this was not a plastic-using civilization. The question, therefore, becomes: how was the plastic destroyed and how did the remains get into the latrine?"

Arniz heard her own words coming back to her in the same deep booming voice, and knew what message they'd intercepted. She sagged back into Tilzon's touch, suddenly exhausted. "Do you not know the meaning of the word *question* when used in this context? It means, we don't know what destroyed the plastic."

"Obstruction is not in your best interest, Harveer. *Sergeant Martin."*

The gun rose again. She knew what the pipe was now; knew it by its effect, even though she'd never seen one before.

Ganes stood, staggered, and put his body in front of the gun. "We don't know *yet*," he panted, still recovering from whatever blow had put him on the ground when Arniz had her attention locked on Dzar's body. "But we'll find it. The more of us you have searching for it, the better the odds."

"You make a valid point." The voice sounded annoyed. *"Sergeant, move them into the anchor so we can have an actual conversation."*

The gun remained pointed at Ganes.

"Sergeant!"

His lip curled and Arniz thought of her department head. Sergeant Martin neither liked nor respected the voice in the ship—but would follow its orders, she acknowledged, as the gun finally lowered, and the Polint, responding to his hand signals, moved out to the edges of the group.

They were being herded, Arniz realized. People herding people. She didn't like it. She didn't like any of this.

"We need to take Dzar to the infirmary," Tilzon said quietly.

Arniz didn't bother moderating her voice. "She's a little past needing the infirmary."

"You need a morgue." It was the red-clad biped. They beckoned two ancillaries over. "Carry her inside."

One of them glanced down at the body. His tongue flicked out, then he spun around and vomited. Arniz thought his name was Nerpenialzic, one of Tyven's but she hadn't bothered to be sure. She met the biped's gaze. This close, she could taste unidentified female in the air. "There's no reason to take the body out of the sun. We'll make her a platform later."

"Will you?" After a long moment, inner eyelids flicked across the black. "We'll see. Now, inside."

Arniz clutched Tilzon's arm as they turned and followed the bulk of the team through the hatch.

Just in front of them, Salitwisi plucked at Ganes' sleeve. "What was that all about? What are we supposed to find? There's nothing *to* find!"

"Either they'll keep killing us until we give them the weapon, or we help them find it."

"Help them find *what*?" When he paused, Arniz could almost hear him thinking about Dzar. "Yes, well, I suppose it won't hurt to look."

•—◆—•

"You still have the duck." Torin ran a fingertip down a blue-green feather.

Binti looked out of her bedroom and frowned at the stuffed waterfowl tucked on a corner shelf in her small living room. She'd picked it up during a drunken fact-finding mission on Abalae, but hadn't been able to remember exactly where. "Why wouldn't I keep it?"

"Because it used to be alive." Torin tapped a nail against the beak. "It's a preserved corpse."

"You have a Silsviss skull on your wall, Gunny."

"Just the skull, not an entire Silsviss. But . . ." She spread her hands. ". . . it's a valid point. Alamber hates it. Hates the duck, that is. He seems fine with the skull."

"Yeah, well, that's partly why I kept it." Binti dropped into a chair and bent over the laces on her trainers. "He doesn't have a lot of buttons I can push. And, anyway, he's not around much when we're on station."

"He's sleeping communally."

"Duh."

Back at the *vertrasir,* Gamar had complained about Alamber liking Torin best. "You okay with that?"

Balanced on one leg, opening up her hip flexors, Binti shot Torin an irritated glare. "We're not a couple. We're teammates and we're friends and we play sometimes. In case you've forgotten, Alamber's di'Taykan; he's after quantity." She changed legs. "And me, I have no interest in being half of a couple or part of a group. I'm whole and happy on my own." She bent, touched her palms to the floor, straightened, and added, "Also, you're not my mother. My mother could run circles around you."

"Too bad speed skipped a generation."

"Oh, you are on, Gunny."

Torin reached the hatch at the end of the passageway first, but Binti passed her on the concourse. They dove into a vertical together, ignoring the handholds rising and falling in the center of the open column and free-falling through the lesser gravity. Also ignoring the profane commentary from two members of Beta Team hanging off straps a level higher.

Briefing Strike Team Alpha in Conference Room H'ata in twenty.

She tongued an acknowledgment and twisted to face Binti. "You get that?"

"I got that."

Turning in the air, Torin grabbed a rising handhold as it passed, and began to climb, Binti right behind her.

Their palms slapped simultaneously against the exit bar. Back out on the concourse, they moved smoothly to either side of a portly Niln who'd frozen in place when they suddenly appeared.

"Knock someone over . . ." Torin hurtled a document cart that beeped at her proximity. ". . . and you lose."

"Unless it's a member of another Strike Team."

"That goes without saying."

Torin reached the hatch back to the living quarters three strides in front. Binti had gotten tangled in a cluster of Katrien clerks emerging from a break room.

"How did you miss them?" she demanded as they ran shoulder to shoulder, the sounds of a cat fight—Katrien was not an attractive language to the Human ear—cut off by the hatch.

"I anticipated their exit."

"How?"

"Gunnery sergeant."

"So you're saying you're just that good?"

"That's what I'm saying."

The clerks from the law offices always finished their first break at 0940 and, as Strike Team Lead, Torin spent more time than Binti on the bureaucratic floors. Enough time to track movement and add it to her mental maps. Apparent omnipotence had become habit.

Nineteen minutes later, they walked into the conference room, the last of the team to arrive.

As Binti circled the table, Torin dropped into the seat beside Ng.

"Running in the halls again, Warden Kerr?"

"Still no treadmills, sir."

A vein throbbed in Ng's temple. "I'll mention it to acquisitions. Again."

"You know, Boss, if you wanted a workout that got your heart racing . . ."

"Warden di'Crikeys."

Alamber sat back and mimed sealing his lips.

Craig leaned closer, a warm, solid weight against Torin's side. Across the table, Ressk had his slate out, Werst had his eyes closed, and Binti murmured in Alamber's twitching ear too quietly to be overheard. Torin wrapped her right hand around her left wrist just above her uniform cuff, and felt settled in a way she hadn't since the day she'd had a conversation with Major Svennson's regrown arm.

Which had been unsettling on a number of levels.

Conference Room H'ata was unattractive and utilitarian and had been built without plastic.

The new normal.

"All right. Three days ago . . ." Ng threw a system map into the air over the center of the table. ". . . an archaeological site on 33X73, a Class 2 Designate, was taken over by a combined Confederation and Primacy force."

"Confederation *and* Primacy? Are you sure?"

That explained the commander's throbbing temple.

"No, Warden Ressk, I decided to brief you on the basis of unsubstantiated rumor."

"We were all thinking it, sir." Torin met and held Ng's gaze. He'd layered calm in a thin veneer over the implications of what he was about to share, and while she had no intention of cracking that calm, neither did she have any intention of allowing his response to stand unchallenged. He could shore himself up with sarcasm if he had to, but he was not to do it at the expense of an honest response from one of her people.

The breath he exhaled was too controlled to be called a sigh. "Yes, we're sure. Most of the scientists are Niln or Katrien, from Alcanton University, but Dr. Harris Ganes is ex-Navy and he got a message out to the Susumi satellite the university has in orbit. Sent it to his old ship, the *CS Nagtucken*. Her captain sent it on to us."

Smart, Torin acknowledged. The Ministry would have buried it in bureaucracy.

"Confederation and Primacy," Ressk said again. "So someone's actually been paying attention to that shit about finding common ground with our enemies."

"Besides both of us being screwed over by the plastic aliens," Werst added.

"Besides that."

"What do mercenaries want with an archaeological site on a Class 2 Designate?" Craig asked, squinting at the map. "That's a pre-Confederation civilization with no possible salvage."

"There's some indication . . ." A message file joined the system map. ". . . that one of the soil scientists found a weapon able to destroy the plastic."

"I can destroy the plastic," Werst muttered as Torin read the message.

"Not even the Krai can eat an entire species," Alamber scoffed.

"And, as the gunny pointed out," Binti added, "no one checked your shit."

"You think they crawled out of my shit?"

She spread her hands. "We didn't check."

"That message doesn't say they found a weapon." Torin sat back and read the message again. "It says they found destroyed plastic where no plastic should be."

"We can assume there was some extreme extrapolation around the word *destroyed* by whoever hired the mercenaries."

"By whoever sent them in looking for a weapon." Torin swept a glance around the table and acknowledged that her team had all reached the same conclusion. "The same *whoever* who sent Major Sujuno to the H'san cemetery world."

"That's a Susumi jump, Warden Kerr." Ng's expression remained entirely neutral. "Space is large, and great minds are not the only ones who think alike."

Alamber's hair flipped dismissively. "Like that time there were two vids about a Human falling in love with a Verfreenitat from the Methane Alliance."

"Those sucked," Binti muttered. "I mean, who wants to use an HE suit as protection?"

Ressk nodded, opened his mouth, and snapped it shut when Torin said, "Critique on your own time, people."

"We have no evidence to link the two incidents," Ng continued as though no one had spoken. "It isn't enough that both involve a search for ancient weapons—the current search is specifically for a weapon to destroy the polynumerous molecular polyhydroxide alcoholydes . . ."

"The plastic," Werst growled.

". . . and the former was a search for weapons to . . ."

"Possibly do the same thing. We don't know how the H'san's weapons were to be used," Torin added when attention turned to her.

"Colonel Hurr assumed revolution," Ressk pointed out.

"Colonel Hurr is Intelligence; he defaulted to the political position. It could be simpler."

Werst nodded. "Plastic aliens fukked us over, let's find a weapon that'll fry their asses."

"Except we don't know where their asses are." Alamber frowned. "Or if they have asses."

"Justice should investigate Anthony Marteau."

Ng stared across the table at Craig. "On what grounds, Warden Ryder? Your dislike isn't enough to void his privacy."

Craig pressed his palms flat against the glass, muscles rolling in his forearms. "He can afford mercenaries."

"If he needs more weapons, he can build them."

"His designs are based on Terran weapons. Centuries old. He's got to be jonesing for new ideas."

"The Justice Department doesn't deal in supposition, Warden Ryder."

"Money. Weapons. Plastic. He also collects plastic."

"Since the revelation, a surprising number of people do."

Werst responded before Craig could. "Are you shitting us?"

"No."

"People are strange."

"Not arguing." Ng glanced down at his slate. "We're developing a permit system."

"Marteau recognized the pistol." At the time, Torin had thought his reasons valid. She still thought his reasons were valid, but Craig clearly didn't, and he had a distance from Marteau's business Torin would never be able to manage. She'd spent too long with a KC-7 in her hands.

"As Warden Ryder reminded us, Per Marteau makes weapons."

"That he intends to sell to the Primacy."

"Say what now?" Alamber demanded, the question bracketed by two sets of snapping teeth.

Ng glared both Krai into a final snap and lowered lips before saying, "We are no longer at war with the Primacy. Per Marteau is breaking no laws. What's more, he was under no obligation to inform us of his business ventures as full disclosure covers only his military contracts. Give me a reason and I'll try to open an investigation, but your job is to deal with situations as they arise . . . and we currently have a situation. This is the one visual Ganes got out."

"Robert Martin." Werst pointed at the closest Human. "He was on Ventris same time I was. Three intake groups ahead."

"And you remember him?" Torin asked as Ng flicked Martin's military ID up off the conference table. There were two possible reasons Martin had been so memorable and, given where he'd ended up, and that neither Werst nor Martin were di'Taykan—Torin suspected the first reason didn't apply.

"Warnings went around about him. He was a bully." Werst drew his lips back off his teeth. "Broke a di'Taykan's arm on Crucible. It was an *accident.* Fukker knew how to work the system. Good shot, followed orders, didn't suck up enough to be noticed. Wanted into the heavies, but psych washed him out. That was the last I heard of him."

Ressk snorted. "Surprised he wasn't a lifer."

"Sounds like the type who makes general," Binti added.

"The other three Humans are Emile Trembley, Brenda Zhang . . ." Ng flicked two more pieces of military ID into the air. ". . . ex-Marines, enlisted for only one contract, nothing special about them except that along with Martin, they're the only Marines in the group." The last Human ID joined the other three. "And there's an eighty-three percent probability the fourth Human is Jana Malinowski, ex-Navy gunner."

Only half of Malinowski had been photographed.

"The Krai," Ng continued, ignoring the interruption, "are ex-Petty Officer Sareer and ex-Lieutenant Beyvek. The di'Taykan are Pyrus di'Himur pink, Mirish di'Yaunah dark blue, Gayun di'Dyon bright blue. Mirish is Gayun's *thytrin* and four of the six ex-Navy personnel visible in Ganes' image served on the *Paylent*. We've requested information on the *Paylent*," he added dryly.

Torin expanded the original. Ganes hadn't captured much of the shuttle. "That looks like an old Navy VTA; there could be a dozen mercs still inside."

"Two dozen if they're friendly," Binti said.

"Yeah, yeah, comrades in arms sharing war stories. Who cares." Alamber leaned in under the nine pieces of military ID and pointed. "Are those Polint? Like you met in the prison?"

"They are." Torin answered. "The biped behind them, dressed in red, is a Druin. Both species are Primacy military." She took a deep breath and turned to Ng. "Confederation and Primacy, working together."

He inclined his head. "As I said."

Alamber had his head tipped sideways, enough light receptors open to make his pale blue eyes look dark. "Not so much working together as being shitheads together. They're all armed. I think the Polint even have swords. Also, not a euphemism." He tipped his head a little further. "Although it could be."

Ng cleared his throat and Torin glared Alamber vertical. "The Navy has informed the local Wardens that the mercenaries' ship is still in orbit around 33X73. The VTA is still on the surface. As Warden Kerr noted, the intercepted message notes only that the scientists found destroyed plastic, not a weapon, so we can assume, given the passage of time, there was no weapon to hand over and the mercenaries are holding the scientists hostage until said weapon is found."

"So this time the Navy has a reason to keep its thumb up its ass," Werst muttered.

"Apparently. And as long as no one does anything stupid, that gives *us* some time. Data's been sent to your slates."

The data included bios of the known mercenaries as well as of the scientists, the geography around the site, and extensive site plans from both Alcanton University and the Ministry for the Preservation of Pre-Confederation Civilizations.

Binti frowned at her screen. "This says there's four other science teams on planet."

"It's a big planet." Ng dropped all images back into the tabletop. "The other four aren't involved. You have fifty-four hours to . . ."

"It's a four-day Susumi jump," Craig protested. Odds were good he'd hooked into *Promise*'s navcom the moment he had a destination and worked out a set of quick and dirty equations. "We can load up and plan in the jump. Time matters in a hostage situation."

"As pleased as I am that you've internalized at least some of your training, Warden Ryder, fifty-four hours."

"That's . . ."

Torin cut Craig off, gaze locked on Ng's face. "That's how long it'll take representatives from the Primacy to arrive. We can't go after their people without them."

"As they can't go after ours without us," he agreed.

"And we've all worked with the Primacy before. Which is why we're here, out of rotation."

The silence that followed didn't last a full three count before it shattered.

"We broke out of prison with the Primacy, Gunny, we didn't work with them."

"And not the whole fukking Primacy either; six . . ."

"Eight."

"Still . . ."

"Yeah, I'm blocking that shit."

"The whole fukking mess was all but over when I rocked up, and *I'm* blocking that shite."

"Before my time, Boss, but this should be Ch'ore's run."

Torin held Ng's gaze while the noise rose and fell.

When the tangle of voices finally quieted, he nodded and said, "Which is why you're here: specialized skills."

Lifted from the battlefield, Confederation and Primacy troops had been imprisoned in a maze of underground tunnels. They'd been kept separate, the entire system automated, and each had assumed the other had put them there. It wasn't until Torin led a group of Marines in an attempted escape and ran into a Primacy group attempting the same thing, that they learned differently. There could have been blood in the tunnels, but they'd chosen to crawl out from under the weight of the war they'd been fighting and see each other as allies and not enemies. They'd still nearly died, but that hadn't been on the

Primacy. The young Primacy leader had proven the qualities that created a good officer weren't confined to the Confederation Marine Corps.

Binti, Werst, and Ressk had been there with her. Binti and Ressk the only other survivors of Sh'quo Company, and Werst the last of his recon unit. One other Human. One other Krai. And three di'Taykan. Although only two of the di'Taykan had made it out. Torin closed her hand into a fist to hide the way her fingers trembled with the effort to not touch the pockets of the dead on the combat vest she wasn't wearing. Seven lives to balance the hundreds she couldn't save.

"Survival isn't exactly a specialized skill," she said over the sound of their names.

"And disingenuous doesn't suit you, Kerr."

Fair point. "Do we know who they're sending?"

"We do not."

"Number?"

"They've been told there's six members on the Strike Team they'll be joining."

"So, seven," Ressk said thoughtfully.

"They'll be in Confederation space and vastly outnumbered," Torin explained when Ng frowned. "Outnumbering us, even minimally, is something they can control. Species?"

"They're refusing to say."

"That's nuts," Alamber snorted. "Why wouldn't they share?"

"They don't want us too prepared," Torin answered. "They have to work with us on this, but they don't trust us. Anyone they send will be expendable."

Ng stared at her, ignoring the reaction from the others around the table. When it died down, he said, "Expendable?"

"If anything goes wrong, they'll be blamed."

"By who?"

She held his gaze. "By those who weren't there when it happened."

"And one of the *serley chrika* will be spying for the Primacy."

"No." Torin turned her attention to Werst. "I expect they'll *all* have been instructed to find out everything they can."

"We would've been," Binti agreed.

Ng sighed. "Their presence will be as much about public opinion as it will be about saving the hostages. We need to assume the Primacy is acting in good faith and that however expendable Warden Kerr believes their operatives to be, that they'll at least know their asses from their elbows. Should they have either. A translation program for Prime, the common language of the Primacy, will be added to your implants. We've provided them with the Federate equivalent." He laid both hands flat on the table. "Given the species of the mercenaries, assume Polint and Druin. I want as much prep as possible completed before they arrive. This is a hostage situation; you can't go in guns blazing."

Under the table, Binti's foot pressed against Torin's. "Unfortunately, sir, that's what we do."

Ng stood and swept a weighted gaze around the table. "Not this time."

"Guess the war really is over," Werst said as the door closed behind the commander. "Never thought we'd be teaming up with the Primacy on purpose."

"At least it's mercenaries," Craig muttered as he tipped his chair back on two legs, balanced his slate on his knee, and began to read, "and not Humans First again."

FOUR

"LOOK, ALL I'M SAYING is that Marteau mentioned selling weapons to the Primacy . . ." Craig leaned in and lowered his voice. ". . . and then there's the Primacy up in our face with weapons."

Torin nodded at a passing Rakva before she answered. "Like the commander said, it's not against the law to sell weapons to the Primacy."

"It should be."

"Not arguing that."

"It's against the law to use those weapons to take hostages."

"And there's no evidence connecting the two," she reminded him, stepping into the lift. They shared a descending handhold and, after a silent exchange, agreed to continue the conversation later. Emphasizing the lack of privacy in the enclosed space, a high-pitched Katrien argument bounced down from above; far enough above, the Katrien themselves were out of sight.

"I had Ressk fossick through Marteau's public accounts and he's doing better than most OutSector colonies," Craig began again when they stepped out onto the level that accessed the least used of the station's eight docking arms. "He'd have hardly noticed the cost of sending mercs to the H'san homeworld looking for weapons and could've sent more mercs off to Threxie with his pocket change."

It took her a moment, then she remembered the planet's log number; 33X73—Threxie. With any luck, the name would be considered part of Craig's unique vocabulary and not stick. The look on Cap's face had been priceless at their first meeting when he'd raised a beer and said, *"So you're the new Wardie."*

"Marteau admitted to Primacy connections," Craig continued, his voice a low growl. "That's three hits on weapons, two on Primacy. When does it stop being coincidence?"

"When we have evidence. He's going to talk about weapons, Craig, he makes weapons." Weapons that had protected the Confederation. Weapons that helped her do her job and bring her people back alive. "I've used guns and mortars, sammies and spikes all stamped with Marteau's name."

"So you give him a free pass, then?"

"No, he's an asshole, but that doesn't make him a criminal mastermind. He has the means, sure, maybe even the opportunity—easy enough to buy that—what's his motivation? Marteau doesn't have to go looking for weapons, so why risk involving the H'san? Or us?" Torin glanced over at the tight line of Craig's jaw, nodded at a group of passing Krai in maintenance uniforms, and when they had the passageway to themselves, asked, "Why has Marteau got your nuts in a knot?"

Craig exhaled, ran both hands back through his hair, stepped out in front of her, and stopped. He didn't look angry. He looked unhappy. "The war is over and, yeah, there's plenty of shit kicking still going on, but Marteau doesn't need to be making new boots. You see weapons as tools, Torin, more efficient than a rock, but essentially the same thing. Other people are going to see them as opportunities."

She touched the Justice Department symbol on his uniform. "And it's our job to stop them."

"Yeah." His hands were warm against her face, his lips warm when he leaned in and kissed her. "And you're good at your job."

As the warmth faded from her skin, she watched him step through the hatch wondering what she'd missed. "Not saying I approve of vigilantes," she muttered as she followed, "but I'm all about taking the opportunity to destroy the plastic aliens."

It made sense that Craig, one of the two Strike Team Wardens not ex-military, would be more concerned about Marteau than either the plastic or the Primacy.

Torin felt the shiver of contact through her boots and released a breath she didn't remember holding. Primacy ships had rammed stations before, and the two accompanying Confederation battleships wouldn't have been able to stop a last-minute acceleration.

The war, she reminded herself, had been over for years.

Compared to the centuries of fighting, there'd been peace for half a heartbeat.

A moment later, interrupting the faint, calming music trickling from the speakers by the air lock, the docking master announced that the clamps had been successfully engaged.

"What was your first clue?" Werst muttered.

"No sirens," Ressk responded.

"Or lockdowns," Binti added.

"And a total lack of explosive decompression," Alamber said without looking up from his slate.

"Vacuum is an unforgiving bitch." Craig moved closer to the air lock, his gaze on the numbers scrolling past on the door. "Stations are all about redundancy."

Commander Ng sighed. "Speaking of redundancy, I summoned Warden Kerr, not the entire team."

"Team," Binti repeated, spreading her hands.

"The whole station knows there's a Primacy ship stopping at Justice on its way to Parliamentary reconstruction meetings," Torin told him when he turned to her. "Primacy 101 has been looping on the station entertainment system for the last twenty-seven hours."

"They're showing the docking feed at Musselman's," Werst said, standing on her left.

"Then why aren't you there?" Ng asked.

Werst grinned. "Rather be here. One of five people with Primacy experience, thought you might need me."

"Of course." Ng brushed invisible dust off the sleeve of his uniform tunic. "And you, Warden di'Crikeys? Having no Primacy experience . . ."

Alamber's fingers skimmed the surface of his slate. "Never met one, so I was curious."

"If you're trying to access the Primacy ship, stop."

His hair flipped out as he looked up and grinned. "Boss already warned me off. Said she'd be annoyed if I started another war."

Ng made one of his noncommittal noises. "And you, Warden Ryder?"

"I was there the last time."

"Your interaction with the Primacy on the prison planet was minimal."

"Yeah, but my minimal was talking, not shooting. That's more talking and less shooting than anyone else on this station. Not to mention . . ." Craig nodded down the arm to the nipple where the *Promise* had been attached. ". . . they're not throwing so much as an admiring eye on my ship without me there."

"The Primacy representative will examine the quarters for their people before attaching, to ensure we've met their specifications."

"And I'll examine them before attaching to ensure they meet *my* specifications."

Torin answered the commander's silent question with her best *nothing to do with me* expression. *Promise* belonged to Craig, not the Justice Department; he had the final word on any refits.

The rebuild, after *Promise* had been damaged by pirates, had given a small salvage ship the capability to add and remove packets as needed—the same capability as the Navy's battleships. The Marine packets, made up of living and training facilities for entire battalions were larger than anything *Promise* might require, but vacuum didn't care about aerodynamics and while the size had been scaled down for the Justice Department packets, the principle remained the same.

Conversations faltered as the air lock lights cycled. Anticipation rose to a measurable force in the corridor, lifting the hair on the back of Torin's neck. They weren't in a prison this time, responses stripped to the bare bones of survival. Nor was this neutral ground. Berbar Station was significantly farther into Confederation space than the Primacy had ever jumped, and her defenses were minimal. They were about to welcome a recent enemy onto a station that housed a branch of their Justice Department and, if taken, would provide valuable information for further attacks. But it wasn't government information Torin cared about, it was the thousands of people—people who not only worked here, they lived here. According to the numbers she'd pulled up while Craig snored into her hip, two hundred and seventy-four children, young of six of the eight resident species, called Berbar home.

She wouldn't let anything happen to them.

The air lock finished cycling.

"What's taking so long?" Alamber's hair had started to twitch. Waiting silently had never been one of his strengths.

Ressk elbowed him in the thigh. "Maybe they're reconsidering."

"Maybe they're nervous," Binti suggested.

"Of us?" Werst snorted. "They're not stupid, that's encouraging."

The telltales on the inner hatch turned green.

The team shifted into defensive positions.

Suddenly surrounded, Ng raised a brow.

"You were a lawyer before Justice tossed us in your lap," Torin explained over the hiss of the releasing seal. "If they exit waving a subpoena, we'll stand down."

They exited waving two feathered antennae above six eyes, in turn above a meter-and-a-half-long, half-a-meter-high body encased in a brown-on-brown-patterned exoskeleton, four upper arms, and a multitude of short, variably jointed legs under chitin flaps. Mandibles clacking together, the Artek sped toward Torin, smelling of cherry candy.

As Ng jerked back, Torin stepped forward. "Firiv'vrak."

The Artek, one of the Primacy's warrior species, slid sideways as it stopped. "I have a voice for your ears this time, Gunnery Sergeant Kerr!" The mandibles clattered once more. "Or should I say Warden Kerr?"

Torin held out her hands, fingers spread, and brushed the ends of the extended antennae. "I think you've earned the right to call me Gunny, Firiv'vrak." She looked up at Alamber's quiet gasp and squared her shoulders as she turned to face the smaller of the two Polint stepping out into the passageway. "Durlin Vertic."

"No longer durlin, Gunny. Like you, I've left the military." The golden-haired Polint twitched at a soft fold of her teal jacket and smiled, only showing her lower teeth. "You're looking good."

"As are you." The last time Torin had seen the young officer, the durlin had been unable to use one of her rear legs due to a deep burn from a spray of molten rock. Torin herself had been starved, scorched, and covered in seeping blisters. Good times. All things considered, the Primacy's first three choices were no surprise and she felt the tension in her shoulders ease as she nodded at the reddish-brown male who pushed up close to Vertic's hindquarters. "Bertecnic."

"Gunny."

With Firiv'vrak greeting Craig and the two Polint moving to Werst and Ressk, Torin turned her attention to the pair of Druin in civilian clothing who followed them. "Durlave Kan Freenim."

"Gunnery Sergeant Kerr." Vertic might no longer be a durlin, but it seemed an ex-durlave kan was as much a contradiction in terms as an ex-gunnery sergeant. "You remember Santav Merinim. Although she's santav no longer."

The ex-santav's inner lids slid across large black eyes. "Gunny."

The Druin were taller than the Krai, but not by much, and Torin's height gave her a clear view of the matching pattern dyed into the upper arc of their hairless heads. When she raised a brow, Freenim nodded. "Yes. We have been *leetinamin*."

"Joined," Merinim added over her shoulder, interrupting her reunion with Binti.

"Congratulations."

Torin had no idea how the Primacy program translated the sentiment, but Freenim flushed, his skin turning from pale ivory to slightly darker ivory, and said, "The pouch will remain empty for a while, but it's good to be with someone who understands."

"Impossible to be with someone who doesn't," Torin agreed as Craig argued about the modifications to the *Promise* with Firiv'vrak. The Artek, as a species, had been active in all branches of the Primacy military, but they loved to fly. Firiv'vrak had been a fighter pilot and from what little Torin overheard, civilian flying just wasn't the same.

The air lock, having cycled through again, reopened and another two Druin emerged followed by a slender Polint with variegated fur—the exposed skin of his face and hands matching the variegation—and, behind him, another Artek.

"Gunnery Sergeant Kerr." Vertic folded over herself until she faced the air lock—Torin had forgotten how flexible the Polint were. "This is Keeleeki'ka."

The second Artek, paler than Firiv'vrak, with raised patterns, paler still, on their carapace, ducked their antennae, and, smelling of licorice, muttered, "We've heard so much about you, Warden Kerr."

"So much about all of you," the variegated Polint added. His upper teeth showed as well as his lower. "For the entire nine days in *terinmun*."

"Seven and a half," Keeleeki'ka corrected.

"Ah. Seemed longer."

"And this," Vertic said sharply enough her voice cut off the buzz of conversation, "is Santav Teffer Dutavar. He *volunteered*."

The efficient management of violence had fairly universal

parameters, and all three of the Younger Races had a variation of *never be first, never be last, never volunteer.*

Given Vertic's emphasis, it seemed those parameters were more universal than Torin thought.

"Still serving?" Vertic had mentioned both name and rank.

Broad shoulders moved inside civilian clothing as though he wanted to shrug it off. Claws showed on all four feet and both hands. "I am."

Had he volunteered to be the military's eyes on? Torin watched a ripple run through the fur on his haunch, black and gray and white and orange all shifting slightly as though he'd thrown off a fly in the enclosed, filtered interior of Berbar Station. "And you, Keeleeki'ka?"

"Not serving now. Never did."

"But you volunteered as well?"

Both antenna flicked down, then up. "No. I was chosen." The licorice scent intensified.

"Good. Good. Now you've all met." The more elaborately dressed of the two Druin frowned at Torin for a moment, flicked his inner eyelids across both large black eyes, scanned the crowd in the corridor, and turned to Commander Ng. The other Druin, tucked close to his right side, clutched what appeared to be the Primacy version of a slate, and looked apprehensive.

"Warden Commander Ng."

Ng nodded. Once. "Representative Haminem."

"I apologize for the invasion . . ."

The other Druin winced.

". . . but when our volunteers saw a crowd of people waiting . . ." Pale, long-fingered hands spread in what Torin assumed was a conciliatory gesture. ". . . they charged past me."

"They were concerned for your safety?"

"Oh no, I'm sure they'd happily toss me out an air lock without a suit." Representative Haminem seemed pleased about the animosity. "Were you not informed? All but two of our volunteers were part of the unveiling of the grand deception on the plastic aliens' prison planet."

"Volunteers?" Torin asked quietly.

Freenim's head tilted toward her. "I assume we could've refused the council's request. I can't prove it, though."

"Unveiling of the grand deception?"

"He's a politician. And he's not wrong about us being happy to toss him out an air lock. He's going on to the restitution meetings." After a short pause filled by the bass rumble of Bertecnic telling Alamber the story of how a tiny crippled Krai had punched him in the nuts, Freenim added, "I'd rather be shot at."

"Who wouldn't." Torin ignored Alamber's offer to examine Bertecnic's nuts with the ease of long practice. Consenting adults. None of her business.

"Warden Ryder . . ."

The noise dropped off. Everyone in the corridor waited to hear what happened next.

". . . you'll accompany the Representative and myself while we examine the packets before installation." As Craig nodded, the commander turned his attention to Torin. "Warden Kerr, IA-3 has been opened for our visitors."

"Yes, sir." She could see the half-dozen warnings he wanted to add flick across his face and appreciated him closing his teeth on all of them.

IA-3, the largest of the level's interrogation rooms, was a right turn and four meters from docking arm eight. The passageway stretched empty in both directions as Torin stepped through the hatch, and she moved to the left so Werst could lead the Primacy team out of the arm. Half the station wanted a look at their recent enemies, but DA8 was used so infrequently, it seemed no one had been able to fake an official reason to be in the area.

She hadn't been told why the powers that be wanted to keep their seven Primacy visitors isolated, although she assumed at least part of the rationale came from there being seven of them in the midst of thousands.

A high-pitched shriek snapped Confederation and Primary both into defensive positions. Torin turned to see a Human child race toward them, a Human male in wide-eyed pursuit.

"Luiza, get back here!"

"Bug!"

"Werst, keep them moving!" Torin would have put odds on a child heading for the Polint, who were furry and looked like they could be ridden. Dropping to one knee, she made a clean interception, the small body slamming into her outstretched arm.

"Bug!"

Torin stood and passed her to the Human male.

He held her struggling body with the ease of long practice. "I'm so sorry, Warden. There's never anyone here, so when she's bouncing off the walls, we bring her down and let her run with no distractions until she's sleepy. When the CCI sounded, we were going to slide into IA-2, but she got away from me." Adjusting his grip, he caught a foot heading toward his crotch—Torin was impressed by his reaction time—and added, "She met a Ciptran on the concourse about three tendays ago and she's been fixated on insectoids ever since."

"Daddy, bug!"

"She met a Ciptran?" The Ciptran, who were built a little like a giant praying mantis, gave lie to the belief that only social species achieved sentience. Torin had never seen two together or one in a good mood.

"It went better than you'd think."

Luiza twisted herself around until she hung nearly upside down. "Daddy! Bug?"

"Artek. My people are called Artek."

Her dark eyes widened as Firiv'vrak settled by Torin's side. "Talks?"

"Yes, child. I talk."

She straightened up so quickly Torin would have worried about whiplash on an adult and put both palms on her father's cheeks. "Down? Please, down."

"Warden?"

Torin glanced behind her, making sure Werst had the rest out of sight in IA-3. Then she glanced down. Firiv'vrak's antennae waved slowly from side to side, but, otherwise, she was completely still, all four arms tucked close to her body, as nonthreatening as a giant insect could look. Torin didn't trust the Primacy, but she trusted her history with this specific Primacy member. "It's safe."

Luiza's father didn't look entirely convinced, but he bent and slowly set his daughter on her feet, a finger hooked behind the crossed straps of her overalls. Luiza threw herself against his hold and got both arms as far around Firiv'vrak as she could reach.

Which put her soft, baby face right up against mandibles Torin had seen crush bone. Muscles jumped in Luiza's father's arm as he fought the urge to yank her back, glancing between his daughter and Torin as though checking for any indication that he should panic.

Luiza giggled as the tufted tips of antennae stroked skin and shifted around to pat Firiv'vrak's carapace when her father reluctantly released her on Torin's nod. She squatted to look at the many legs, but didn't touch, then straightened, leaned forward, and licked along one of the swirls of darker brown.

Torin caught one word in five, but, given the disgust, it wasn't hard to work out Luiza's negative reaction to Firiv'vrak not tasting like she smelled.

"Luiza! We don't spit on the deck!"

Amusement intensified Firiv'vrak's cherry candy scent.

Embarrassment having displaced a good portion of his unease, Luiza's father snatched her up and set her back on his hip. "All right, you've met the Artek. Now we have to let the Warden get back to work."

Not for the first time, Torin noticed the same tone that worked on second lieutenants worked on small children.

"I know we're not supposed to be here, Warden, but if we could deal with the consequences later . . ." He pressed a cheek against his daughter's hair, looking resigned.

Firiv'vrak replied before Torin could. "It was a pleasure to meet a child."

"And your pleasure has been noted." Although Torin had no intention of reporting the security breach, the CCVs had recorded not only Luiza and her father's presence in a restricted area, but unregulated contact with the Primacy. "You need to head for the perimeter hatch. Now." There'd be Wardens waiting on the other side.

"Before it looks like we're lingering."

"A little late for that. Feel free to blame me."

"Blame you?"

"Tell them I said I was expanding the parameters of the peace."

"And that'll . . ." His eyes widened. "Holy shit, you're . . ."

A small hand covered his mouth. "Daddy! Swears!"

He pulled her hand away and shook his head. "Come on, baby girl. We need to go."

A head of dark curls and a small hand appeared over her father's shoulder as he hurried away. "Bye, bug!"

"Artek," he corrected firmly.

"Bye, Arkek!"

"Close enough," Firiv'vrak allowed. They stood silently for a moment, then Firiv'vrak rose and pivoted on her lower legs. "You are known in your Confederation," she said as she dropped back to the deck. "We are also known for exposing the manipulations of the plastic aliens. There were many tests to prove we told the truth, that the images were real, that the manipulation occurred."

It sounded as though the translator paused before deciding on *tests.* Torin wondered if there was a way to find out what other words had been considered.

"The young of many mammals fear us," Firiv'vrak continued, as though she hadn't just implied that the years after the prison planet had been less than pleasant. "It was good to see one who doesn't."

Luiza hadn't cared that the Artek were Primacy. She didn't yet know the concept of enemy. "Changing perceptions, one two-year-old at a time."

Antenna brushed against the back of Torin's hand. "Indeed."

"Because this is the first joint venture between our two people . . ."

"Second," Werst muttered.

Ng ignored him. ". . . we, the Confederation and the Primacy . . ."

"Does he think we've forgotten who . . . *Chreen!*"

Standing at the side of the room, Torin appreciated Ressk silencing Werst's running commentary before she had to.

". . . have agreed the mission to 33X73 must be documented."

"Documented not only in case this attempt at cooperation fails," Representative Haminem added, "but also should it succeed."

At the back of the room, Vertic crossed her arms, one front foot clawing at the floor. "You believe we have a better chance of success if we're being watched?"

"Yes." Haminem moved a hand up and down in front of his narrow chest. A nod by any other name. "We do. Impartial witnesses bring out best behaviors."

Not in Torin's experience. "This wasn't in the briefing packet," she pointed out, not liking where things were going.

Haminem's inner eyelid flickered. "Politics. And," he continued, "there was an extended discussion about who to embed—those in favor of impartiality on one side and those in favor of experience on the other. There is, of course, no such thing as impartial experience that I've found in my own extended experience."

If his phrasing had been intended to be funny, no one laughed.

"It's in the briefing packet now," Ng said bluntly. "Credentials and clearly stated parameters." He squared his shoulders and met Torin's gaze. "In the end, given the dangers involved, experience won."

"Sir?" She really didn't like where things were going.

Right on cue, the hatch opened.

"So, ex-Gunnery Sergeant Kerr, we are meeting again." The Katrien paused just far enough inside the hatch to allow her camera operator to enter behind her. Confederation law required cameras large enough to be easily seen in order to prevent any perceived invasion of privacy. As lights on top of the camera indicated recording was in progress without the required signed permissions from everyone in the room, it appeared privacy applied as little to the Strike Teams as it did to the military.

Fluffing silver-tipped dark fur, the highlights too artfully natural to be real, Presit a Tur durValintrisy of Sector Central News and a reoccurring stone in Torin's boot, waved a small hand that looked like a black latex glove emerging from the cuff of a thick fur coat, and declared, "We are not being at all surprised to be finding ex-Gunnery Sergeant Kerr are leading the first combined Confederation and Primacy law enforcement exercise. It are sure to be an eventful trip." She paused, the finger she raised for silence hidden from the camera by her fur. At a nod from her camera operator, she adjusted the dark glasses that protected sensitive eyes from the light and turned to sweep a disdainful gaze over the assembled company. "That are being enough for now, this room are being too depressing to be shooting in. The government are clearly buying ugly gray paint in bulk. We are going to be getting individual interviews when we are having a more attractive background."

Torin took a step closer to the commander. "Sir?"

"You will be the Combined Strike Team Lead while resolving the hostage situation. As lead, the material the Primacy has provided on their people on 33X73 has been added to your briefing packet."

"Yes, sir. But Presit . . ."

Ng cut her off, pitching his voice under the dull roar of Presit working the room and the room reacting. "Not my decision, Warden. But better the devil you know."

She released the breath she'd been holding a little too quickly for it to be called a sigh. "You'd think so."

The Primacy had been able to identify the three Polint. The black, Camaderiz, was ex-military. He'd served the minimum time with no distinction, but he knew how to fight. The bay, Netrovooens, and Tehaven, the variegated, would know how to fight as well, but they hadn't been trained.

"Not trained by the military," Freenim amended, as they went over the information. "That doesn't mean they don't know what they're doing. The Polint are strong, and fast; and those blades they're carrying? They're civilian weapons."

Torin paused, pouch of coffee halfway to her mouth. "You arm your civilians?"

He laughed. "They arm themselves. The Polint will have been using them since they were small, the blades growing in size as they do."

"All the Polint?" Torin asked him, brows up.

"Statistically unlikely, but all the Polint I've known. Mind you, all the Polint I've known have been military so . . ." He spread his hands.

Pacifist Polint weren't like to join up. Torin frowned down at her slate. "They didn't give us much."

"There wasn't a lot of time."

"Single names?"

"Werst? Ressk? Like the Krai," he continued before she could speak, "formal names among the Polint include lineage details. The odds are high that's more information than my government wants the Confederation to have. Or it's possible they were only able to find out the day names of Netrovooens and Tehaven in the time available and left Camaderiz short as well so it looked like a choice."

Torin tossed her slate down onto the galley table. "If Justice needs full names, they can ask them when they're brought in."

"Prisoners."

The Primacy didn't take prisoners.

"Our orders are to keep fatalities to a minimum," Torin reminded him.

Freenim spread one, long-fingered, pale hand out on the table's surface. "Primacy prisoners will be returned to the Primacy."

"That's for politicians to decide. Not us." She drained her coffee and added. "Our job, as a team, is to rescue the hostages. What about Druin names?"

He blinked, but after a moment followed Torin's train of thought back to the earlier part of the conversation. "Descriptive." Hand now against his chest, he added, "Freenim of Murglin on Shepten. That's city and planet. If I was still on my home planet, I'd introduce myself as Freenim of Thoi in Murglin. Community and city."

"Except that you made those places up."

He grinned. "Of course I did."

Torin tapped a fingernail against the screen of her slate. "There's no information about the Druin in red."

"I believe your people have a saying that covers the lack." When Torin raised a brow, he raised his pouch of coffee in salute. "Space is big."

• ——◆—— •

"Is this all of them?"

"It is." When the Human called Martin nodded toward the half of the common room crowded with scientists and their ancillaries, Arniz hissed.

Salitwisi slapped his tail against the back of her legs. She ignored him the way the murdering *yerspit* had ignored her.

The Krai, the voice out of the shuttle, crossed to stand beside the map table. The Druin in red positioned herself behind his left shoulder. "I'll keep this brief." His nostril ridges were nearly closed. He didn't feel safe. Perceptive—Arniz wanted to pummel him. He was in charge, which made him as much at fault as the Human who'd killed Dzar. "I have a buyer," he said, "who will pay a great deal for a weapon able to destroy the plastic aliens. Who found the debris?"

Arniz hissed and pushed between Tilzon and her ancillary to the front of the group. "He killed her." She jabbed a finger toward Martin.

The Krai's focus drifted past her. "One death to keep the rest safe."

It sounded like a quote to Arniz. "One death for no reason!" she snapped. "And it wasn't debris. It was molecular evidence and that's all it was. There's nothing here for you. Go back to where you came from!"

A few voices murmured agreement behind her.

Salitwisi grabbed a handful of her overalls and yanked her back. "Don't antagonize them! They've already proven they're willing to kill us."

"You should listen," Martin sneered. "Could be the smartest thing your lizard friend ever said."

Part of Arniz acknowledged that was possible; the greater part wanted to scream insults and accusations at the Human. Or perhaps just scream.

The Druin's hand in its red glove lay on the Krai's shoulder like a splash of blood. His focus had locked onto Arniz. "Did you search for a weapon?"

"No!"

Before she could add that they hadn't searched because the weapon didn't exist, his lips drew back off his teeth. "If you haven't searched, you have no idea of what's out there, do you?"

• —◆— •

Torin wasn't surprised to find Presit waiting for her as she stepped through the hatch into DA8, heading for the *Promise*.

"I am seeing you are alone. Craig are having finally come to his senses and found a mate who are less likely to be getting him killed?"

"Craig is already on board." Torin had no doubt that Presit not only knew where everyone was but had worked out how best to ambush them for her "story".

"You are not being with him."

"Like you, I have duties to perform."

"Are there being trouble in paradise?"

"What? No."

"That are being too bad. He are almost not entirely useless with a brush. What?" she demanded when Torin stopped and stared down at her own reflection in Presit's mirrored glasses. "I are merely determining where the lines are being drawn before we are all being locked up together in the pitiless vacuum of space."

"We're in a station. We're already locked up together in the pitiless vacuum of space," Torin added when Presit waved an imperious demand for more information.

Black lips drew back off small, pointy teeth. "Oh, yes, now I are remembering how pedantic you are being." She sighed and a fraction of the pretension fell away. "I are wanting to tell you that regardless of what I and my family are owing you for the return of Jammers' body, I are going to 33X73 to be reporting on the first Confederation/Primacy joint venture. In spite of what certain politicians are believing, I are going to be entirely impartial."

"You always are."

Presit cocked her head and, given the position, Torin assumed she was being studied from behind the glasses. "I are pleased you are finally admitting to what are being my superior reporting skills."

"I've never had a problem with your reporting skills."

"And I are never having had a problem with your mate."

She'd said it so sweetly, Torin couldn't stop the snort of laughter.

Dalan a Tar canSalvais, Presit's camera operator, waited at the air lock with Alamber who had the fingers of one hand buried in fur as he scratched behind one of Dalan's ears. Tongue protruding slightly, Dalan sagged against Alamber's leg.

"You are being in public," Presit snapped. "I are not caring what you are doing if I are not needing you, but you are to be remembering that your behavior are reflecting on me."

Dalan pulled in his tongue, straightened, and pointedly glanced up and down the empty arm. "And the crowds are going wild," he muttered, looked over the top of his glasses at Torin, and winked. "Was recording all through the war, Gunnery Sergeant, was having been with the crew recording on Horlong 8. I are knowing when to be keeping my ass out of your way."

Horlong 8 had been a disaster, due as much to bad officers as the Primacy. She was impressed he'd gotten out alive. "Then I'm glad to have you along, Dalan a Tar canSalvais."

"I are just Dalan." His muzzle was grizzled, his ears notched. There was a chance, if only a small one, that he could temper Presit's . . . enthusiasm.

Torin noted Alamber's expression and waved the two Katrien into the air lock. Whatever he had to say, he didn't want an audience. As Presit's voice rose, greeting Craig and giving the impression she hadn't seen him for years rather than hours, Torin beckoned the young di'Taykan closer. "What is it?"

"There's seven Primacy on board, Boss. Two Artek, two Druin, three Polint."

"I can count."

His light receptors flicked open, then closed. "And only six of us. We're outnumbered, and the Polint are . . . large."

"I'm not surprised you noticed." She studied his face while he worried at the old piercing scars in his lower lip. "Is this about adding another di'Taykan to the team?"

He sighed. "Let it go, Boss. This is about ratios. If anything happens, they have numbers, size, and appendages on their side. I mean, three Polint, six more legs, two Artek, four more arms and a fuk of a lot of legs. And the cherry one, she's a pilot. Me and Craig, we can fight, but . . ." His hair continued to flick back and forth as his voice trailed off.

Alamber had no history with the Primacy aside from what he'd seen on vids. He'd never served, didn't have the undercurrents of *at least you understand where the fuk I'm coming from*—common ground no matter how deep or dark the currents. He hadn't been on the prison planet, didn't have the personal history of fighting beside most of their new team members.

"I can't speak for the two I don't know, but you have my word the rest are no danger to us. And don't forget . . ." They winced in unison as a high-pitched voice demanded to know what was wrong with the ship's air filters. ". . . we have Presit."

His hair stilled. "She's on our side?"

"There are no sides, Alamber." Torin touched two fingers to the inside of his wrist. Without another di'Taykan on board, the necessary physical contact had to come from his non-di'Taykan teammates. "But we've built up more resistance to her."

"I remember you."

Presit preened. "I are hard to forget."

"Yes." Settled on her haunches, Vertic frowned. "Your dialect is causing my translator some difficulty."

"Your translator isn't alone," Torin said, stepping out of the air lock into *Promise*'s control room. "Their dialect is a pain in everyone's ass. Alamber, show the Katrien to their packet."

"Sure thing, Boss."

Alamber stepped forward. Presit held up a hand. "The air filters are not being up to the number of species on board."

Torin smiled. If Alamber, with the Taykan's sensitivity to scent, could handle the combination, so could Presit. "Shed less."

"You are always being so amusing. And I are not liking what you are having done with the place." Presit curled a lip at the seats then extended her dislike to the lockers holding the HE suits. She'd shared *Promise*'s original cabin with Craig on the way to the prison planet and, given how that original cabin had been laid out, Torin carefully avoided thinking about the logistics. "That being said, I are preferring to remain here. If the harnesses are fitting the Krai, they are fitting me."

"Presit . . ."

She huffed at Craig's tone.

". . . I'd prefer it if you strap down in your quarters until we're clear. For safety's sake."

"Fine. For you." She smoothed her ruff and glared at Torin. "But you are not hiding things from me this time, I are telling you that now."

The last time, things had been hidden with Presit's agreement for Presit's own good. But Presit always edited her own story.

Torin waited until only Craig and Vertic remained, and raised a brow.

Vertic softened her shoulders and curled her claws in toward her palms, her posture deliberately nonthreatening. "I wanted a chance to speak to you alone before we begin the actual mission."

Between preparation and overt observation, there'd been no chance on the station. Torin fell into parade rest and waited.

"I can't leave," Craig began.

"That's not what I meant by alone," Vertic interrupted. "You should hear this as well." Her mane flattened slightly as she drew in a deep breath. "I realize I have neither experience nor training in police work, and I therefore will have no trouble with your command, Gunnery Sergeant Kerr."

"Thank you, Durlin."

Her broad mouth curled up at the corners. "If we don't refer to my previous rank, my people will be less likely to respond to it by default."

"Of course." Torin had missed the sense of security a good officer provided, the knowledge that the big picture would remain in focus while she concentrated on the details. Vertic had been an officer Torin had been happy to serve under during their combined escape and she'd matured during the intervening years, now exuding an air of confident authority.

Although *exuding* might be the wrong word given the overwhelmed air filters.

"Also," her mane rose as she continued, "I checked into Samtan Teffer Dutavar's service records, and he's proven himself to be level-headed when it counts. Not always a given with our males."

Torin would bet big that Vertic had checked while trying to work out why he'd volunteered. And bet bigger that she hadn't found out.

"Keeleeki'ka is a Sekric'teen, from Neesemin'c, the Artek homeworld. The Sekric'teen are a sizable collective who opposed the war from the beginning. Their scent is . . ." She licked her lips. ". . . tart. The exposure of the plastic aliens strengthened their position both politically and popularly, and they insisted one of theirs join the team. All that's in the briefing packet. What isn't included is the certain knowledge that if we don't return, the Sekric'teen will rip into the government, and, with the media on their side, that's the last thing our government wants. If we do return, they'll want the story told their way, and the government's not likely to enjoy that either."

As a general rule, Torin had little sympathy for governments she'd spent years fighting against. "Damned if they do, damned if they don't."

"Unfortunately, that makes Keeleeki'ka more important than she seems." Vertic shifted in place when Torin indicated she should continue. "I'd have preferred someone less inflexible in their view of how things work."

"But her lot were right," Craig pointed out.

"That's what they're inflexible about."

Torin grinned at the dry disapproval in her tone. "Is there a chance Dutavar is here to keep an eye on her?"

"It's possible," Vertic allowed, rising to her feet. "But I doubt it. Our military seldom involves itself in politics. Now, as these seats most certainly aren't configured for my body—nor did I expect you to rebuild the core parts of your ship for my benefit," she added when Craig opened his mouth. "I'll strap in with the others. I look forward to working with you again, Gunny. Captain Ryder."

"Captain?" Craig asked quietly, watching Vertic's signal move toward the quarters designed for the Polint.

"Your ship."

"I like her. So you're good with the rank thing? Just like that, then?"

"Why wouldn't I be?" Torin dropped into the copilot's seat. "You've always ranked me on the *Promise*."

"Not my rank." He turned toward her, gray-blue eyes searching her face. "Her rank."

"My Strike Team."

She had no idea what he was searching for or if he'd found it when he turned his attention back to the board. "All right, then."

• —◆— •

"You fought in the war."

"We all fought in the war, one way or another."

Arniz turned away from the screen and frowned up at Emile Trembley—who stood far too close, who smelled of warm, damp, hairless, male mammal, and who wouldn't understand the data scrolling past as the last of the soil analysis finished even if an entire deceased plastic alien as well as a functioning example of the weapon that killed it were found. This was the first time he'd been assigned to watch her. The black Polint, Camaderiz, had remained silent. Arniz had thought him sullen until she realized that without the translation program on Martin's slate, they had no language in common. Mirish di'Yaunah, the di'Taykan with the deep blue hair and eyes, had sprawled gracefully in a chair far too small for her, her complaints about the heat and humidity repetitive and dull. Trembley, however, liked to talk. "I didn't fight," she reminded him.

"I wasn't talking about *you*. You were hiding at your university while *we* kept you safe." The emphasis might have been a Human way of speaking, but she suspected Trembley was young. Firsters always spoke in a cascading series of absolutes.

"Hiding?" she snapped. "I wasn't hiding. I was teaching and learning; knowledge being a component of what you were fighting to protect. So, to whom are you referring when you say *we?* I can't read your mind, you know."

Trembley's mouth twisted up into what Arniz assumed—from the little she knew of Humans—was a smile, although it showed no sign of amusement. "Lucky for you, lizard. *We* . . ." He doubled down on the emphasis. ". . . are the Younger Races." His chin rose. "And *we* have decided to keep fighting."

"All of you?" She thought of Ganes, who'd stepped forward and been struck when Dzar was murdered. No, not all of them.

"Not yet. *Some* people are blind to their oppression. But *soon*."

Possibly younger than Dzar.

This was what came of giving children weapons.

Before she could ask who or what the Younger Races were continuing to fight, the scanner trilled. She squinted at the screen and cut the power. "That's it. No further signs of anything that might have once been plastic and no sign at all of what might have been responsible for the state of the sample we found previously."

"No weapon?" Trembley scowled at her equipment.

Arniz snorted. If bad temper could improve performance, she'd have long since finished. Dzar's murderers would be gone and good riddance to them.

"You *had* to have missed something."

She sighed. "Follow me. Watch where you put those ridiculous boots." He was used to following orders; she'd give him that. He stayed between the perimeter pins delineating an animal shelter and the foundation of a tower—neither visible to the uneducated eye. She stopped him well back from the edge of the latrine and pointed.

"It's a hole," he said.

"Yes, it is. Had I not been catering to the violent enthusiasms of an invading force . . . you lot," she added when he looked confused, "we'd have performed a CPT, taken samples of varying ASTM dimensions, we'd have done gas tests in the bore holes, and, because I *had* an ancillary who needed the experience—until you murdered her—we may have used the FFP."

"I didn't *murder* anyone!"

"And yet she's still dead."

The acrid undertone of his scent spiked. "How many of *us* have died?"

"Oh, for . . ." Arniz ignored the boundary stake, knocked flying by her tail. "That's a faulty comparison. The two things aren't at all connected. The point is . . ." She cut off his response, unwilling to put up with his disconnect. ". . . this is a hole. We're on a first-year mapping expedition on a Class 2 Designate and we have no business digging a hole."

"Bullshit. You have a *digger*."

"Yes, we do. That doesn't mean we intended to use it. Nevertheless, in order to find an imaginary weapon, I not only had the digger excavate the latrine, but six centimeters of undisturbed soil outside of the dimensions of the latrine. I've done a full . . . well, not a full given the volume, . . . but a relevant analysis of every bit of soil taken out of the hole at the molecular level. There is no weapon here. Nor are there bodies, plastic or otherwise. Or parts of bodies. Or anything but very potent urea and evidence of fecal matter that supports the possibility of a predominately carnivorous species pre-destruction, some small amount of solid debris consistent with what we know about their civilization—which is less than we would have known by now had you not shown up and interrupted our work and starting killing people—and trace amounts of plastic in a single specific location."

He stared into the hole, at her, and said, "Thrown out by a non-plastic-using civilization." As though that one data point justified everything.

"Yes, all right, fine. But only three years ago, this planet was a Class 1 Designate. Class 1 allows orbital scans only. Class 2 allows scientific study at selected sites and a complete restriction on anything leaving the planet. I've distilled those definitions to the essentials, by the way. There are terabytes of rules the Ministry for the Preservation of Pre-Confederation Civilizations requires us to follow, and not following them to the letter will destroy careers. This is only the third year on-site research has been permitted, and this is the first year there's been a dig at this particular site. There are things we've been able to surmise—that they were a non-plastic-using civilization being one of them. Unless the four sites currently being examined are, by some outlandish coincidence, four extremely large, historic recreations and we entirely missed the actual pre-destruction population centers."

"What?"

"Highly unlikely, of course, because the orbital scans are really very thorough."

Trembley shifted his weight from foot to foot, compacting the soil, dark brows drawn in. "Then why did you even *mention* it?"

"Every possible hypothesis needs to be considered." Arniz retrieved the dislodged boundary stake and tried to push it back into place. An insect with long, delicate legs, thorax a shade darker than the soil, scurried away. After a moment, Trembley muttered under his breath and replaced the stake for her. "I don't see why you need a weapon to destroy the plastic aliens." She pulled the stake back out and shifted it two centimeters to the right, waiting pointedly until he threw his weight against it again. "By their own admission, they've completed their experiment and left the Confederation. Granted, their behavior was unforgivable, you lot would know about that, but they're gone."

"Yeah, you just keep your head in the sand with the *rest* of the Elder Races. Oh, no . . ." His voice rose as he straightened and picked up a peculiar accent. She had no idea who he thought he was imitating. ". . . *we're* too socially evolved to be bothered again."

"That's not . . ."

He cut her off. "When they come back, we'll be *ready*. Times are changing."

"Into what?"

If Humans came with a neck pouch, he'd have inflated it. "Into times where the Younger Races *won't* be the cannon fodder anymore."

"I understand fodder. What's a cannon?"

"It's a . . ." He glanced around with his strange, brown-on-white eyes, as though the answer was on the plateau with them. ". . . it's something the sergeant says. It's a weapon, I guess."

"You guess?" Anger pushed beyond background noise, she rounded on him. "You guess? You don't know why Dzar was murdered?"

His scent spiked.

His palm felt like a stone, slamming into her chest.

"Shut up, lizard!"

Staring at the sky, catching her breath, that seemed like a good idea.

"You weigh her death against the *millions* of us who died in your war and you know what—you *owe* us."

Ideology.

Or possibly rhetoric.

When, Arniz wondered, had it become us against them?

"And this was the only place the plastic residue was found?" Yurrisk slapped the edge of the map table, making the image flare.

If she hadn't been sitting in a chair one of the ancillaries had dragged from the end of the anchor where the rest of the expedition was once again confined, she'd have lashed her tail. "It's the only place the soil has been analyzed that completely."

His lip rose. "You're saying no more plastic has been found because no one has looked for it."

"I'm not . . ." Arniz thought about it for a moment. "All right, fine, I am saying that. Essentially."

"Then I want all your scanners calibrated to search for plastic residue."

"I thought you wanted the weapon?" Ganes' mocking question drew Yurrisk's attention. Arniz was all in favor of mockery, but Martin had proven himself willing to kill and Yurrisk willing to allow it. Ganes needed to be careful. He was the only one who knew how to keep the tech functional. "Why would you expect to find a weapon with the residue? Do you think your mercenaries dropped their KCs every time they shot a Primacy soldier?"

All three Polint growled at the spill of words from the translation program in Martin's slate and Arniz realized Ganes was attempting to divide and . . . well, not conquer, but divide at least. Clever. She may have underestimated him.

"You were Navy. Have you forgotten your training?" he continued as Yurrisk's nostril ridges closed and his lips drew back. Arniz thought she saw the Krai's hands tremble. "I always thought the Navy preferred you to keep hold of your weapons."

"His Navy did. Our Navy also." Qurn, the Druin in red, moved to Yurrisk's side, her shoulder against his, the contact leaving him no room to move his arms. "But when it comes down to it, Dr. Ganes, no one cares what you think." Her voice managed to be both precise and melodic. Arniz had no idea how. "We currently have no search parameters for the weapon. If your scanners find more plastic residue, more points of reference, we'll have identified the layer of history we'll need to excavate. Or, specifically . . ." Her narrow lips arranged themselves into what Arniz assumed was a smile. ". . . that you will need to excavate. Stop assuming we're all uneducated and have no idea of what we're doing. It's annoying."

Maybe not a smile, then.

Tail up, Salitwisi squeezed between the black and the variegated Polint. "And you want to use all the scanners?"

"Yes." Yurrisk had calmed while Qurn spoke. "All the scanners. And all of you out there . . ." He expanded the image of the site and called up a grid pattern. ". . . scanning."

Arniz frowned as she considered the changes that would have to be made. "The results will be scientific garbage."

"But sufficient for my needs?"

She wondered what would happen if she said no. Would he know she was lying? Would someone else die to convince her to tell the truth? Had Dzar died so she'd consider that before speaking? "For your needs, yes, it should be sufficient."

"*Harveer* Arniz does not speak for all of us!" Salitwisi's tail rose higher, the tip tracing small arcs in the air.

Yurrisk showed teeth. "Be quiet."

"But . . ."

Martin stepped away from the wall where he'd been leaning, arms crossed. "We don't need you, lizard."

Salitwisi's tail dropped so fast Arniz thought he might have cramped his ass. In response to Martin's signal, Camaderiz shoved him back into a huddle of scientists with enough force he took Lows to the floor with him. A warning, Arniz assumed, more overt than usual. The other expedition members had been locked in the anchor during the detailing of the latrine and whispered conversations while curled together in the nest at night had included complaints about minor bullying and wasted food, but no overt physical abuse. Yurrisk's people seemed to have little midground between childish petulance and death.

"When you find residue . . ."

Attention drawn back to the common room, Arniz cut him off. "If."

"When," Yurrisk repeated.

"Science doesn't work that way."

"Commander? Why not go right to the ruins in the jungle?" Pyrus, the pink-haired di'Taykan, leaned over the map table and expanded the view. "Wouldn't they have stored the weapon in a building?"

"Very likely," Arniz answered, not caring she hadn't been the one asked. "But without knowing where to look, you could be in there for years. We've found plastic residue on the plateau, we need more data points, it makes sense to look for it here first."

Yurrisk stared at her.

She stared back. His eyes were a deep green, the same shade as the darker parts of his mottling. She'd expected them to be cold. Calculating. The sort of eyes that could see the death of an innocent and not care. They weren't. They were haunted. She couldn't tell how much of the present he actually saw. The disconnect made the scales on her neck itch.

"No tricks," he said at last.

"No tricks." Arniz pointed at Ganes. "He'll have to do the calibrating."

"Why him?"

She blinked. "Dr. Ganes is our engineer. It's his job."

Yurrisk stared a moment longer, blinked in turn, focused, then nodded. "Get it done."

Having moved to stand behind him, Qurn kept her attention on her slate.

"That made no sense at all," Salitwisi hissed after Arniz was sent back to the others and Ganes escorted to the equipment lockup. "What are you playing at?"

"They're not leaving without the weapon," Arniz told him.

"There is no weapon!"

"Given what the plastic has done . . ." She thought of Trembley's millions who'd died. ". . . they won't believe that. We're stuck with them, and more of us will join Dzar if they think we're being deliberately obstructive."

He cradled his left wrist against his body. "But you said . . ."

"I know what I said. The longer we can keep them on the plateau, out in the open, the better the odds that the Ministry satellite will register an unscheduled shuttle, take a closer look, and realize something is wrong."

"If any of those useless bureaucrats even looks at the data," he sniffed.

"How long did it take them to show up when the Mictok opened that tomb?" When Salitwisi began to smile—nothing cheered him up faster than a rival's misfortune—Arniz added, "Spread the word. No one turns this into a teaching moment. Let them believe they'll get results."

His tail tip flicked back and forth. "And the Druin thinks she's educated."

"Why does the jungle just *stop*?"

Arniz approved of curiosity, of a willingness to learn, and this was a teaching moment that had nothing to do with a nonexistent weapon.

"It doesn't just stop." She indicated Trembley should turn and look at the tree line. "We assume the pre-destruction peoples continuously worked to keep the jungle from encroaching on the western city limits. With nothing to stop it post-destruction, it went over the wall and moved east. Over the centuries, it engulfed three quarters of the city before the drier, shallower soil here on the plateau slowed it, but it hasn't *stopped*. Eventually, it'll reach the edge of the cliff." The dark line of jungle visible on the other side of the broad crevasse, vines and creepers tumbling down the rock, supported this hypothesis. "Some of my colleagues believe that the ruins on the far west of the city, those first covered, may have been preserved with little deterioration. I, personally, subscribe to the belief that said ruins have been pulled down by the weight of the cumulative years of vegetation. You can't trust vegetation."

He shifted dark glasses taken from one of the larger Katrien down his nose to frown over them at her. "Why not?"

"It's too ephemeral."

"I don't know what that *means*."

Smarter than he looked to admit that, she acknowledged. "It changes too quickly. You can count on soil."

"If the west side is *better*, why are you here?"

"It's not *better*. And we're here because our license from the Ministry extends to the tree line and no further. Had this season gone well, next season the university might have been permitted to breach the canopy."

"So this season *hasn't* gone well?"

She stared up at him. He flushed.

"Trembley!"

She hadn't heard Martin approach. From the way his scent spiked, neither had Trembley.

"Enough talk. Get the furballs back to work."

"Yes, sir!" Trembley jogged toward the high-pitched sound of a Katrien argument, boots thumping out a bass line against the packed dirt.

Arniz had placed her chair in full sun, close enough to where three ancillaries worked to be available if needed, far enough away from the hastily constructed cluster of terminals to avoid another conversation during which Salitwisi declared he could read the results as well as she could. When the sun disappeared behind a Martin-shaped shadow, Arniz tasted the air. The big Human wasn't happy.

"What do you think you're doing?"

"Sitting. I'm old."

"What were you doing with Trembley?"

"Teaching. You shot my ancillary, I'm making do."

"You know nothing he needs to learn." He knew how to show his teeth, she'd give the *verbin kur* that.

She watched him rejoin the tall, pale female Human by the edge of the grid and wondered why she couldn't remember her name. She thought it might have something to do with melons. Had the female Human been an alkali soil, she'd have been able to remember her unfavorable physio-chemical properties, so there was nothing wrong with her memory. Now Zhang, on the other hand, that was a name she remembered. The word felt good in her mouth. Zhang's first name, however . . .

"There's something not right about this."

"There's nothing wrong," she snapped at Ganes, who'd snuck up on her other side. "They're a different species with a different naming structure."

He blinked, twice, and raised both hands. "What are you talking about?"

"What are *you* talking about?" Shells, she sounded like Trembley. And how could Ganes know what she'd been thinking about. "Of course it's not right," she muttered. "Dzar's dead, and we're being kept from our work by her murderers."

"The Humans are holding themselves separate."

Arniz took another look around. With the Katrien back at work, Trembley had returned to Martin's side, his head down as Martin's mouth moved. The other two were listening passively, hands on their weapons. She couldn't see Yurrisk and Qurn, but Beyvek and Sareer, the other two Krai, were by the shuttle, the three di'Taykan were watching three ancillaries running scanners over on the northern edge of the grid, and the three Polint were racing in from the edge of the cliff. Tehaven, the smallest, with the variegated pelt, seemed to be winning. "Holding themselves separate from the Primacy?" she asked. "That's to be expected, isn't it? They're all ex-military and they spent centuries slaughtering each other."

"No, separate from everyone."

"I don't see it." No one seemed to like each other very much, but they were working together. Where *working* could be defined as maintaining a threatening environment.

"You're used to academic infighting," Ganes told her. Unnecessarily. She'd been there. "I spent years working as part of an integrated team—Humans, di'Taykan, and Krai. Trust me. These Humans are not integrated."

The Polint tended to remain close to Martin because of the translation program on his slate, but she doubted that's what Ganes meant by integration. When he moved closer, she saw the darker blotches of bruising on his arms and a scab at the corner of his mouth. "Martin?"

For a moment she thought he'd pretend to misunderstand the question, and in all honesty it had been a bit anomalous, then he shook his head. "No, the Polint."

"Why?"

"Multiple Niln, multiple Katrien, and I'm alone. They were emphasizing that."

"And Martin was encouraging them." It was exactly what she'd expected of the *yerspit*.

"No, surprisingly he wasn't. They ignore me when he's around."

Martin had a big hand cupped around the back of Trembley's neck and seemed to be shaking him gently. Like an elder with a hatchling. Arniz felt a bit ill thinking of Martin performing any kind of a parental function, but she couldn't deny what she'd seen.

From the way Ganes stared, he'd seen it, too.

• ——◆—— •

The Polint quarters had been the largest packet added onto *Promise*, a five-by-six-meter rectangle with food storage and preparation along one long bulkhead and sleeping mats laid out along the other. With the mats rolled up, it was the only cabin large enough to hold all thirteen of them.

With a single exception, Torin noted, the two teams maintained a careful separation. Alamber, having already decidedly lost a game of *you show me yours, I'll show you mine,* had draped himself over Bertecnic's back, and the big Polint seemed pleased by the contact. Other than that, the Primacy had gathered to the left and the Confederation to the right, nearer the hatch. Presit and Dalan stood by the far wall, opposite Torin, camera ready but not on, clearly waiting for something notable to happen.

Torin was impressed. Five years ago, Presit wouldn't have waited.

The low buzz of conversation barely rose above the hum of the Susumi engines, the agreement holding that Federate and Prime alone would be used and that every word spoken would be translated.

Expression carefully neutral, Torin watched Vertic speaking to Dutavar. Her expanded briefing packet had compared Polint social structure to that of bees. A female chose a pod of three to five males, the most favored male eventually becoming fertile, the others remaining in close support. Torin wondered if Vertic was assessing Dutavar under biological parameters, recognized it was none of her business, and let it go.

"We're ready, Dur . . . Vertic." Freenim's inner eyelids flickered. Torin appreciated the problem he had dropping the rank. Before Vertic could respond, had she intended to respond, 33X73 appeared in the center of the room.

"It are being a hard light mapping feature!" Presit gestured and Dalan, who looked bored even considering fur and mirrored glasses, raised the camera, the recording light on.

"We took into account that this would be our gathering place, and asked to have it installed rather than use a flat image." Freenim reached up, touched the control panel set into the bulkhead, and adjusted both size and brightness. "Although we weren't permitted to sync our slates to your ship."

"Fukking right you weren't," Craig muttered. "Need more slates slaved to my ship like I need a third armpit. My slate's plenty."

"So we were informed."

Freenim shot her a side-eye. The Taykan had no whites either and Torin was used to interpreting expression around the absence. Damned right her slate had full access to *Promise*'s systems. Craig knew; he chose to ignore it.

She walked to the planetary image, then around it, then indicated the half-dozen red dots in orbit. "The satellites belong to the Ministry for the Preservation of Pre-Confederation Civilizations and are entirely useless for defense. Or offense. Or early warning. I'm sure they have a function, but it has nothing to do with us. There's one communication satellite used by all the scientific teams. It has Susumi access, but minimal bandwidth because the Ministry cheaped out and has nothing to do with us either. We'll run communications through the *Promise*. The only ship in orbit belongs to the mercenaries as the universities dropped their teams off at the beginning of the season and won't return until the season ends. That's a little over eight ten-days from now, planetary day at 29 and a half hours give or take a minute or two we're not going to worry about." For the moment, she ignored the planet itself. "Were I one of the mercenaries in this situation, I'd have set up an orbital alert."

"If you were a merc, Boss, known space would be fukked."

"*You'd* have set up a perimeter alert?" Ressk asked over the laughter.

"Valid point," Torin acknowledged. "I'd have an orbital alert set."

"You are having given this some thought, Warden Kerr." Presit's teeth showed. "Should we be being grateful you are being on our side?"

Torin raised a brow toward the camera, then turned her attention back to the room as a whole. "We'll look for an alert when we're close enough. For now, let's discuss how much we'll be able to see on the ground. Craig."

"Through that much atmosphere . . ."

The map table showed it at six hundred and twenty-two kilometers with a high concentration of both moisture and particulates over the site in question.

". . . our scanners'll light up where the warm bodies are, but that's it."

"Wait!" Presit held up a hand, fur ruffling around her wrist. "You are saying the military are not having scanners that are being able to find the Primacy on the ground through any atmosphere our military can breathe?"

"No idea. This ship doesn't have military scanners."

"And are that not being just a little bit shortsighted?"

"This isn't a military operation." Torin answered before Craig could.

"I are just saying, it are being strange to me that the military are having better equipment to be killing people than you are having to be saving them."

"The Justice Department and the military are continuing to define their levels of cooperation," Vertic said smoothly, having moved to Torin's side. "Just as the Primacy and the Confederation continue to work on theirs. Every new operation takes time to reach full efficiency."

Presit combed her claws through her whiskers. "I are assuming Primacy officers are having training in dealing with the press?"

"Of course." Her tone was so neutral, it was almost a threat. "If you'd continue, Gunny."

"Sir." She couldn't prevent the involuntary response. Craig glanced over and frowned. "It'll be boots on dirt, people. Unfortunately, this . . ." She expanded the map. ". . . is all the dirt there is. That's the anchor, that's an occupied landing pad, that's a cliff, and that's a lot of jungle."

"They'll have eyes on every square centimeter of open ground," Craig added. "I'll have to put us down in the trees."

Freenim leaned in, shifted the perspective, and touched the map where it showed a break in the vegetation. "Here?"

"There," Craig acknowledged. "Tight, but doable."

"Eight point six klicks out." Torin pulled up the relevant data. "On the way in, you could drop a recon team here, at five point three."

"Would it be worth the extra burn for just over three kilometers?"

"It's a jungle," Torin told him. "Triple the time, minimum, to cover any distance."

"Uh, Gunny . . ."

"Unless you're Krai."

"Your shuttle is Taykan built, and their stealth tech was adopted by your military." Freenim expanded a side bar showing average temperature and humidity. "Under these conditions, five point three is sufficiently distant for horizontal flight to remain unheard on the plateau."

Craig frowned. "And you know this how, mate?"

"It was part of my job." He nodded at Torin. "As the opposing knowledge was part of hers. Infantry are vulnerable to air attacks."

Bertecnic stood, tumbling Alamber to the deck. "What if we burned out a base camp closer . . . will they see the smoke?"

"We aren't burning out a base camp," Ressk snarled, nostril ridges closing.

By his side, Werst had begun to growl, the sound rising from deep in his chest. He'd been raised by the space port, on concrete and wire catwalks, while Ressk'd had a more traditional Krai upbringing, but trees were a part of their species identity.

Both male Polint scraped the deck with their front claws. Muscle quivered under golden fur as Vertic held herself still.

Alamber retreated to a safer position behind Binti.

Firiv'vrak chittered out a string of consonants the translator ignored while she and Keeleeki'ka—who'd been avoiding each other—tucked their legs in close and settled to the floor smelling of wet dog and cinnamon.

Dalan no longer looked bored.

No one in the cabin was armed. Even knives had been surrendered.

Not that it would matter.

"Enough!" Torin snapped, voice filling the empty spaces, impossible to ignore. "Relevant emotional context aside, we aren't burning out a landing site because that *would* attract the attention of the Ministry satellites and if we want to get all hostages out of this situation alive, the last thing we need is a swarm of bureaucrats descending to complain about interference with biodiversity on a Class 2 Designate."

In the long moment of silence that followed, everyone in the cabin considered the possibility.

Krai nostril ridges slowly opened. The sound of claws against metal stopped.

"Bureaucrats," Werst grunted, like the word was profanity. "We'd have to rescue them, too."

Bertecnic's tail flicked from side to side. "No matter how little we'd want to."

"Common enemy?" Freenim said quietly beside her.

Torin set 33X73 spinning. "Whatever works."

FIVE

THERE WERE STORIES OF SHIPS emerging from Susumi space, the crew dead of old age having traveled for a lifetime yet still arriving moments after they left. There were stories of ships jumping in and out, traveling no distance at all, their crews unchanged, centuries having passed since their departure. Susumi engineers declared both stories false. An error in the jump equation would lead *only* to an error in destination.

Light-years off course, unable to jump home.

A few thousand kilometers off course, attempting to share space with another solid object.

Torin disliked the engineers' qualifier.

The jump to 33X73 would take four days. They'd arrive twelve seconds after they left. She had no idea how it worked, but she trusted the math. It was that or stay home.

Strike Team Alpha had never needed the simulators the Corps used in Susumi space. With full knowledge of individual skills and the way six individuals fit together into a whole, they could plan over pouches of beer in the galley and train in the area opened up when Torin and Werst surrendered their personal quarters. But all three Polint wouldn't fit into the galley, and the "gym"—even at double the minimum—had never been able to hold more than four bipeds. Fortunately, the Polint quarters were large enough to knock off a few of the rough edges.

Torin watched Bertecnic take six running strides and stop abruptly a meter from the bulkhead. Werst used the momentum to launch himself off Bertecnic's back, hit the bulkhead, flip around, grab the first of a dozen hanging ropes, and start back the way they'd come, avoiding Dutavar, who'd risen up on his hind legs and braced his palms flat against the ceiling. The Polint were too top-heavy for him to hold the position, but Dutavar was flexible enough to twist and land facing his one ninety, making him harder to escape than Bertecnic. Ressk still hadn't managed it. Werst was trying for best two out of three.

Dutavar grabbed an ankle as Werst slapped the target.

"Tie," Torin called, and the developing argument became a loud replay of the run.

"More civilized than the first time," Vertic observed.

The first run had very clear delineations between *us* and *them.* As well as a clear imprint of Werst's teeth in Dutavar's foreleg after having been pinned under a hundred and fifty kilograms of angry Polint. Krai bone being one of the hardest substances in known space, Werst's ribs hadn't broken—which was why Torin had begun with this foursome. She trusted training and experience to stop Werst and Ressk before biting became biting and chewing, even when the Polint used their size against them. The Corps had strict policies against eating allies.

Turned out the Primacy held a similar position against disemboweling. Claws remained sheathed.

"They should be able to work together by the time we arrive," Torin allowed. Shared military backgrounds allowed for shortcuts—even if those backgrounds had involved shooting at each other.

"You don't sound happy about it."

"I'd be happier if we could get a look at each other's weapons."

The weapons had been stowed behind a time lock; inaccessible until they came out of Susumi. Torin didn't like it, but she understood. After four days, misunderstandings likely to lead to violence would have been dealt with nonfatally.

"At least we know how much damage each other's weapons can do."

"That's very comforting."

Vertic made a noncommittal noise and said, "How did your interview with the reporter go?"

Torin assumed Vertic had already debriefed Freenim. "Presit seems to think we should be angry."

"All of us? Or specifically you and the durlave kan?"

"Specifically."

"Because the two of you were essentially the same rank?"

"That's what I assume. Werst! Don't make me into your goddamned playground monitor; it's Ressk's turn."

"And are you? Angry," she added when Torin turned toward her.

Torin curled her right hand into a fist to keep from touching the weight of cylinders in a vest she wasn't wearing. "Not at Freenim."

"You're not stupidly patriotic," Torin said, stepping off the treadmill after ten kilometers and reaching for a towel. She'd been watching Dutavar use the resistance bands, waiting until exertion had worn off some of his prickly defensiveness—he'd share the gym, but he wouldn't talk. "I can spot that kind of jingoistic crap a kilometer away. It's not species specific."

He grunted and leaned into the maximum resistance, patches of his fur dark with sweat. He smelled better than a Dornagain. Probably better than she did right now, Torin admitted.

"I understand why your military wants one of their own here." She tossed her wrist weights in the bin. "You'll be reporting back on our preparedness. They'll want information on weaknesses that can be exploited should war begin again. The things a civilian wouldn't notice."

He thought he was giving nothing away, but she'd learned his tells watching him train. He'd have lied to her had they had this conversation back on the station. Physical honesty was a lot harder to fake.

"If a weapon to destroy the plastic exists, the more realistic among your superiors want you to get as much information on it as possible—pictures, scans, schematics. I guarantee someone highly placed and political suggested you grab it and run."

His tail flicked once. Dismissively.

"They likely pointed out that there'll be six Polint on 33X73 and six Polint can easily overcome the minimal opposition present because, of course, the other three, the three currently working for Robert Martin, will, in the end, choose to fight on the side that benefits their own species."

His shoulders rose and his rhythm faltered.

Torin crossed to the cooler and pulled out a pouch of water. "We have those types on our side as well. Some day, if you're very good, I may tell you about General Morris." She preferred room temperature water, but she'd forgotten to get a pouch out before she started her run, so she drank it cold. Ex-Gunnery Sergeants didn't make mistakes. "What I don't understand," she continued, leaning against the wall, "is why you *volunteered* for this shit job. You can't fully integrate into the team because you're serving military. You have to remain an objective observer. When you get back, no matter how much information you give them, your superiors are going to want more, and you won't be able to fully reintegrate into your old unit because you've been behind enemy lines on your own. What were you up to? Not to mention . . ." She took another swallow of water. ". . . no one entirely trusts volunteers."

The muscles in his back were so tense, he was going to hurt himself if he kept working the bands.

"If you're trying to impress Vertic . . ."

"No!" Dutavar jerked toward her, breathing heavily, bands at full extension. "Our ship was in Susumi before I met her. My presence here has nothing to do with Vertic!"

Alien gender politics; Torin knew better than to get involved. "But it's personal, isn't it? Tehaven, down on 33X73, he shares your markings."

Dutavar's lip curled and, arms trembling, he slowly let the bands slide back into the bulkhead. "We're the same color, so we must be connected? Is every human with brown hair and eyes personally connected to you, Warden Kerr? Is your universe so small?"

She shrugged. "Brown on brown's the default in my part of the universe. Your particular pattern—the orange in the black, white, and gray variegation—Vertic tells me that's rare. She says it only occurs in one family line. Rare enough that a variegated Polint there with Martin and a variegated Polint here with me is unlikely to be coincidence. Do not," she snapped, "claw the deck padding."

"Sorry, Dur . . ." Teeth cracked against each other as he shut his mouth around the Primacy rank, ears down, mane flat.

Not the first time he'd been told. Torin shifted until her posture became more gunnery sergeant than Warden and met his eyes. "Well?"

Claws emerged and disappeared again. His mane rose. A muscle jumped in his jaw.

Torin waited.

"He's my brother," Dutavar growled reluctantly. "The youngest. He's a damned fool and desperate to prove himself. Our mother wants him home alive. My superiors owed her a favor."

"You didn't exactly volunteer."

"Not as far as my mother's concerned, no."

"And your orders?"

"Everything you said. Watching. Reporting. Get the weapon if I can, get specs if I can't."

"And your actual orders?"

"To bring my brother home alive."

Torin threw Dutavar a clean towel. "I hope she knows you can't guarantee his safety."

"Do you *have* a mother?" he asked, rubbing at the moisture on his chest.

"Fair point."

• ——◆—— •

Craig shifted his knee on Keeleeki'ka's back, his weight pressing her against the floor. "Stay down!"

"We're on a ship in the between." Her antennae flicked back and forth, a touch against his thigh and away. "A ship you control. Where would I go where you couldn't find me?" Antennae relaxed into a sullen curve, she clicked her outer mandibles. "I only wanted to learn about you. About your worlds. About your Confederation."

"In my control room?" A new scent overpowered the licorice and lifted the hair on the back of Craig's neck. On ship or station, even considering the amount of ceramic, the smell of heated metal did not evoke a neutral reaction. Vacuum was unforgiving.

"This is the heart of your story." Keeleeki'ka unfolded her upper arms, fanning out the multiple slender digits, reaching for nothing. "We only know what the council tells us about you and we know they lie."

Civilian salvage operators, even ex-civilian salvage operators who'd become Wardens, weren't big fans of the government. Craig eased his weight up. "Don't run."

Releasing his white-knuckled grip on the edge of the shell, Craig shifted his weight to his other leg, and straightened. His knee had barely cleared the carapace when the Artek slid out from under him and scuttled toward the control room hatch.

"*Promise*. Lock two." The bolts slid home. Craig folded his arms and glared. "Not until we're done, Keelee."

Pivoting 180 degrees on her rearmost legs, Keeleeki'ka backed into the closest corner. "The others were your enemy, not me!"

"You're the drongo who stuck her mandibles in where they didn't belong!"

"I didn't bite your ship!"

"I never said you did!"

"Keelee is not my name!"

"So I should click for forty minutes when I talk to you?"

"Yes!"

"Why didn't you ask if you had questions?"

Keeleeki'ka's wedge-shaped head swung from side to side. "Why would I believe you would answer? Knowledge is power. Why would you give power to me?"

"Knowledge also keeps you from making stupid mistakes. Like messing around in my control room!"

Craig?

He tongued his implant. "Torin."

I just got a ping that you locked the hatch. Everything all right?

It hadn't been that long ago that he'd been unable to handle having anyone else on his ship. Limited air. Limited supplies. Unlimited sweat and a bung brain. He could hear the memory of those bad old days in Torin's voice. "Everything's aces. Keeleeki'ka's up here giving me an ear-bashing, didn't want to be interrupted."

Remember what Vertic told us.

"That she's not the officer you're looking for?"

What? No, that Keeleeki'ka is more important . . .

"Than she appears." Craig blew out a deep breath and ran both hands back through his hair. "Memory's not that bad, luv."

If she's willing to talk, see if you can get anything useful out of her.

"About?"

Anything. Except how to kill her. Torin sighed and he knew exactly the expression she'd be wearing. **I know how to do that. Anything else. Knowledge is power.**

"So I've heard."

Record everything.

"Teach grandpa to suck eggs." With thirteen members of the Primacy Torin's responsibility—however much she changed the definition of responsibility when referring to the six on the ship and the seven on 33X73—she'd be a fool not to want any conversation between the two halves of her team recorded. And Torin was no fool;

any foolishness she'd brought into the Corps had been trained out of her. Replaced with responses more useful to war. He'd hoped he'd replaced a few more of those than it seemed as if he had.

Craig!

Keeleeki'ka had swiveled all eye-stalks in his direction.

"Sorry. Thinking."

I said, or I could listen through your implant.

"Or you could continue turning recent enemies into a cohesive unit because you're just that good. We're not fighters, Keelee and I. We'll be fine. No drama." He tongued off his implant before Torin could reply. The pilot's chair creaked a protest when he dropped into it and, as the familiar support wrapped around him, the stiff ache in the line of his shoulders relaxed. "All right, you want to find out about us, fine, we keep it fair. You and me. For every question I answer, you answer one of mine."

Keeleeki'ka scuttled closer, upper body slightly raised, the weak points on her undershell exposed. Craig couldn't decide if it was trust or carelessness. She smelled of acetone. There was a fair go that meant she was feeling smug. "We're in your territory; you ask first."

"Okay." He couldn't think of a damned thing. The smell of acetone grew stronger. "Okay. What makes the Sekric'teen different from the rest of the Artek."

"Ah, a good question." The smell changed to cedar shavings. Approval. "We hold the origin of our people."

"You're historians?"

Her arms and antennae waved in counterpoint. "Yes. When the council agreed one of us would go, I was chosen to hold this story."

"Can you fight?"

"I hold the story."

Yeah, that was helpful. He swung his feet up onto the board and crossed his legs, right heel in the paint-free divot he refused to have repaired. "Your turn."

"I believe you asked two questions, so my first is this: Who is grandpa and why must he learn to suck eggs?"

• —◆— •

"Historian?" Eyes on the two Druin working their way through the Krai's climbing lines, Firiv'vrak rolled her antennae. "Inflexible, pedantic, hypercritical, unimaginative, bombastic, supercilious . . ."

Torin was impressed by the translator's vocabulary.

". . . opinionated pains in the thorax. The Sekric'teen had the lowest percentage of military involvement across our entire species. I knew nymphs cut off entirely because they bucked tradition and volunteered. On behalf of the Artek, I apologize to Captain Ryder for the lecture on our glorious history he's no doubt having regurgitated on him as we speak."

"We only know what the people who put the translation program together want us to know."

Craig kissed her shoulder and wrapped an arm around her waist, pulling her closer. "Your brain goes sexy places post-coital."

"The translator had no problem with supercilious, but paused before spitting out plastic aliens."

"So the Primacy calls the little fukkers something else, no surprise."

"You're right. It isn't. But why can't we know what the Primacy calls them?"

He yawned, her hair moving with the force of his eventual exhale. "Why do we need to?"

"No reason." She pushed his arm up under her breasts, and cradled it with her own. Whatever had been bothering him earlier seemed to have passed before they had to talk about it. "Don't change."

"Wasn't planning on it," he murmured sleepily against the top of her head. "You neither, my paranoid *preciosa*."

"Spanish?" It had been a common second language among Craig's Human friends on the salvage station.

"Spanish is sexy."

"Craig . . ."

"No. You can listen to the Q&A tomorrow. Go to sleep."

That night she dreamed of Staff Sergeant Harnett, her hands bloody as she threw herself at him over a wall of the starved Marines he'd enslaved. She twisted a muscle in her back as she jerked awake, but managed to keep from punching Craig in the throat.

". . . and what you have to realize is that my people are essential in keeping the history of the Artek alive and it was because of our intimate knowledge of that history that we were able to notice our leaders were not acting as Artek always had, so we knew there had to be outside influences . . ."

Craig shook his head as he silenced the recording. "Her people spent centuries holding onto traditions with all eight to sixteen fingers and arguing that something was wrong. A lot of it's devolved to rhetoric at this point, but being right has put them in an interesting position. A chunk of the Primacy want to elevate them to a kind of priesthood. And a bigger chunk resents the hell out of them for being right."

"She told you that?"

"Didn't need to, did she?"

No, Torin acknowledged, she didn't. Get past the superficial physical and cultural differences and sentience was a one trick pony. And Craig could read everything he needed to know about a mark across a poker table between one card and the next. "What did she learn about us?"

"Bit of history, bit of politics. Then it got personal." He grinned as Torin's eyebrow rose. "She thinks it's disgusting that mammalian embryos are internal parasites, and was appalled we didn't use artificial wombs. Apparently, they're a popular option in the Primacy."

"Get the import license, we'll make a fortune. Since you don't need to listen to this . . ." She fished her shoes out of their compartment in the bulkhead. ". . . I'll review it on the treadmill. Later, we'll compare it to Presit's inevitable interview and look for discrepancies. If someone's been sent to sabotage the mission and prove the Primacy and the Confederation can't work together, Keeleeki'ka is our wild card."

"Her people were against the war."

"Thus, wild card." Her voice trailed off under the weight of Craig's regard. "Look, I'd like us to be part of a new alliance, but I have to consider other possibilities."

"And if it's beer and bikkies all the way to the bottom?"

She stopped to brush his hair back on the way to the hatch. "Then we'll have a party. But everyone's safer if I assume death and destruction. And I need to keep my people safe."

"Two thirds of this file is redacted." Werst looked up from his slate and glared at Torin. "What the fuk happened on the *Paylent*?"

"My file is as redacted as yours," Torin told him. "We're all reading the same thing."

They were all back in the Polint quarters, going over the briefing packets on the mercenaries together. Experience had taught Torin that one of the best ways to avoid conflict was to make sure everyone involved had the same information and be damned sure they interpreted it the same way. "Petty Officer Sareer, Lieutenant Beyvek, Lieutenant Gayun di'Dizon, and Seaman Pyrus di'Himur all in the image Commander Ganes got out, all among the fifteen survivors of the destruction of the *Paylent*."

"It was a cruiser," Binti said softly. "That's three hundred enlisted, thirty officers."

"That's three hundred and fifteen dead," Alamber added, his hair flat against his head.

Ressk snorted. "No surprise they didn't re-up after that."

"Any of you lot able to fill in the details?" Werst growled across the room.

"My people don't . . ." Keeleeki'ka began.

Firiv'vrak cut her off. "Your people don't fight. They all know that. Shut up. It wasn't a boarding I was a part of, nor a ship whose honors I know."

Werst's lip lifted. "Three fifteen dead isn't hon . . ."

"Enough." Torin cut him off. "We're not refighting it now."

"We . . ." Vertic's gesture included Bertecnic and Dutavar. ". . . were ground troops. We'd no more know about it than you would. Freenim?"

"No." He glanced at Merinim and tapped the back of his left hand. She returned the gesture. "Neither of us have heard of it."

"I guess you lot killed so many . . ."

"Werst!"

He took a deep breath and exhaled slowly. "Sorry, Gunny." Looked across the room, nostril ridges open. "Sorry. War's a fukking waste." Another deep breath. "A lot of Krai in the Navy."

Ressk reached over and wrapped a foot around his ankle.

"Fourteen survivors mentioned a Commander Yurrisk, helmsman. Krai. All of them said they wouldn't have made it without him." Torin knew she was holding her slate too tightly. "Although what he actually did has also been redacted."

"Heroes." Vertic held her hand out, palm up. "Villains." And turned it over. "Those who give the orders seldom want to know exactly how we carry them out."

"Fukking right." Werst lifted his pouch of sah, and everyone save Keeleeki'ka, Alamber, and Craig joined in the salute.

Torin had to force her swallow of coffee down before she continued. "Three years ago, an Aggressive Class minesweeper, the *DeCaal* was registered to Commander Yurrisk, who'd left the Navy with a medical discharge."

"Details of the discharge redacted," Ressk muttered. "A lot of detail redacted given the original file length. Poor fukker."

"His registered crew, Sateer, Beyvek, di'Dizon, di'Himur, and Mirish di'Yaunah, di'Dizon's *thytrin.* Seems safe to assume that when Commander Ganes took his picture, Commander Yurrisk was in the shuttle or still in the *DeCaal* in orbit. Robert Martin, Brenda Zhang, and Emile Trembley have all been keeping a reasonably low profile since they left the Corps. Trembley's only been out for a year and there's nothing in the public records. Zhang was cautioned ten tendays ago by security on an OutSector station after a public screaming match with a Katrien."

"Cautioned or deafened, Gunny?"

Presit had declared she was perfectly capable of reading briefing packets without help, borrowed two bottles of Alamber's nail polish, and disappeared into her quarters.

"Martin," Torin continued when the laughter quieted, "has been fired from fifteen jobs in the last eighteen months. And Jana Malinowski, who saw more combat than the other three put together, has been arrested six times for fighting and has consistently skipped out on her court-appointed therapy."

"Who would do such a thing, Boss?"

"Damned if I know. All four were basic infantry, no specialties, and Martin did legitimately make sergeant. Field promotion, just before his contract ended."

Keeleeki'ka waved both antennae and all four arms. "I have a question. If Robert Martin is a sergeant and Yurrisk is a commander, why do you give Commander Yurrisk his rank but not Robert Martin?"

"Because for whatever reason they're on Threxie, Commander Yurrisk is broken and Robert Martin is an asshole. What?" Craig tossed his empty coffee pouch in the recycler and reached for another. "More than just a pretty face. I can read between the lines, and I know how Torin thinks."

"Poor fukker," Ressk repeated.

"You're very lucky," Vertic told him.

Craig smiled tightly. "Yes, I am."

Alamber's eyes darkened and he glanced from Craig to Torin. When he opened his mouth, Torin glared it shut again. "Vertic, if you could cover the Polint."

Vertic had no more personal information on any of the three than what had been in the briefing packets, but she went over the different fighting styles. "And while Camaderiz may be the only one with military training," she concluded, "do not discount the other two. They can point a gun and pull a trigger and, if they have enough ammunition, no one will care about a lack of precision shooting."

"What about the blades?" Ressk asked.

Her nostrils flared. "Don't get close enough for them to bring the blades into play."

There were a few snickers, but no one in the room doubted she was serious.

"If it comes to close quarter fighting, one on one," Torin said, holding the attention of everyone in the room, "the Polint will fight the Polint. No arguments. And, while we're on the topic of Polint fighting Polint, *Santav Teffer* Dutavar has information to share."

His ears flattened.

"Or would you rather I did it?"

Arms folded, mane up, Dutavar glared at Torin.

Torin raised a hand and cut Vertic off. Then she waited.

"It's no one's business," he growled.

"You know better."

He did. She could see his mother's instructions fighting the abilities that had kept him in the military longer than most Polint males. His lips curled back off the ivory slabs of his teeth. "Tehaven is my brother."

"Are we supposed to be surprised?" Alamber asked as Dutavar swept a challenging scowl around the room. "You look like copies."

"Netrovooens has the nearly same coloring as Bertecnic," Dutavar snarled. "Do you assume they're brothers?"

Alamber grinned. "Even with crappy slate resolution, I can tell Netro's not a copy of Bertecnic. Not by, as it were, a long shot."

When Vertic tried to silence the resulting commentary, Torin caught her eye, shook her head, and mouthed, *let it go*. Credit where it was due, Vertic seemed to understand the reasoning.

"Integration seems to be going well." Freenim had crossed to Torin's side during the initial flurry of speculation.

Torin opened another coffee. "It's a di'Taykan thing."

"I remember."

"Still nothing on the Druin in red?"

"Surprisingly, no." He blinked, inner eyelid sliding across the black. "Our government seems to have forgotten to load a full Druin population census onto our slates."

Torin toasted him with the coffee.

"But Merinim says she dresses well."

The bulkhead outside the Polint quarters rang under Bertecnic's fist. Dutavar grabbed his shoulder and hauled him back with enough force his front feet came off the deck.

Torin moved toward them, mouth open to call for backup, and staggered sideways as Vertic raced passed her, the width of the passage not adequate for an adult Human and a running Polint. When she reached the males, Vertic grabbed an ear in each hand, yanking them first apart and then into their quarters. Bertecnic's tail flicked out of the way at the last second as the hatch slammed closed behind them.

The silence was definitive.

And quickly broken.

"Do not be asking them about the Ner. They are getting emotional," Presit added as Torin turned toward her.

The Ner rode the Polint into battle. Torin had faced them at the siege of Simunthitir and while they were small, they were vicious fighters and good shots—the latter remarkable as the Polint ran like cats. "The durlan and Bertecnic lost theirs before the prison planet, in the battle that destroyed Sh'quo Company," she pointed out. "They were over it then."

Presit combed her claws through her whiskers, right side, then left. "Then, yes, but now Dutavar are still being military and are having to leave his Ner behind. He are still being unhappy and that are reminding Bertecnic about being unhappy. They are not willing to be sharing being unhappy." She snorted dismissively. "Competitive grieving."

"Maybe they'll bond over it."

"Is that being what the young are calling it?"

* * *

"We can't hide in the trees and pick them off one at a time; we'll risk the hostages' lives." Torin folded her arms. "We can't land on the plateau and swarm them; we'll risk the hostages' lives. We can't bomb them from orbit; we'll . . ."

". . . risk the hostages' lives." Vertic didn't join the chorus, but everyone else did.

"And we aren't carrying the ordinance anyway. There's a shitload more we can't do," Torin continued. "I know it. You know it. We're done discussing it. Bottom line, doing the job means getting all the hostages out alive. Recon team will drop in this clearing here . . ." She tapped the image. ". . . Craig lands the shuttle here."

"You and me on recon, Gunny?"

She wanted to put boots on the ground. Wanted to look up and see sky. Wanted to get from point A to point B over hostile terrain, with someone she trusted watching her back. Wanted her eyes on the mercenaries instead of on briefings. No one would argue her dropping with Werst—they were the only two members of the extended team with time in reconnaissance.

"You and Ressk," she said, expanding the image until they could see through the canopy. "You can take the path of least resistance. Move faster."

Werst shook his head, nostril ridges closing. "Ressk . . ."

"Was a Marine, like the rest of us. And he has better range scores than you do." Torin held Werst's gaze. "I convinced Commander Ng your bonding wouldn't affect your performance in the field. Calling me a liar?"

"No, Gunny."

"Good." Ressk would likely have more to say about it, but she was done. "Get the lay of the land, max out the DLs, then head back. Top speed. DLs, look and listen," she added in response to Vertic's silent question. "Surveillance."

"One of our people should go as well." Dutavar's mane lifted as Torin turned toward him. He was challenging? He was showing respect? She could not get the hang of the mane movements. "This is a joint mission."

"Artek are frequently used in reconnaissance," Freenim said. "They're adaptable to a number of environments and hard to kill. As long as the three Polint remain unaware the Wardens have been joined by the Primacy, they'll assume local insect life."

"Which are large." Alamber stared unhappily at hard light images of sixteen insects that ranged from half a meter to a meter long.

"Only nine are on this continent . . ." Torin deleted seven images. ". . . only four of those are poisonous, though not to all of us, and the med kit will have finished the inoculates in plenty of time."

"Yeah, not exactly comforting, Boss."

"Insects and snakes." Merinim spun another screen of images, bright colors flashing. "This one has wings. It's a snake. With wings."

"Vestigial. It can't get its entire body off the ground."

"Since it's big enough to swallow me whole, not really comforting, Gunny."

"The Artek . . ."

"Are not trained for this," Firiv'varic broke in. "Pilot." Her antennae flicked toward Keeleeki'ka. "Civilian. You point me at a target, I can fight, but don't expect me to know what I'm doing scuttling through the undergrowth on my own." The scent of heated milk momentarily overwhelmed the air filters. "And don't expect *anything* of her."

Keeleeki'ka snapped her mandibles. "I'm here to witness."

"Us die?"

"Enough." Torin swept an uncompromising gaze over her team. "We rescue the hostages, we attempt to capture rather than kill the mercenaries, we do it together. And we do not question every word out of my fukking mouth. Is that clear?"

A low rumble of *yes, Gunny* and *clear, Gunny* ran through the room. Not exactly resounding, but Torin would take it. Craig, leaning on the bulkhead by the hatch, winked.

"I'd still rather avoid the jungle." Vertic spread her hands, claws emerging. "We're good climbers, but we're heavy. Our people are more comfortable on rocks than trees. If we drop into the cleft . . ." She expanded the edge of the plateau. A waterfall poured out of the jungle one point seven three kilometers in from the advancing tree line. The anchor was barely visible. ". . . it's a two-point-two–kilometer climb."

"We'd still have to land far enough away the mercs won't hear the shuttle." Torin adjusted the angle. "And there's jungle at the bottom of the cleft."

"The river . . ."

"No room to land upstream, so we're shit out of luck on a nice silent float." Craig shrugged. "Trust me, I looked. Unless we want to announce our presence with authority and land on the plateau, it's hack through approximately five klicks of jungle and then climb the cliff, or hack through eight klicks of jungle, skip the cliff."

Vertic looked ready to keep arguing, visibly stopped herself, and nodded.

Torin threw up a meteorological report before the Primacy officer could change her mind. "It'll be hot and humid under the trees, but they have three di'Taykan to our one and, as they're not in uniform, they don't have access to environmental controls. That'll be to our advantage. There's an impressive amount of pollen, so no one forget broad spectrum antiallergens before leaving the VTA." She stepped away from the image and folded her hands behind her back. "All but one of the hostages are Niln or Katrien, so they'll be easy to identify."

"And small enough to be considered a food source by most of what's in the jungle," Ressk noted.

Bertecnic snorted. "So are you."

"Not my idea to be here. They came willingly."

Merinim brought up the information on the scientists. "Their license is only for the plateau. They're not to explore past the tree line. Nondisruptive sensor sweeps only."

Ressk spread his hands. "Because most of what's in the jungle will kill them."

"But the plateau is safe."

"Except for the mercenaries."

"I'd rather be shot at than swallowed."

Everyone turned to stare at Binti.

"What?" she demanded. "I hate snakes."

• ——◆—— •

"So you are being the Strike Team's sniper? You are providing long-distance cover for the dangerous jobs they do, that are being correct?"

"It is."

Presit shifted under the intensity of Binti Mashona's regard. "I are asking you to stop doing that."

"Doing what?"

She was wanting to say, *don't be paying so much attention to me*, but she are not being able.

* * *

"I are hearing congratulations are in order. For your bonding," Presit added before either Krai could be voicing the question she could see on their faces. "I are imagining it are a great assistance to be having your bonded with you on a mission like this."

"Can't see why," Werst grunted. "He's useless in a fight."

"And he couldn't use polymorphism to create a set from an array."

"What the fuk does that mean?"

"That you're proving my point."

"Sex is good, though."

Ressk's nostril ridges fluttered open. "Very."

"Sure, it's strange having all these other people on the team, but on the other hand . . ." Alamber's eyes darkened. ". . . there's all these other people on the team. Have you looked under Bertecnic?"

"Are you serious? Freenim will pouch when we decide it's time to have children, and I don't want to discuss it. We're here to assist in the rescue of innocent hostages not talk about . . . that."

"So you are being a pilot? Of small fighter craft?"

"I was."

Presit sneezed at the sudden overpowering scent of cherry candy.

"Torin told everyone to cooperate fully." Craig drew the brush carefully over Presit's shoulder.

She frowned at the amount of fur in the air. "I am being very sure she did."

• ——◆—— •

"You two need to talk."

Torin let her head fall into her hands. "Alamber . . ."

"Something's got Craig's nuts in a knot, Boss."

"If you'll stop sounding like Werst, I'll talk to him after Vertic and I go over the order of the march."

"You're marching? It's jungle. How do you march through all that . . . stuff?"

"Go away."

* * *

"Exiting Susumi in three . . ." Craig held both hands over the board. "Two."

Torin drummed her fingers on her knee. In order to enter the system unnoticed, the *Promise* was nearly a million kilometers away from the buoy and regulated safety. Torin trusted Craig significantly more than any Susumi engineer the Navy had ever used, but solid objects cared sweet fuk all about trust.

"One."

The stars reappeared.

"Yay." Binti twirled a finger, unsnapped her harness, and stood. "We survived again. Second seat's all yours, Firiv'vrak."

"Got bonzor eyes like Binti's on board, might as well use them," Craig had explained, back when they'd all started working together.

Neither of them had acknowledged that the extra heartbeat Binti's eyesight might give them would make no difference if they exited Susumi bearing down on an asteroid, or a planet, or another ship. The only people with more superstitions about survival than Marines were civilian salvage operators.

As the Artek were unable to sit in chairs designed for bipedal species, Firiv'vrak rose nearly vertical on her rearmost legs and shuffled sideways into the narrow space between the copilot's chair and the board, both chair and board supporting part of her weight. She didn't look comfortable, but the growing scent of cherry candy said she was happy.

"I need that light at 521." Both hands busy, Craig pointed with an elbow. "Slide it left."

Firiv'vrak had piloted an unfamiliar VTA off the prison planet and close enough to the *Promise* to run a gangway between air locks. When she'd asked to learn the *Promise*'s controls, Craig, who'd been on that VTA, could find no good reason to refuse.

"Shitload of bad reasons," he'd admitted. *"I'm not going to hand her the keys and tell her to be back by 26:30, but she's an ace pilot and I'm not going to insult her either."*

"That shuttle was made of plastic aliens. You could argue they were flying themselves and that she had nothing to do with it."

"You can argue it, I won't. She stinks of burning hair when she's unhappy."

The Artek didn't perceive color like a biocular species and, although

Confederation numbers were only a combination of ten symbols, they were ten symbols newly learned. She'd never be able to fly the *Promise*, which made keeping her happy a minimal risk scenario.

They passed the planetoid at the farthest edge of the system, close enough to its gravity well to pop proximity numbers out above the control panel.

Torin stood and leaned over Craig's chair to get a better look. "Two hundred thousand klicks? Cutting it close."

"Plenty of room. It's a warning, not an alert. I'm using the planet to mask our emergent point if the mercs have a sweep going."

"What are the odds?" Binti asked. She'd moved to stand against the HE lockers along the back bulkhead instead of taking another seat. "If Ganes hadn't got word out, no one would know they were there. They're in the clear if they're gone before the university's supply ship returns."

"Bet your life on it?" Craig asked.

She smiled. "I'd bet yours."

Firiv'vrak waved an arm, both antennae swaying in counterpoint to delicate fingers. "I'd bet. Four of a kind beats a full house. Royal flush beats a straight flush."

When Torin glanced over her shoulder, Binti shrugged. "Four days in Susumi, Gunny." Her smile slid into a smirk. "Couldn't spend it all planning a rescue, reading briefings, and fluffing Presit."

"Warden Mashona owes me a week's wages and a duck."

• —◆— •

"Let me see that." Arniz pushed Salitwisi away from the monitor with her tail.

When one of the recalibrated scanners got a hit just inside the point where the leading growth of the jungle met the plateau, Yurrisk had most of her equipment moved to where he could watch both the monitors and the excavation of another latrine.

Arniz suspected he didn't trust them. "These aren't similar readings," she said, splitting the screen and pulling up the original data. "They're identical readings."

"Which means?" Yurrisk leaned forward, one hand shading his eyes as he squinted at the screen.

"Eventually it'll mean that a great many people will be spending a lot of time trying to work out why, but, for now, with next to no information, I think I can safely say that identical molecular readings of what could be plastic residue in multiple latrines . . ."

"Latrines used by a non-plastic-using civilization," Salitwisi added.

". . . means, colloquially speaking, it's the same stuff. There . . ." Arniz waved a hand in the general direction of the first unsanctioned hole in the ground. ". . . and here."

"There's more residue in the second latrine." Yurrisk expanded the second reading.

"I'm aware." Roughly, eleven times more. Her tail twitched with the effort of not hauling his hand off her equipment. He'd ruined the joy of discovery, the *feecont*. She should be teasing answers out of Dzar, guiding her to an application after years of study, not reciting the *de con talbin* to her spirit every night. "There's more of everything in the second latrine," she snapped. "It's five times the size and most likely communal. Which we'd know for certain if we hadn't abandoned science for a treasure hunt."

"How do you even *know* it's a latrine?" Trembley asked, frowning at the pile behind the digger. "It looks like dirt."

"Darker dirt," she told him, softening her tone. "Before my complicity in the destruction of an irreplaceable archaeological site . . ." She scowled at Yurrisk. ". . . I could have shown you the differences in vegetation caused by the nutrients available in subsurface rot."

"But the rot . . ." Trembley began.

"Enough." Yurrisk cut him off. "Is the residue at the same historical level?"

Arniz sighed and turned to face the Krai, arms folded over her field overalls. "What part of latrine do you not understand? The contents of latrines rot. We're not plucking data out of the stratification of bedrock here, it's shit and piss and whatever they—whoever the predestruction they were—used to wipe themselves clean. Easily identified by a high localized concentration of urea and sulfides, it's why we start our ancillaries—like the one you murdered—on them."

"Bodies get dropped into latrines." Martin had suddenly appeared by Trembley's side. For a big man, he could move quickly and quietly, and he clearly wasn't happy about Trembley being part of this discussion. Possibly because there was half a chance the young Human was still intellectually flexible enough to learn, to move from archaeology 101 to taking a second look at the morality of murder.

"I bow to your greater knowledge of what happens to a body destroyed by violence." She used the tone she'd perfected for her department head: so completely devoid of sarcasm, it verged on insult. "But that's not the point. Anything we find in a latrine now will be among the last things that were ever put in there. So, technically, the answer is yes. It doesn't matter that we found the two bits of residue at different depths, if a weapon ever existed, it existed immediately pre-destruction."

Martin peered over at the screen, although, given the angle of the sun, she doubted he could see the results from where he stood. "You should've known that from the initial data."

"The initial data was a statistical anomaly. Now it isn't." He frowned, but to her eye he seemed more thoughtful than angry. Of course, he hadn't seemed angry when he'd shot Dzar, so what did she know about Human expression. "This latrine . . ." She waved a hand at the excavation. ". . . was a trench and not a particularly deep one compared to some I've seen. The first latrine was a hole in the ground; again, not particularly deep in comparison. Any structures built over them have disappeared. It's entirely possible the structures were wood in order to make them easy to move when the latrines filled. There's a chance the latrines were used in a specific rotation to fertilize a nutritionally poor soil, but given the destruction you're responsible for there's little chance of discovering . . ."

"Destruction I'm responsible for?" His nostril ridges closed, his cheeks flushed a darker green, Yurrisk stepped forward, only to be brought up short by Qurn's back. Arniz hadn't seen her move to intercept.

"Get to the point, *Harveer*." Arms folded, Qurn looked near the end of her patience.

Arniz sighed and waved emphatically enough to take in the entire plateau. "You're *looking* at the historical level of the residue; well, out to the foundation stones of the city wall, at least. GeoPhys has turned up no evidence of intact levels belowground—no real surprise given the depth of soil out here—so it's reasonable to assume there's no hidden weapon."

Yurrisk laid his hand on Qurn's shoulder but neither moved her nor moved around her. "Then explain the plastic."

"If *we* can't come up with a weapon capable of destroying the plastic aliens," Trembley said before Arniz could respond, "how could a civilization that shits in holes?"

"Maybe they ate them." Sareer, the most identifiable of the Krai due to the nine small rings piercing the outside curve of her ear, joined the group at the monitor. When all attention turned on her, she gripped her weapon like a security blanket. Arniz wondered who around the monitors she thought was dangerous. "The vids say Krai can eat the plastic aliens, and that would mean the aliens would get shit out." Her voice trailed off under the weight of Yurrisk's gaze. "Because this is a latrine. Right, Commander?"

His expression softened into amusement. "According to the *harveer*, yes, it's a latrine."

"Then maybe they were like us."

And continued softening into sadness. "Then we should pity them."

"I are having a theory!" Tyven hurried over, trailing her bonded. Arniz hadn't realized she'd been close enough to listen in, but the Katrien had excellent hearing. Yurrisk's features snapped back into what she'd started to think of as his crazy face as the two Katrien joined them. "Perhaps the foods of the pre-destruction are recombining in the digestive tract to be resembling plastic residue."

"It doesn't resemble plastic residue," Arniz sighed. "It is plastic residue."

"But how are you knowing for certain?"

"Science."

Tyven's shoulders slumped and even her fur seemed to flatten. "I are not being able to argue with that."

Sareer's nostril ridges closed as Yurrisk ground his teeth. Muscles tensed, eyes flicking between Yurrisk and Qurn, she took a deep breath and said, "If the people here were a kind of Krai, we should think about where we'd hide the weapon."

"Parallel evolution!" Blood rose into Trembley's ears when everyone turned to stare. "I saw a vid," he mumbled, scuffing a foot in the dirt.

"It's possible," Arniz allowed, absently waving away the perpetual cloud of tiny blue insects. Trembley's surprise at being right was an expression familiar to every teacher in known space. "We don't know what the builders on this world looked like because we haven't yet found remains and none of the art in the ruins at the Mictok dig is representational."

Tyven nodded. "It are all being geometric in that part of the world. Repeated patterns. We are hoping to be eventually finding art on the ruins in the jungle."

"Then today's your lucky day." Yurrisk spread his arms. "If the weapon's not out here, it's in the jungle. In the ruins of the buildings."

"It's not fukking rocket science," Martin added.

"No, it's not; it's archaeology and you don't have the faintest understanding . . ." His hand closed around her throat, warm and mammal moist and so tight she could barely pull air past it.

"I don't think *you* understand what's happening here." His breath lapped against her face. "If I find you've been deliberately delaying us in the futile hope of rescue, I'm going to bury you in one of your latrines."

"Why would I delay?" she gasped as he released her and she dropped to the ground. Out of the corner of her eye, she saw Trembley's boots move closer. "I want you gone."

"Martin." Yurrisk's voice held the Human in place. "This isn't helping to find the weapon. Move them into the jungle. Now."

"*Harveer* Tilzonicazic is going to be unbearable," Salitwisi muttered as he helped her to her feet.

Arniz huffed out a reluctant laugh. As a botanist, Tilzon had wanted into the jungle from the moment they'd landed.

• —◆— •

The only ship in orbit around 33X73 . . .

"Threxie."

Torin rolled her eyes. "I'm not calling it that."

. . . was an Aggressive class minesweeper, decommissioned by the Navy when they brought in the Avengers class. The exterior looked rough. It had no external packets attached, an extensive repair to one of the two shuttle bays had sealed that bay shut, and a large Krai symbol had been applied just back of the main air lock.

Leaning on Craig's chair, thumb rubbing against a familiar fold in the duct tape, Torin could see the sigils for out and tree as well as the emphatic curl that identified a command. "Get out of my tree?"

"Yeah, that's the polite translation, Gunny." Werst shoved past Binti and leaned against the side of the second seat, where Ressk ran the minesweeper's ID. "It means, *fuk off, this is mine*."

"It means . . ." Ressk paused, stared out at the ship, and shook his head. "Never mind. Close enough. It's the *DeCaal*, all right. I'm impressed Commander Yurrisk found a minesweeper in one piece."

"Shoved to the back of the yard and forgotten when it was decommissioned." Craig adjusted the scanners as he brought them closer to the *DeCaal*. "Navy yard at Ventris crammed the new in on top of the old. Strippers ran twenty-seven/ten, but I never saw much headway."

"I'm impressed your government allows the sale of old warships to civilians. Particularly given their emphasis that the Confederation is made up of peaceful peoples." The translator perfectly reproduced Freenim's sarcastic tone. It was a major improvement over the program Torin had used in the Corps. But then, no one sent in the Marines if they wanted to have a conversation.

The entirety of Strike Team Alpha plus both Druin, Vertic, and Bertecnic filled the seats and all the empty floor space in the control room. Even Torin, who'd spent most of her adult life crammed into various types of armored transport with one more Marine than the transport could comfortably hold found it crowded. She admired how well Craig was holding it together. Not long ago, he'd reacted badly to the thought of sharing the *Promise* with her.

Presit had banked half a dozen shots of the *DeCaal* when they'd first come into range and then left, muttering about overwhelmed air filters. The odds were high she'd also disliked not being the center of attention.

"What's wrong with the government unloading old ships?" Craig spun the pilot's chair around to face the room. "Weapons are removed."

"And the peripherals?" Standing behind Merinim's chair, although their asses were narrow enough they could have shared, Freenim folded his arms. "Weapons are easy enough to replace if the mountings and the firing systems remain."

"Not the kind of weapons you'd hang off a minesweeper." Torin thought of the cases stolen from MI and made a mental note to check thefts in other sectors when they returned to the station. Weapons were, after all, still being made.

"A decommissioned ship isn't in the yard because it's full of holes. It's stripped but solid. That kind of ship's not cheap." When Torin, Craig, and both Krai turned to look at him, Alamber's hair flicked back and forth. "You guys have got to work on mastering that lateral slide from violence to economics. Big Bill had a sideline selling previous owner vessels—whether the previous owner agreed or not—and a spaceworthy ship that size? Very much not cheap. Add the cost of getting it up and running, plus the major outlay of replacing the Susumi engine—which I can guarantee was removed in the yard—and Commander Yurrisk isn't going to have enough left to remount guns. Not unless he started out rolling in it."

Ressk thumbed his slate. "He didn't. His line's not wealthy. Not even close."

"You know this commander?" Freenim sounded like he didn't trust coincidence.

"Sure. We all know each other." Torin cleared her throat and Ressk rolled his eyes at the warning. "Fine. No. I've got his Navy records."

"In my humble and yet informed opinion," Alamber declared, ignoring the rising tension, "over the last three years Commander Yurrisk has spent everything he has to keep that ship flying. Explains why he's threatening scientists."

"Does it?" Vertic's hands closed over a seat back, her claws dimpling the padding.

Alamber sighed expansively. "He has expenses. Someone's paying him for applied violence. Destroyed plastic aliens." He raised a finger. "Destroyed by what?" Another finger. "A weapon." A third finger. "I'll pay you the money you need to bring it to me, and you can keep your ship flying. If he's as damaged as you lot think, he might not have any other options. Most people paying to ship cargo through Susumi want proven stability."

"Makes the commander highly motivated to find the alleged weapon," Torin said after a moment's silence.

"And if there is no weapon?" Vertic asked.

"Given the plastic, I doubt they'll convince him of that."

"And if there is a weapon?"

"We'll confiscate it if he has it, but our mission is to save the hostages."

Her claws scraped against the floor. "Of course."

The green light flashed on the upper left corner of the board. "Scan's done. No life signs. The commander must've been lurking in the VTA when Ganes got his shot, then." A second light flashed as Craig shut off the first. "And there's SFA in the way of security."

"If Alamber's right about the cost . . ." Torin began.

"Trust me, Boss, I'm right."

". . . orbital security would be low on the list of systems to restore. War's over. Who's going to blow him up?"

"You should." Vertic moved closer to the screen, pushing between the seats. Golden hair rubbed off her sides, tumbled through the air, and was sucked toward the filters. "Destroy the ship to keep the mercenaries from returning and running."

"Not allowed," Werst grunted.

She reared a few centimeters, the crest of her mane brushing the ceiling. "You're not permitted to prevent the enemy's escape?"

"Not by wholesale destruction," Torin answered before Werst could voice his opinion of that particular restriction. "We're not soldiers; we're Wardens, dealing with civilians, governed by different regulations."

"Dealing with assholes, governed by bureaucrats," Werst muttered.

Torin raised a brow. He ducked his head, nostril ridges closing. "R&D is developing an explosive cartridge we can attach to the hull, by the engine." She switched her attention back to Vertic. "If the engine is started before the cartridge is removed, or there's an attempt to remove it without the right codes, they'll be dead in space. The *ship* will be dead in space," Torin clarified the idiom when both Druins' inner eyelids flicked. She had no idea if, in this particular instance, the motion meant approval, disapproval, or dry air.

"But they're still developing this cartridge," Freenim said thoughtfully.

Werst snorted. "We'll have it any day now."

"You don't currently have it, though. That's my point. Without a way to secure the ship within the restrictions of your Justice Department, how do you stop the enemy from escaping?"

"We tag their ship so we can find them later." Torin nudged Craig, and he tossed the specs of the tags into the air above the board—the spirit of transparency albeit not the letter as none of their Primacy companions could read Federate and only Firiv'vrak had bothered to learn numerical symbols.

The look Freenim shot her suggested he was well aware of the subtext. "So you don't *stop* them from escaping."

"We do our best to stop them from reaching their VTA and taking off."

"But not from leaving orbit."

"We find them later. We find them," Torin repeated when Freenim spread his hands. "Craig, set the tag and let's get dirtside. Those hostages aren't getting any younger."

"On it."

"Wait!"

Hands on the tagging system, Craig glanced up at Torin who nodded as Firiv'vrak appeared at the hatch, scuttled over the lip, and disappeared between the seats. Vertic shifted from foot to foot to foot and ended up on her haunches, forelegs in the air as Firiv'vrak forced her way past and up between the last two seats, trailing the scent of grapefruit and pepper. "I feel that reading!"

"Feel?" Torin tossed a silent question at Freenim. Who gave the minimal shrug, NCO to NCO, that said he had no idea what Firiv'vrak was talking about.

"I feel it here!" She waved her antennae as she slid into place between the copilot's seat and the board.

"You feel it in your antennae?" Space was big and stranger things had happened, Torin acknowledged. "What reading do you feel?" She didn't recognize the string of numbers Firiv'vrak isolated on the board, but Craig did.

"It's background radiation, Firiv."

"No." Unfolding her arm, she expanded the status line. "None of you were Navy, so you wouldn't recognize it, but this is a Primacy wave. It's used on exploration vessels, specifically to identify the composition of asteroids, of missiles, of debris. My best guess—they've tucked pieces of scientific equipment into the holes left when the sensor arrays were removed. Equipment that's not specifically military, so it's easier to buy and doesn't attract the wrong kind of attention. If a solid item comes close enough to the *DeCaal*, it'll send the analyses down to a slate—probably the Druin's. No offense, Durlan, but I can't see Polint males thinking of this."

Vertic waved it off. "None taken."

Warned, the mercs would, at best, use the hostages as shields. "So we don't tag."

"Could we neutralize the wave?" Alamber asked.

Half of Firiv'vrak's eyes stalks turned toward him. "No idea. I'm a pilot."

"How close is too close?" Craig demanded his eyes on the proximity readings.

Torin rested her hand on the rigid curve of his shoulder. If they were close enough for the wave to identify the composition of the *Promise* . . .

Firiv'vrak extended her antennae, tips quivering. "I haven't felt an interruption in the wave."

"That could just mean you haven't felt it." Craig's hands hovered over propulsion. "Would you know if we triggered it?"

"Yes. No. Probably."

"Two to one odds we haven't triggered it, then." Torin pitched her voice closer to command. "Let's pick up the pace, people."

"Have you scanned for Primacy life signs on board?"

Craig twisted out from under Torin's hand to stare at Vertic. "No. Why?"

"Because there are Primacy among the mercenaries." Her gesture involved both arms, and still managed to include only her own people. "I'm sure your military has ensured you have the required algorithms on board. Why haven't you used them?"

"We never have," Torin said before Craig could answer.

Habit.

As Wardens, they'd only ever been sent out after the Younger Races.

Habit got people killed.

"Will a scan interact with the wave?"

"No." Firiv'vrak's cherry candy scent had returned. "As I said, science, not security."

"Do it, then." As Craig set up the scan, Torin noticed Keeleeki'ka and Dutavar had followed Firiv'vrak into the control room. The full, combined team only fit because Bertecnic had opened the hatch to the head and backed in. "You could all watch this on your slates."

"Not like being here, Boss." Alamber had perched on the back of his chair, boots on the seat, head higher than the tallest of the Polint, risking Craig kicking his ass if the boot prints marred the pleather.

"Suck it up." She straightened. "Everyone without a tactical reason to be in the control room, gear up and get to the VTA. Werst, you're on Presit detail. Freenim . . ."

"I'll see that everything's ready, Warden."

Vertic moved back to the open area behind the chairs, clearly not considering herself to be a part of *everyone*. It took Torin a moment to realize *she* hadn't included Vertic as part of everyone either. It was one thing to ignore her rank and another to get past the certain knowledge that gunnery sergeants didn't give orders to officers.

"No Primacy life signs." Craig leaned back in his chair. "If Alamber could neutralize the wave, they'd likely see it as an equipment failure, given the state of the ship as a whole."

"He's smart enough," Torin allowed. "But we don't have the time."

"I'd feel better if we could tag."

"I'd feel better if we got to the hostages while they're still alive. Put her where you want to leave her, and lock her up, Craig. We'll meet you at the VTA."

"Not like you can leave without me." He reached up and squeezed her hand. "I want you or Werst in the second seat, in case we run into trouble from atmospheric disturbances or surface-to-air missiles on the way down." He grinned at her expression. "Your paranoia is catching."

"Glad to hear that."

Binti waited for her at the hatch to the VTA. "You think they know we're coming, Gunny?"

"Only one way to find out."

"I are well aware there are space restrictions, this are being a very small and very uncomfortable looking vehicle." Presit's voice didn't so much drift out the hatch as jab through the opening. "But I are not seeing how those restrictions are applying to me. I are half your size, I are therefore allowed twice the space."

"We could drop Presit by the anchor," Binti suggested. "They'd surrender in an hour. Two max."

"Tempting."

• —◆— •

Of the scientists, only Dr. Ganes was large enough to use one of the two laser cutters Martin brought out of the shuttle.

"Sized for Humans." Tilzon patted the power pack a little sadly. Arniz had watched her wield a cutter on other digs and was just as glad she wouldn't be flinging a beam around.

"Looks like you're up, Doc." Martin tossed one of them at Ganes, who caught it, nearly dropped it, and straightened as Martin laughed. "Too bad you're not human-sized, lizard."

"Truly." Tilzon sighed. "I'm feeling the lack."

"And the second cutter?" Yurrisk asked. Arniz recognized the tone. He wasn't asking for clarification; he knew the answer and wanted to know if Martin did, too.

"We'll keep it charged and ready to trade out."

"We haven't time to let it sit idle. We need to be gone, with the weapon, before their supply shuttle returns. One of you will use it."

Martin folded his arms. "One of us?"

"They're Human-sized, Sergeant."

As expected, Yurrisk won the staring contest. Arniz doubted anyone could look into his eyes for very long.

"Malinowski!" Martin held out the cutter as she trotted over. "You're taking down trees."

"Fuk you, Sarge."

"Yeah, well, next time, don't complain about being bored."

Arniz didn't understand why Martin didn't use the Polint. They were certainly large enough and looked significantly more bored than Malinowski had, constantly shifting from foot to foot. They were even carrying large, heavy-bladed knives. Almost machetes.

"Let's go!"

Martin, Arniz noted, yelled for the same reason Salitwisi did. To attract attention to himself.

The three di'Taykan, hair clamped close to their heads, looked as unhappy about following Martin past the mesh of vegetation that marked the edge of the jungle as Arniz felt. As she understood it, they disliked both heat and humidity. Well, in that case, they'd certainly picked the wrong dig to terrorize.

Yurrisk seemed sympathetic. "No reason you can't sit while you guard the scientists. The shade under the trees has to be better than standing in direct sunlight."

"Guard the scientists doing what?" Salitwisi demanded.

Turning slowly, Yurrisk waved toward the jungle. "Guard the scientists as they clear debris away from the road."

Salitwisi's tail rose. "We have ancillaries for that."

"They'll be joining you."

The road had been obvious on the orbital surveys. Before the flacid tail-wobblers had arrived, Salitwisi and his ancillaries had begun to slowly examine the point where the road met the city wall in hope of identifying a gate house. Or, at the very least, a gate even though the surveys showed no break in the wall along the cliff. A meter and a half wide, the road seemed narrow out on the plateau and ridiculously wide in under the trees. Yurrisk had insisted they clear it completely, shouting, *"Because important items are kept in buildings by the sides of roads! What use is a weapon if you can't get to it!"* Then he'd sat down heavily on a rock and waved Qurn off.

Sareer and Beyvek swarmed up into the trees, swinging from branch to branch using both hands and feet. Arniz could hear them yelling out directions to Ganes and whichever human had joined him. They'd left their weapons leaning against one of the thicker trunks and looked so happy it was difficult to remember they'd been a part of Dzar's murder and the . . .

It wasn't kidnapping, not really, they hadn't been taken anywhere. Arniz supposed they were being held hostage, but no one had actually tried to leave, so *held* wasn't entirely accurate. Science performed under duress?

Manual labor performed under duress.

"What part about 'drop the debris off the path' do you not understand?"

Arniz peered up at Gayun, visibly wilting in the higher humidity under the trees, brilliant blue hair flat against his head. "I was thinking. You should try it some time."

He raised his weapon. She touched the air with her tongue and smelled the sweat staining his clothes. "You're lucky you got old. Too many p . . . p . . . people are dead."

"Well, the old can't work very fast. If you helped, you'd be out of the heat faster."

Salitwisi, Tilzon, and the four closest grad students froze in place, branches and vines dangling from their hands, a slender tree resting on two sets of shoulders.

The end of Gayun's weapon jabbed a circular bruise into her chest. Arniz staggered back, arms flailing.

"*I've* got this, Gayun." The warm press of Trembley's hand on her back kept her on her feet. "Find a *breeze* before you melt."

Gayun's eyes darkened, then he muttered something Arniz didn't catch and headed back toward the plateau. She turned just far enough to watch him go, but not so far she lost Trembley's support.

"If he wants, he can shoot you from out there." Trembley gave her a shove forward. "Get to work. All of you, get to work!"

Although the ends had been cauterized by the cutter, the pieces of vine were heavy and wet with sap. It seeped through the periderm all along the length, viscous and unpleasant and sticky when dry. Lifting and carrying and lifting again, Arniz could hear insects over the cutting and gathering and muttering, although it sounded as though the birds had moved deeper into the jungle away from their intrusion. The air, only four meters from the plateau, tasted of moisture held in the folds of foliage, and of rot, flora and fauna falling to decompose and become a rich humus.

Three hours later, only six meters of the road had been cleared. It was farther than Arniz had thought they'd get even for a variable definition of *cleared* that included abandoned stumps and roots that buckled slabs of stone up at enough of an angle edges had been visible through the mats of vine. She closed her fingers around a gauzy insect trying to drink from the surface of her eyes and listened to Yurrisk as she pulled pieces of the body out of the dried sap on her palm.

"You're cutting organics!"

"Yeah, but really fukking tough organics." Zhang wiped sap off the side of the cutter, continuing the motion to scrub a damp, green stripe off her cheek. She didn't seem particularly bothered by Yurrisk's rising anger, but Arniz assumed she'd seen battle and that had to put a crazy Krai into perspective. "Everything gets bigger and thicker the farther in we go, and I've got three percent power left."

"Two five here," Ganes panted.

Martin looked ahead into the trees and back along the road as he said, "That's it for today."

"That is not it!"

Martin shrugged. "It'll take six hours to charge the cutters back in the VTA. It'll be dark in less than five."

Arniz flicked the last piece of insect away and sank to the ground.

"Are you being all right?" Tyven, fur in random sap-stiffened clumps, squatted beside her. "You are having collapsed."

"I sat quickly." She reached out to pat the geophysicist on the arm and snatched her hand back as she thought better of it. "I'm fine. I'm tired."

"I are . . ."

Yurrisk grabbed Tyven's shoulder, hauled her to her feet, and shook her. "Where are the ruins?"

She twisted in his grip. "There!"

"Where?"

"There!"

It would have been funny had Tyven not been flapping back and forth, toes digging into the ground trying to keep her balance. Arniz tasted the air, the prevailing scents of dying vegetation and the bitter residue of the cutter overwhelmed by anger and fear. Both emotions coming from Yurrisk.

"Are you not seeing that ridge?"

Arniz could see only vegetation, insects, and shadows, but the Krai had eyesight almost as good as the Katrien.

"It's a fallen tree."

"No! It are being an intact wall."

He stopped shaking her. "A building."

"A wall," she repeated. "There are being no more than a meter remaining. Maybe less. Maybe more. Organic cover are being deceptive."

He shook her again. "I want buildings! Storage! They should be by the road!"

"How are you knowing? We are knowing almost nothing about those who are being pre-destruction!"

"Because this is a city, and not the best part of a city either, given communal latrines."

"Everyone shits," Arniz muttered. And Yurrisk needed to make up his *retunin* mind. One minute yelling and shaking, the next all quiet reason even while maintaining his grip on Tyven's shoulder as though expecting her to hold him up.

"Cities, especially walled cities, have limited space," he continued. "They're not going to waste that limited space by setting buildings back from the road."

Tyven waved her arms. "We are not knowing how many people this city are holding. There are maybe being small numbers and big properties."

"The amount of work a city this size requires wouldn't be practical for small numbers."

"We are not knowing what is being practical for the pre-destruction people. We are having nothing to extrapolate from. We are only beginning to study . . ."

"You be taking your hands off her!"

Arniz turned in time to see Lows hit the ground, Gayun's foot in his back, boot almost covered by fur.

"Lows!" Tyven fought to get to her bonded. Yurrisk had to hold on with both hands.

"I need buildings!"

And there was the yelling and shaking again.

"It are not working that way!"

Arniz hissed as she rose to her feet, cutting in before Yurrisk could respond. "According to the survey, the first intact buildings are five hundred meters in from the plateau. We've gone six. Barely. Why don't you join your people on the high road to the ruins . . ." Sareer and Beyvek hadn't touched ground all day. ". . . the three of you can spend the time looking for your weapon instead of landscaping. I thought Krai were all about climbing."

Breathing heavily, hands opening and closing, Yurrisk glared down at her for a moment, then turned and vomited.

"I thought the Krai are not doing that?" Tyven muttered.

"What's going on here?" Martin demanded, punctuating his arrival by crushing a round-bodied insect under his boot.

"None of your b . . . b . . . business, Sergeant." Gayun stepped between the two as Qurn hurried past and handed Yurrisk a bottle of water.

"My business is keeping the peace, through force if necessary." Martin sneered at the di'Taykan. "Looks like this is my business, *Lieutenant*."

"Blades," Yurrisk said, spat, and stood. "We need heavy cutting blades."

"You are really not understanding archeolo . . ."

The force of Martin's kick threw both Tyven and Arniz back. Spores exploded under Arniz's head as she slammed into a round gray-green fungus. Trying to catch her breath, Tyven began to cough.

"The Polint have b . . . b . . . blades," Gayun pointed back toward the plateau. "They're more than just b . . . b . . . big knives, that's for sure."

Martin's eyes narrowed. "Cutting brush was not what the Polint were contracted to do."

Yurrisk's nostril ridges closed. "Then I suggest you convince them to amend the contract."

•—◆—•

“Boss, that medical discharge Yurrisk got?” Alamber passed her his slate. “I took a look at the details.”

“Wasn’t that redacted?”

“Sure, but the information’s still there if you dig deep enough and medical files all use the same encryptions. You manage to get into one file, you can get into all of them. Long story short, Krai bone still surrounds a delicious creamy center and Yurrisk’s got whipped.”

Vertic leaned forward and grunted. The VTA had been modified to hold the Polint, but the strapping had been designed for safety not comfort. “What’s he talking about?”

“Brain injury.” Torin frowned down at the screen. “Transient vertigo and acrophobia on top of what has to be a really messed-up head space given all the redaction.” Bodies could be rebuilt. Brains couldn’t always be rewired. “Werst, could he . . .”

“No. Live with ground dwellers, sure, but he’d never be able to live on a Krai planet, with Krai.”

“They’re arboreal,” Torin explained before Vertic could ask. “Natural forests where possible, concrete towers where it isn’t.”

“I’m not sure how perceiving the world as swaying when it isn’t would prevent him from living arboreally—trees sway in the Confederation, don’t they?”

“They sway when they sway,” Werst told her. “Not when they don’t. If they zig when you zag.” He slapped his palms together. “Impact.”

“I see.” Vertic tucked her thumbs behind the straps over her chest. “I see how fear of heights could be a problem, if the translator has defined acrophobia correctly, but he’s on a ship. A ship that goes into space.”

“He has both feet on the deck.”

“What if he looks out a window?”

“Both feet on the deck,” Werst repeated. “That’s what matters.”

Vertic shook her head. “That’s . . .”

“That’s what matters to the Krai, Durlan. If he loses his ship, he loses the only home he’s comfortable in.”

“So Yurrisk’s hiring mercs and holding scientists hostage because he picked up a brain injury while serving and psych dropped the ball. He can’t live with his own kind, so he crossed the line in order to keep his ship, the one thing he has left, flying.” Binti spread her hands. “You sure this isn’t a job for Veterans’ Affairs, Gunny?”

"Depends on how many hostages we find alive."

"Hands and feet inside the ride, kids." Craig cut off any response. "Atmosphere in three, two, one . . ."

•—◆—•

"Brenda!"

"What?" Trembley, dripping sweat, leaned into her line of sight.

Arniz straightened, trying to work the kinks out of her back. They wouldn't be able to keep going much longer. It got dark under the trees first and the road was as much a suggestion as a guide. "Brenda Zhang. It's the shorter, female Human's name."

Trembley shrugged and slapped at an insect. "Well, yeah. So?"

"I have a theory that it's harder to kill people who know your name."

"Unless you kill them *because* they know your name. What?" Her expression pushed him back a step. "It's just another *theory*. We're here for the weapon. Nobody's going to get killed."

"Else. Nobody else is going to get killed," she added when he looked confused.

His brows drew in and he looked back along the path, as though he thought he'd be able to see the anchor where Dzar still lay in a stasis pod in the infirmary, out of the sun. "That's not . . ." He frowned.

She waited.

"Get back to work!" he snarled instead of telling her what it wasn't.

Arniz knew guilt when she saw it. Guilt for watching beings smaller and older than he was work themselves to exhaustion. Maybe guilt for allowing Dzar to die and doing nothing to stop it, but maybe not. Age had taught her it was easier to acknowledge smaller guilt, however angrily, than larger.

Stomach growling, she staggered as she dragged a branch to the side of the road, tripped on the mossy edge of a slab, and pitched forward into the tree where Sareer and Beyvek had leaned their weapons. Hands out in full view, she backed away as quickly as she was able. No telling what Martin would do if he saw her near the guns, but her options weren't good. He enjoyed hurting people.

"Are you being all right, *Harveer*?" Magyr, one of Tyven's ancillaries steadied her as she stumbled.

"No. I'm tired and hungry and bruised." Arniz didn't know Magyr well. They hadn't been here long enough, but she seemed intelligent and not particularly entitled which was the minimum Arniz required.

She patted the young Katrien on the arm with the back of her hand to keep from clumping yet more fur together. "But then I expect everyone is."

She trudged back to the debris she'd been clearing, paused when she realized Magyr hadn't followed, and turned to find the ancillary still standing by the tree. Staring down at the weapons. Probably wondering why the horrible things existed now the war was over. Arniz certainly was. She'd never given them a second thought until she'd suddenly seen how quickly they could end a life. And yes, so could a rail car, she acknowledged, forcing her back to bend, reaching for another armload of cut vines, but a rail car would have been harder to acquire and significantly less portable. Although Arniz had no doubt that, if given a chance, Martin would happily use one to . . .

She tightened her grip on the vines when Magyr picked up one of the guns. Dropped them when the ancillary turned, holding the weapon in both hands. She heard a shout; it sounded Human, and she froze, unable to move either toward Magyr or away. Her mouth formed the words *put it down,* but she couldn't draw in enough of the heavy, humid air to make a sound.

The shunk, shunk of edged steel cutting through organics stopped. Dragging her gaze from Magyr, Arniz saw Martin beckon Cameradiz over and ask for his blade. Cameradiz laughed, said something short in his own language, and handed it over, watching as Martin walked toward Magyr, the heavy blade, flecked green, held loosely in front of him.

In the distance, birds and insects maintained the background noise. On the road, silence.

"All of you!" Lips drawn back off her teeth, Magyr waved the weapon in a wobbly arc. "All of you who are not being us! You are going back to your shuttle and you are leaving! Now!"

"I have the shot, Commander."

One of the di'Taykan. Arniz shuddered.

"No," Yurrisk said as Martin drew closer. "Let the sergeant handle this. He's here to keep us safe."

Barrel of the weapon visibly shaking, Magyr pointed it at Martin as he stepped into a beam of sunlight.

One step. Two.

Kept it pointed at him as he left the sunlight behind.

Three steps. Four.

Arniz closed her eyes. But she still saw Martin swing, Magyr fall. She could smell the blood. Hear the screaming.

When she opened her eyes, one frantic heartbeat later, none of that had happened.

Standing close enough Magyr could lean forward and poke him, Martin put down the blade.

"You are leaving!" She held her ground, her ears flat against her head. "You are all of you to be leaving! Now!"

He said nothing, merely took another step. Over the wet snap of crushed flora, Arniz could hear Magyr's breathing grow faster, shallower. In a parody of gentleness, Martin cupped one hand around the back of her head and cradled her jaw with the other, his dirty fingers pale against the dark fur.

At first, Arniz thought a branch had broken. Then Magyr fell, staring back over her own shoulder, her glasses resting skewed on her muzzle.

Martin caught the gun and glared up into the canopy. "Next, I'll deal with the idiot who left a weapon . . ." His voice rose, growing louder and angrier. ". . . lying around like it didn't fukking matter."

Someone whimpered. Arniz didn't know who. She was pretty sure it wasn't her. Soon the whimper would become a wail. Then, if not stopped, a chorus of wailing.

A Niln cried out as they hit the ground. Arniz didn't . . . couldn't turn to see who it had been, but the whimpering stopped.

"I don't appreciate arbitrary killing of our civilian workforce, Sergeant." Silence followed Yurrisk's comment. Arniz could hear a bird in the distance and Salitwisi breathing beside her. "We won't find the weapon if we have to search the entire jungle by ourselves."

"And we may need the combined expertise of the entire scientific party," Qurn pointed out.

Martin ignored her, his gaze locked on Yurrisk. "You'd rather I let it pull the trigger? They don't do that, though, do they? Pull triggers. That's why they have us."

"Perhaps you should have thought of that before you killed her."

"My job is to keep you safe."

Arniz thought Martin sounded like he was delivering the punchline of a joke.

Apparently, Yurrisk didn't. "I know. Thank you. But I will deal with my crew members' carelessness. That's not your place." His sigh when he looked down at Magyr's body sounded weary. Arniz fought the urge to throw a rock at the back of his head. "Leave her here until we finish for the day."

Tyven pulled himself out of the weeping ball of fur he'd curled into with his bonded. "*Sivern contra!* You can't . . ."

"I can." Yurrisk turned, nostril ridges closed, lips off his teeth. "It's not the dead you have to worry about. Never the dead. The dead are beyond pain, theirs and yours. The dead can be left where they lie because your fight is with the living, and the living never stop. Everyone, get back to work!"

Eyes locked on Yurrisk, Martin slowly leaned the weapon back against the tree and just as slowly picked up Camaderiz's blade. "You heard the Commander," he said. "Back to work. Now!"

One by one, in fits and starts, grief and shock in control of their movements, they went back to work.

Arniz bent to pick up another armload of chopped vegetation, heard Qurn's quiet voice, and shuffled closer, bent over, old and harmless.

". . . sell the weapon, can also repair the head on deck two. Or, unless you're very attached to that field kitchen, we could rebuild the galley."

Even allowing for species differences, her tone spoke of more than friendship. Primacy she may be, but Qurn was a part of Yurrisk's crew. She hadn't arrived with Martin. Arniz dropped her armload off the path and huffed in annoyance as she bent for another. If not for the two dead, an ache in her heart to accompany the ache in her back and arms, the whole experience—mercenaries, ship captains, ancient weapons, destroyed plastic aliens, inter-species romance—had all the plot points of a bad vid.

All they needed now was a daring rescue.

SIX

THE SCANNERS HAD GIVEN THEM the exact size of the clearing, even accounting for the shifting foliage around the edges, but it looked a lot smaller approaching at the VTA's slowest atmospheric speed. Which was slow only in comparison to the speed used escaping a gravity well.

"Drop team, sound off."

"Ready, Gunny."

"Ready, Gunny."

"I are wanting to be having a better visual than two Wardens disappearing."

"I want people to stop being assholes; we're both doomed to disappointment. Stay away from the hatch." Boots magged against the lateral movements of the shuttle fighting gravity, Torin touched the controls. "Hatch opening in three, two, one." As each layer of the deck slid back, she felt the individual vibrations, vibrations distinct from the steady churn of the engines. All military VTA had an atmospheric hatch, no air lock, just a hole in the hull for dropping supplies or personnel. Its existence had freaked the hell out of every vacuum jockey Torin had ever known. Firiv'vrak included.

"I are needing to be . . ."

"I don't care." Torin had allowed Presit into the drop compartment only after having extracted a promise that she'd stay secured against the rear bulkhead, well out of the way. Then she'd explained Confederation transparency laws to the astounded Primacy contingent. "Drop team advance to drop position."

Werst and Ressk shuffled forward, toes of their drop sleeves out over the rush of green.

They had better gear than usual this mission. Whether it was because the budgetary committee had finally surrendered to the Strike Teams' constant demands for military grade equipment or they were showing off for the Primacy, Torin neither knew nor cared.

"Drop in three, two, one."

With seconds to react if something went wrong—and *seconds* was a generous estimate—Torin's perception slowed as Ressk followed Werst with barely a centimeter between them. When they cleared the underside of the ship, both drop sleeves engaged. The moment Torin got the orange—di'Taykan shuttle—her fingers moved to the hatch controls.

Silver-tipped fur appeared at the edge of her peripheral vision.

She flexed her knees as the ship bucked.

Reached for Presit's hand.

Missed.

A claw painted copper dug a line through the end of her finger.

She unmagged her boots and dove out the hatch following Presit's scream.

Presit was fluff and attitude. Torin, plus the emergency gear and weapons worn by the hatch controller—the Corps believed in redundancies and Torin continued to believe in the Corps—weighed significantly more. "Craig!" Teeth clenched to keep a face full of wind from interfering with her jaw mic. "Secure the hatch!"

Hatch secure . . . The fuk are you doing outside the ship?

"Presit." Her tone filled in the details.

The skills involved in dropping from a moving shuttle, attempting to prevent the messy death of a fellow Marine, weren't easily forgotten. She hoped.

Son of a . . .

She hooked two fingers around Presit's ankle, adjusted her grip, and flipped the reporter over her shoulder. Presit's fingers and toes wrapped around the straps crossing Torin's back as she slapped her other hand against her chest to activate the drop sleeve controls. "Continue to landing site."

Torin!

"That's an order." The drop sleeve, stripped down antigravity tech, pushed against her ribs as it slowed their descent. Werst and Ressk had been set to make a feet-first landing, sleeves over their boots, but emergency gear had to take a second body into account. Three Marines had come out of Torin's training group with cracked ribs. Another had secured his sleeve incorrectly and ended up with a broken larynx when it had shifted suddenly up his chest. No one had died, so the Corps had counted it a win.

Arms and legs spread, Torin ignored Presit's shrieking, and looked down.

They'd gone past the clearing and were about to make contact with the top of the canopy.

"Shit."

Arms tight against her sides, legs folded back at the knee, eyes squinted nearly shut, Torin made herself as small a target as possible.

The drop sleeve had been designed to slow descent; on long drops with plenty of time to counter gravity, the sleeve brought everything to a stop half a meter above the ground. On shorter drops, when momentum remained too high for a safe stop, the sleeve added a vertical bounce to bleed energy off. Designed for open ground, even if the opening was a single Marine wide, trees made the experience significantly more . . .

The flexible end of a branch slapped against her cheek.

. . . painful.

The world turned green.

Orange and yellow exploded past her, shrieking.

A new line of pressure across her pelvis. Suddenly gone.

Her shoulder scraped against . . . something harder than her shoulder was all Torin knew.

Half a meter above the ground, they were two meters deep in bracken.

Torin!

The charge burned out and she dropped, the crushed vegetation almost comfortable. Under the unblinking gaze of a glossy brown beetle as big as her fist, she checked the medical readout on her cuff while carefully working muscles and joints.

"Bruises. Small lacerations. I'm fine. Presit?" She shifted left, the straps digging in as Presit hung on. "Presit, let go. We're down."

Just as Torin had begun to think she might be stuck with the

Katrien permanently attached, she felt the pressure shift, heard a soft grunt and the wet crunch of broken vegetation as Presit hit the ground beside her.

Her eyes wide, she panted, tongue protruding from her open mouth.

"Are you hurt?" Torin rose to one knee, working the drop sleeve release. "Presit! Are you hurt?"

"I are not having . . ." She shook her head. Waved both hands. Sat up. "You are . . . You were jumping . . . You are being crazy!"

"You're welcome. Are you hurt?"

Clouds of fur danced briefly in the damp air as she stroked herself. "No. I are not hurt."

Gunny!

"Werst, you and Ressk continue to the plateau. No change in plans. Craig, we'll get a ping off the shuttle and join you."

Understood, Gunny. No change in plans.

Torin . . .

She waited.

The beetle buzzed, then closed its wings against its back and lumbered away.

Understood, Torin. No change in plans. Craig didn't sound happy, but right now she didn't need him to be happy. She needed him to follow orders. **We're going to talk about this when you fossick in.**

The cut on her cheek dribbled fresh blood into the corner of her mouth. She licked it clean. "Roger, that." Of course, they were going to talk about it. She'd jumped out of a moving shuttle; a little above and beyond what Craig had begun to see as SOP. Heartbeat beginning to slow, she took a deep breath. "Going silent in case mercs are monitoring." The Confederation mercenaries, sergeant and above, had implants. Vertic had been certain the three Polint did not. The Druin was a wild card. It was unlikely they'd pick up the signal, but not impossible. "Kerr out."

Do not get eaten by snakes. Ryder out.

She tongued the mic off as she unfastened the drop sleeve, hit the self-destruct, and tossed it aside. Her helmet scanner showed a multitude of life signs in the immediate area, but most were small and all were invisible.

"Your kind are being all about destroying things," Presit muttered, poking at the rectangle of ash on the bracken.

"Sleeves are single use. We don't leave gear behind for the enemy."

She picked a piece of bruised vegetation out of her fur. "It are a good thing I are not having a button that are doing the same."

Torin frowned. That had almost sounded like an apology. "You sure you're all right?"

"I are just having plummeted to certain death that are suddenly not so certain." Blinking rapidly, she combed her claws through her whiskers, right side, left side, right again. "I are as all right as I are able to be at the moment. Although I are having lost my glasses."

"Light's not exactly bright down here."

Presit sniffed dismissively. "I suppose it are dim enough for me to be managing."

"Good, because we've got a three-kilometer walk to the V . . . Fukking hell!" Torin flicked her hand, sent the insect feeding inside the cut on her finger flying, then quickly sealed her cuffs and collar, activating the electrostatic charge just in time for multiple insects to fry before they reached bare skin. The rest dropped off her uniform, disappeared into the bracken, then reappeared, heading for flesh.

Torin rolled up onto her feet, bit back profanity when the bruising across her pelvis complained, and hit the charge again.

"Be bending down now!" Presit commanded, rising onto her toes. Torin bent and felt claws, cool against her cheek. "This one . . ." Presit held up an insect, multiple legs wriggling, mandibles cutting the air, then, when she was sure Torin had seen it, tossed it aside. ". . . was almost reaching the cut on your face. Now we are being even."

"Even?"

"You are stopping me from dying on impact. I are stopping you from being eaten alive."

"That's . . ."

"And I are willing to seal it for you."

As a declaration, it was definitively Presit. Torin was impressed at how quickly she'd recovered her aplomb and dialed back a smile too broad for their relationship. "You can seal it after we get out of here."

Presit shrugged. "As you are wishing."

Using her KC to push back the two-and-a-half-meter-high fronds, Torin led the way out of the bracken along the path the VTA had taken. While she didn't have Binti's level of training in camouflaging her movements, she could have left the edges of the bracken undisturbed, masking their arrival from unfriendlies at ground level, but the numbers of insects flushed out by her presence made her skin crawl and she moved at her best speed. The unfriendlies were five klicks away. The insects occupied ground zero. Threat assessment 101. "Presit, move."

"Apparently, I are not tasting good. That are being so sad. Don't be waiting on me."

Torin took her at her word. She'd been standing on the spongy ground a meter away from the bracken long enough to seal the cut on her finger when the Katrien finally emerged.

Presit held out a hand, and Torin tossed her the sealant, bending to minimize the difference in their heights. "They are touching my feet . . ." Presit matched the angle of her head to the angle of the canister as she sprayed. ". . . but are not climbing on. Also, there are being a snake on the branch behind you."

"It's a vine." Torin tapped her fingers against her cheek. The sealant held, the wound beneath it numbed.

"Then it are being a vine with a mouth."

About as wide as her thumb, the snake was a greenish brown that perfectly matched the surrounding vines, and appeared to have no teeth. Or eyes. Torin moved a little further away and watched it settle; head, or end with the mouth at any rate, turned toward her.

"You are sinking."

"I know." She lifted one foot then the other as her boots broke through layers of decay on the jungle floor. Multiple insects scurried away—a huge improvement on the direction the bracken insects had taken—and the smell of rot rose around them. A row of buildings, all but one a single story high and all nearly covered in vines, were visible through the trees. If Presit had waited another few meters to fall, their landing would have been significantly rougher.

"I are not liking it here."

"For the first time ever, we're in complete agreement." Helmet scanner down, Torin drummed her fingers on her weapon, tuned out Presit's list of exactly what she didn't like, and waited for . . . "That's the signal. Let's . . ." She stared at the mesh of vegetation between the trees. ". . . go."

"If I are assuming you are not also carrying a flame thrower, then I are expecting we are to be taking the high road."

Torin glanced up. At around the three-meter mark, she could see sizable empty spaces amid the foliage, perfectly illustrating why Werst and Ressk had gone out on recon. They could move faster through the trees than they could at ground level.

"We are not being Krai, still being living in trees, but my people are being arboreal not so long ago in the evolutionary scheme of things." Less prehensile than the Krai, Katrien toes had the advantage of claws. "As I are understanding it, your people are having been arboreal too, way back." Presit waved a small, black hand, the humidity plastering her fur around her wrist. "I hear it are like *seekindying dae hurricna;* you are never forgetting."

"Yeah, I heard that, too." She'd look up a translation later. Weapon secured across her back, Torin centered the signal from the shuttle and jumped for the nearest low-hanging branch. At the edge of her peripheral vision, she saw the eyeless snake drop to the ground and reminded herself not to grab any vines.

• —◆— •

"Look at it this way, you're contributing valuable anecdotal research." Grateful for the chance to rest, Arniz slowly pulled an insect out of a bleeding scrape on Trembley's arm, maintaining a constant pressure so as not to leave the head and mandibles in the wound. "The predestruction inhabitants must have tasted Human; they're ignoring the rest of us."

"Yeah." Trembley winced as she closed the tweezers carefully around the narrow body of another insect. "I noticed and . . . Fukking *hell*! That hurts!"

"I thought Marines were tough." She smacked him on the thigh with her free hand. "Hold still. You enjoyed it a lot less when I had to cut pieces out." The insects smelled of copper when crushed although, Arniz allowed, the scent could have come from the bulging blood sacs distorting the shape of the rearmost legs. "If I survive this . . ."

"Survive *what*?"

She blinked at him, a slow slide of her inner lids, and his skin darkened as blood rushed to his face. The little she knew of hairless mammals suggested shame. Good. "If I survive this," she repeated, "I'll petition to take a specimen or two back to the university and have the entomologists our budget wouldn't extend to include, find out what, exactly, they're feeding on."

"Me. They're feeding on *me*."

"In a less general sense." Gripping Trembley's wrist, she turned his arm from side to side, tongue tasting the air as she leaned in close. "I think that's got them all."

"You *think*?"

"I also think that next time you're bitten by an organism you haven't previously been exposed to, you should do more than scratch." Maintaining her grip, she applied a generous layer of sealant. "I'm sure Marine training must have covered what to do when killing people on a world new to your species."

"We didn't *just* kill people," he muttered sulkily, examining his arm.

The edges of the scrape were pink and puffy, and Arniz hoped it was from trauma. Alternately, she hoped the broad-spectrum antibiotics in the sealant could deal with it. Every member of the expedition had had emergency medical training, but that training had been focused on getting the injured to the autodoc alive. "We have Human parameters loaded, thanks to the presence of Dr. Ganes. If Yurrisk had allowed me to take you to the infirmary . . ."

"Yeah, right." Trembley flexed his hand and rolled the muscles of his arm under the seal. "If you're in the *anchor*, you could send for help."

"From the infirmary? With you right there? Swelling, but functional? That makes as little sense as the rest of this," she sighed, reassembling the first aid kit.

"What part of wanting a weapon to defeat the *plastic* doesn't make sense? Look, we're going to find the weapon and then we're boots up. The commander already has a buyer. He'll sell it, pay Sergeant Martin, who'll pay us. You can raise the alarm or keep doing whatever you were doing. I don't get what's so *hard* to understand."

Arniz sat back, bracing herself on her tail. "And if there isn't a weapon?"

"But the plastic *residue* . . ."

She should never have sent that message. Should have buried it in the day's report as an unexplained anomaly. Dzar and Mygar would still be alive if she had. She wasn't responsible for their deaths, she carried too many years to carry unnecessary guilt as well, but she had put the circumstances leading to their death in motion. She watched Salitwisi scan the area around a stone corner protruding from the undergrowth, Mirish standing guard, weapon aimed, as though Salitwisi with a scanner was dangerous. The di'Taykan's deep blue hair flicked back and forth, two of the gossamer-winged insects riding the air currents above it. Ignoring the extremely high probability that the absence of underground chambers out on the plateau meant there'd be no underground chambers in the jungle—both locations part of the same city, on the same shallow soil, the jungle arriving at this point thousands of years after the city had fallen—Yurrisk had decided the ruins along the path were to be scanned.

Why kidnap experts and not use them? Arniz wondered. Guns did not supersede education. "Fine. Yes. There's plastic residue, but your conclusions are unsupported. What if there isn't a weapon?"

He shrugged. "Well, we have to be *sure*, don't we?"

It sounded as though it made sense. In a crazy *let's extrapolate from nothing at all* kind of a way. She sighed. It wasn't an argument she could win. "Evidence suggests the insects are attracted to open wounds."

Trembley nodded. "Standard operating procedure: if physical integrity is breached, leaks are to be sealed immediately."

She blinked.

He poked her leg. "That was like a *joke*."

"I see." She didn't. He reminded her more and more of her firsters. Young and uninformed, needing guidance to meet his potential, guidance he wouldn't get from the violent . . .

"What the fukking hell!"

Martin. Coffee delivered too hot or an absence of ginger biscuits deserved an extended burst of profanity at high volume—although he'd killed two people without raising his voice.

"Fukking fuk fuk, God fukking damn it!"

He'd was raising it now.

And flailing.

Bits of broken bracken filled the air around him.

And tearing off his clothing.

Humans looked unfinished without scales or fur.

"They're in my fukking dick!"

Dick?

Oh. Penis.

Even in the dim light under the canopy, that looked painful. With luck, it was actually excruciating.

She could taste his fear on the air. Even better.

A whimper drew Arniz's attention back to Trembley. He surged up onto his feet, whites showing all around his eyes, a Human feature she hadn't previously been aware of, and stood on first one foot, then the other. She reached out and swept an insect off his lower leg. "It appears they're drawn to openings, not only open wounds. I suggest you seal your clothing."

"How?" Still shifting from foot to foot, he plucked at his trousers. "These aren't combats!"

"There's tape in the anchor." When he stared at her, not understanding, she sighed. "Specifically for this. The Ministry didn't tell us what flavor the locals would prefer. This isn't our first dig and tape doesn't malfunc . . ." She sighed as Trembley raced past, back along the cleared road toward the anchor.

"Where do you think you're going?"

Trembley didn't slow at Yurrisk's question. "Penis bugs! Need *tape*!"

"You don't . . ." Before he could raise his voice, Qurn touched his shoulder and pointed at Martin. With her gloves off, her long ivory fingers looked like bone.

Arniz froze as Yurrisk strode past her toward Martin. When he ignored her, she relaxed, stiffening again as Qurn shot her a suspicious glare. She shot one back.

"You're not being paid to play with yourself." Yurrisk's voice had dropped to a low growl.

The crack of a crushed insect between thumb and forefinger sounded before Martin raised his head.

"And," Yurrisk added, arms folded, "you don't get paid at all if we don't find that weapon."

Martin's expression dropped Arniz's tail tip to the ground. Did Yurrisk not recognize how close he stood to death or did he not care? Was this something the military trained away?

"I think he needs a moment." Glove back on, Qurn touched Yurrisk's shoulder. She saw the danger, even if Yurrisk didn't. "The faster Sergeant Martin deals with his infestation, the faster he can go back to work."

"Infestation. Of course." Yurrisk's shoulders rose and fell as he drew in a deep breath. He swayed right, then right again as he drew in another. Arniz hoped he'd choke on an insect, but no such luck. "Then deal with it."

"Ganes!"

Arniz twitched at the aggressive bark. Tossing aside an armload of cut bracken, Ganes met Martin's gaze, apparently unaffected. He raised an eyebrow as Martin stomped naked toward him.

Everyone stopped what they were doing and watched. From the road, from the ruins . . .

"Why haven't the little fukkers gone for you? You have some kind of science anti-bug charge in your clothes?"

"My clothes are impregnated with a species nonspecific repellent. And . . ." Ganes raised a hand, a broad band of familiar gray tape around wrist and cuff. "Insects are a given on most planets. I'm not impressed by your preparation for this mission, Serg . . ." He rocked back, lessening the impact. Straightened, wiped blood off his mouth, sweat off his forehead, and smiled, teeth a defiant red-stained slash across his face. "There's tape in the anchor. You wouldn't fit into my clothes."

Arniz had never liked him so much.

"I would," Zhang snapped. "Let's go!"

"No. I . . ."

"You got them in your dick, Sarge, you want to consider where me and Malinowski will get them?" She grabbed Ganes by the wrist and dragged him past Yurrisk, breaking into a run. "We'll be back."

"You can't . . ." Yurrisk began, but Qurn shook her head. "Fine. Malinowski, you'll . . ."

Malinowski crashed through the underbrush, running diagonally toward the road, leaving Lows unguarded.

Martin opened his mouth, closed it again, gathered his clothes, and walked toward the anchor, clothes in one hand, KC swinging from the other. One of the Polint, Arniz thought it might have been Tehaven, barked out an observation the rest of them found very funny.

"New worlds, new life-forms," Yurrisk muttered. "Steering clear of both was one of the reasons I joined the Navy."

"Penis bugs are a frequent occurrence in the Confederation?" Yurrisk jerked around toward Qurn so quickly he lost his balance and had to grab for her with both hands even as she braced hers against his chest. "Okay?"

He wet his lips, swallowed, and nodded. Unaware, or uncaring of his audience, he leaned in, rested his forehead against Qurn's and said, "Penis bugs?"

"A good reason to join the Navy."

Arniz couldn't see Qurn's face, but what she could see of Yurrisk's expression made her wonder how he could have hired the piece of trash that was Sergeant Martin.

"All right." One hand still holding Qurn's arm, Yurrisk turned toward the Polint and made a slashing motion with the other. "Keep cutting!"

Without Martin's slate, they wouldn't have understood the words, but they seemed to understand the motion and, after a short exchange, began swinging their heavy blades again.

"Mirish!"

The di'Taykan, now leaning against the slender trunk of young tree, raised her head. Mouth open, she seemed to be panting slightly, the ends of her deep blue hair stroking a lime-green coil on her shoulder. Clearly unhappy in the heat and humidity, her weapon hung by her side. Not that it mattered how she held her weapon; Magyr's death had taught Arniz these people were weapons in and of themselves.

". . . keep an eye on Dr. Lows as well as *Harveer* Salitwisi."

"Sir."

"The insects nest in the bracken." Qurn held up her slate and, had Arniz the energy to spare, she'd have crept closer to try and get a look at it. "Keeping the Humans out of the bracken should solve most of the problem."

"The road goes through the bracken," Yurrisk replied. "And if we leave the road . . ."

"Commander!" Sareer dropped down from the canopy, weapon across her back. She'd seemed quieter since Magyr's death. Arniz wanted to think it was because the adventure of pushing people around, of taking what was wanted, had consequences. Sareer hadn't known Dzar, but Magyr had been more than an unfamiliar face in a crowd, and it had been Sareer's weapon picked up from where it had been leaning on the tree. Arniz wanted to think shame had quieted her, but she didn't know enough about the Krai. Trees were important to them; perhaps it was nothing more than Sareer having a spiritual moment in the canopy. "We've seen the ruins, Commander. They're mostly low stone buildings, with narrow rectangular doors and triangular windows—like they couldn't make up their minds. We saw two or three buildings with an intact second story and half a dozen more with broken walls and missing roofs." When Yurrisk held out his hand, she passed over her slate. "I came back with pictures. Beyvek's still mapping. Weird layout, though; a lot of empty spaces for a city."

"There were empty spaces on the plateau as well," Arniz said thoughtfully, forgetting she wasn't part of the conversation. "We theorized the pre-destruction inhabitants might have produced food inside the wall. For themselves—or for livestock, given the distinct probability of them being carnivorous. Soil analysis will identify dedicated agricultural use and, with the soil this shallow, they could have used a rotation of manure, theirs and their livestock's, as well as agriculture to build up organic matter."

Off in the underbrush, Salitwisi snorted. Loudly. "Simpler theory; all structures made of wood have long rotted away, leaving gaps."

Arniz waved a piece of cut bracken at him, the anger at Magyr's death finally finding a safe outlet. Architecture blinded Salitwisi to smaller possibilities. "I'd *planned* on testing for that as well."

"Not to mention that any clay structures," Salitwisi continued, ignoring her, "would have eventually washed away in the rainy season."

She whirled around to fully face him. "And where would they have found clay?"

"From the river basin. At the bottom of the cliff. There had to have been a reason they build the road pointing directly at it."

"There may have been a hundred reasons. We'll never know them all!"

"But next year, when our charter includes the river basin . . ."

"Are you hallucinating?"

"I know people at the Ministry, and I'll be speaking to them in the off season."

"Oh, you'll be . . ."

"Be quiet!"

The whole jungle fell silent at Yurrisk's roar. Insects, birds—even branches stopped rubbing together.

Heart pounding, she turned to see Yurrisk and Qurn and Sareer staring. Yurrisk's nostril ridges were closed. Behind him, Sareer's were open. Arniz wished she'd paid more attention at the xenopsychology workshop her department had been forced to attend. Xenoanthropology? She didn't remember, she hadn't been paying attention, but it looked as if the two Krai were experiencing opposing reactions.

"We are on a timetable here," Yurrisk growled. He pivoted on the ball of his foot, and Sareer's nostril ridges slammed shut as he faced her. "How much farther?"

"One point four seven kilometers, sir."

Not by the path she'd taken, Arniz assumed. On the ground.

"When Martin and the rest of the Humans return, you'll guide them around the bracken and then back on course."

"Yes, sir." She glanced around and frowned. "Where have they gone?"

"They're back at the anchor taping up their genitals." This time when they turned to stare, Arniz shrugged.

"Snake."

"I see it."

It slid off the edge of the branch and just before gravity won, thin arms lifted out from its sides, spreading triangular wings.

"Flying snake."

"Yep." Werst watched it glide to a lower branch, admiring the way it used its tail as a rudder. Did snakes have tails? he wondered. How did anyone tell where the snake ended and the tail started? And shoulders—if it had arms, did it have shoulders? And how did it taste?

"Not the forests we're used to." Ressk capped his canteen and secured it, his shoulder a warm weight against Werst's side.

"Definitely not what I'm used to." The city around the spaceport had been an artificial forest on a good day. On a bad day, it had been a prison, a broken promise, a kick in the nuts.

"Yeah. Not going to your pity party." Ressk's hand closed around his wrist and squeezed. "You've been up a lot of trees since then, and this . . ." He waved his free hand. ". . . isn't like any of them."

Werst plucked a long, narrow leaf and chewed it thoughtfully. "Reminds me a bit of Dunjub. Took the squad out to find the Primacy camp, and had to figure out the best way to bring the company up through old-growth forest."

"And has your experience told you the best way to get a Polint through that?" Ressk shifted his feet, tightened his grip, leaned out, and peered into the underbrush nearly three meters below.

"Yeah. Behind a flamethrower." He rolled his shoulders and cracked his back. "Be a lot easier if we could land on the anchor again. Hostages complicate the shit out of things."

"That's why they sent the best."

"Us and the other guys." His nostril ridges flared. "Mammals moving on the ground."

They looked together. If they hadn't been tracking by scent, they wouldn't have known anything was there.

"Light on their feet."

Ressk drew in a deep breath. "Omnivore. Noting location."

They tapped it into the map forming on their visors at the same time and Werst checked that the anchor's beacon remained centered. "Running silent from here in. Stay close."

"On your six."

Werst fell forward, caught the branch the snake had dropped from, swung up, grabbed another with his left foot, and released, picking up speed as he moved. Surrounded by the movement of leaves and branches, of insects and birds, of Ressk following behind, he had a job to do and a team he could count on. If he found one of the mercs alone and circumstances forced him to beat the crap out of him—where *beat the crap* meant subdue in an entirely lawful manner—the day would be damned near perfect.

•—◆—•

"Maybe next time I tell you to stay strapped to the rear bulkhead," Torin snarled, focus finally shattered by Presit's constant complaints, "you'll damned well stay strapped to the rear bulkhead!"

"There, that are not being so hard, are it? Are you feeling better?"

Torin stopped and slowly turned. They'd left the trees when Presit had nearly fallen, recent arboreal ancestors no longer able to compensate for a lack of physical conditioning. She'd done well, considering, and they'd covered over half the distance to the VTA before they'd had to drop to the ground, but from the moment her feet hit dirt, Presit hadn't shut up. She was tired. Her hands and feet were sweaty. Her eyes were watering. It was hot. They were taking too long. She was hungry. Leaves were sticking to her fur.

She looked tired and hot and there *were* leaves sticking to her fur, but they'd just spent the last ninety-seven minutes moving two point six kilometers through jungle, so what did she expect?

Presit spread her hands, unconcerned by the knife Torin held. Or her expression. "You are wanting to have said that since we are hitting the ground. I are watching your movements becoming tighter and tighter as you are holding it in. Now you are not having to hold it in."

Her movements had become tighter because the Confederation had laws about silencing the press. Even temporarily. On the other hand, she did feel better. "We're nearly there. Less than half a kilometer." Torin recentered the signal from the VTA and slipped between two tangles of vine, walking along a fallen pillar, carefully avoiding the open doorway of a nearly intact structure. As long as Presit stayed on her heels, she'd be fine.

"I are also needing to thank you for saving my life."

"You're welcome."

"There are being a large beetle on your back. It are not doing anything, but . . ."

"Presit, be quiet."

"I are just thinking you are wanting to know."

"Please, be quiet."

The next two hundred meters were blissfully silent.

"I are smelling smoke. And considering how you are smelling, that are not an insignificant amount of smoke."

"Craig dropped a VTA into a jungle clearing." Torin could smell it, too, although given the humidity, she expected it was stronger closer to the ground. And stronger to those species with a better sense of smell than Humans. Which was most of them. She bent her knees, twisted to duck under a branch, and twisted again to avoid crushing a fungus. Those who'd fought on Rinartic knew spores were no one's friend, and nightmares about vigorous growth in moist places had been common.

"I are not able to go on."

"We're less than half a kilometer away. I could throw you that far."

"You are being so full of shit, it are being no wonder your eyes are brown. I are taking three steps for every one of yours. I are having walked three times as far."

Torin had shortened her stride and slowed as much as possible. Without Presit, she'd have been back at the VTA long enough to have showered and begun familiarizing herself with the Primacy's weapons. That said, even with environmental controls keeping her from overheating, her face was slick with sweat around the sealant on her cheek, her right hip ached with every step, and given the way her shoulders and arms felt, she clearly need to increase her resistance training.

Presit had every right to sound exhausted. "You've done a good job keeping up," Torin said as she turned. "Given . . ."

The knife in her hand was not a throwing knife. Torin threw it anyway, yanking Presit toward her with her other hand.

The snake's body writhed, blood and viscera painting loops on the ground.

"When something that size goes after something your size . . ." Torin reclaimed her knife and lifted the head on the blade. Clear fluid oozed from the revealed fangs. ". . . assume poison."

Shed fur danced through a beam of sunlight slanting through the canopy. Presit opened her mouth, closed it again, and finally managed, "This are being a Class 2 Designate. You are not being permitted to kill the indigenous species."

"You're welcome."

Several of the fist-sized beetles had already appeared. One approached the snake and took an audible bite. When the snake didn't respond, the rest advanced. They were an efficient cleanup crew, and Torin assumed the beetle she'd seen after impact had been an advance scout, waiting to see if she'd been too injured to fight back.

"I are not needing to be standing here and watching that, Gunnery Sergeant Kerr!" After falling out of the VTA and traveling three kilometers of jungle, both in the trees and on the ground while wrapped in heat and humidity, it seemed the bloody efficiency of beetles devouring a snake had pushed her to the edge. Presit sounded close to breaking. "You are taking me back to the shuttle now!"

"So you can fall out of it again?"

If Presit lost it in front of Torin, given the relationship they had, she might never get it back.

Torin raised a brow when Presit glared at her.

"You are not being funny, Gunnery Sergeant Kerr." Her lips curled back off her teeth. "If I are having this on camera, people are reacting to you entirely differently." She combed her claws through her ruff with one hand, and pointed imperiously with the other. "If you are going to cut a path, I are suggesting you begin cutting."

"You made terrible time, Gunny."

Presit shrieked and jumped back as Firiv'vrak rose up out the underbrush.

His hand cradling her jaw, Craig brushed against the sealant on her cheek with his thumb. "You. Medical. On the knocker."

"I'm fine." Torin pressed a kiss into his palm, pulled his hand away from her face, and tugged him into step beside her as they crossed the clearing. Charred foliage crumbled under their boots. Freenim's sitrep—perimeter established, weapons familiarizing begun—allowed her the time to reassure Craig who hadn't developed the kind of high-level compartmentalization that years in and out of a war zone required. "I've a few cuts and bruises, but I need a hot shower, not medical. Neither of which we brought down with us." Sonic showers were better than nothing, but only just.

"You jumped out of the VTA." He sounded so neutral he was clearly still dealing with his reaction.

"I was the SO at the hatch. I had a drop sleeve."

"You jumped out of the VTA."

"Presit fell. I went after her." Dalan had somehow managed to convince Bertecnic to carry Presit across the clearing. Walking alongside, he had his camera up to record her report of their three-kilometer walk. Presit managed to look exhausted and revitalized simultaneously. Dalan had brought her a spare pair of glasses, and she wore her clumped and matted fur like a fashion statement.

"By jumping out of the fukking VTA!"

Torin sighed, stopped, and turned toward him. "Craig . . ."

"Tell me you won't do it again." He grabbed her shoulder, snatched his hand back when she flinched. "I've learned to cope with people shooting at you . . ." His eyes crinkled at the corners although his smile looked reluctant. ". . . you seem to enjoy it, you freak, but I know what happens when a body leaves a ship unexpectedly. It's never good. I'm not blind, Torin. You're walking like you're a hundred and forty."

"But I only feel like I'm a hundred and thirty-nine." He didn't smile. She could feel the heat of his hand on her waist, through her uniform. "Compromise, after I give Durlan Vertic . . ." No. ". . . after I record my after action, I'll strip down and you can scan me."

"If anything's broken, you're heading up to medical."

She knew her body. She knew it whole. She knew it bleeding. She knew it broken. Reminding Craig of that wouldn't help; he needed to know she was all right now. Nor did he like being reminded of the injuries war had written on her. "Fine. If anything's broken, you can take me up to medical."

The rigid line of his shoulders relaxed. "Good."

"Warden Kerr!"

Torin turned, fully aware she'd be on camera.

Presit waved from Bertecnic's back, as though Torin might have had no idea who'd called her. "I are agreeing with Warden Ryder on this one." She laughed, a soft chuckle intended to let her viewers know she'd just gone through very trying circumstances, but was bearing up. "If it are up to me, we would not be doing such a thing again."

"If you'd stayed secured to the rear bulkhead, as requested, we wouldn't have done it this time."

Her ears flattened. "I should be knowing you are somehow making this my fault!"

Vertic stood as Torin stepped into the shuttle, both hands tugging down the front of her tunic. "Repor . . . I mean, what happened."

With only the energy screen activated, the air in the shuttle was as humid as outside it, but Torin suddenly found it easier to breathe. "Civilian went out the hatch, sir. I was wearing the rescue drop; I went after her. We landed safely and made our way back to the shuttle. I have additional information on the local wildlife."

Her mane rose. "Dangerous?"

"So far, only to Humans."

"You're Human."

Craig made a sound they both ignored.

"Fortunately, the discovery included minimal interaction."

"I see." Her lips twitched. "And your injuries?"

"Cuts and bruises. A little dehydrated." Torin paused to accept the pouch of water Craig pushed into her hand. "Presit's exhausted, but uninjured. Given the amount of exercise, she'll stiffen up later." Which might keep her out of trouble for a while, but Torin doubted it.

Vertic's expression suggested her thoughts ran with Torin's. Her mane settled and she nodded. "Well done, Gunny."

"Thank you, sir."

The golden eyes met hers. "Don't do it again."

"I'll do my best."

As she followed Craig to the rear of the shuttle, she noticed his shoulders had tensed up although she wasn't sure why. Nothing she'd told Vertic should have been a surprise to him. As soon as he was satisfied she hadn't been hurt, she'd find Freenim and they'd discuss how to use this new information to help keep their people safe.

• ——◆—— •

Werst shifted his weight as the branch bowed beneath him, froze in place behind a screening spray of leaves, and held up his fist. Ressk dropped silently down by his side.

Up ahead, several buildings rose up out of the underbrush, the canopy significantly thinner around them in spite of a number of mature trees rising directly up from the stone. He'd seen these particular trees once or twice as they'd traveled. Tall and slender above the tentacle-like sprawl of their roots, they contributed few lateral branches of any size to the high road and seemed to only grow where the buildings were solid and barely covered by foliage. Ressk had speculated that those buildings had been built of a harder stone that fed different minerals into the soil. Different soil, different trees.

Werst didn't care. He hadn't stopped because of the trees, but because of what sat on one of the buildings.

The male Krai either wore a couple of different pieces of combat uniform, or the camouflage function had split the difference between winter and urban before it crapped out. Not female, so not Petty Officer Sareer. Not Commander Yurrisk, too high off the ground. Had to be Lieutenant Beyvek. They'd all been Navy and that might explain the combats—trust Navy not to be able to work basic tech. Crouched on a peak of decorative carving like he didn't have an enemy in known space, Beyvek pulled something wriggling and banded out of a hole and dropped it into his mouth. Once. Twice. Three times. A nest of juvenile snakes maybe.

Werst plucked a large blue-and-yellow slug off a broad leaf by his hand, hunger aroused by watching another eat.

Easy shot, had he been allowed to take it.

Not significantly harder to take Beyvek out silently. The lieutenant's weapon was a good meter away. Come up behind him, slit his throat, drag his body off, let his crew assume he'd been taken by a local predator. One less *serley* mercenary to deal with later.

But those weren't his orders.

It was, unfortunately, easier to kill an opponent than to disable one and silently remove them from the fight.

It was easier to be a Marine than a Warden. The skill sets overlapped, but the morality had gotten a lot more complicated. Case in point, after the *Paylent*, Commander Yurrisk and his crew hadn't been given the support they were entitled to. That didn't make what they were doing right, but Werst carried enough of his service with him to make it . . . not *there but for the grace of Turrist go I*, but understandable.

He took one last look and shifted enough to meet Ressk's gaze, signaling he should leave a DL at this location. With it secured and activated, they faded back, curving to the left where the foliage was thickest. From now on, they'd have to move slowly and carefully, using every bit of cover.

Humans had to be trained to look up.

Krai didn't.

• ⸻◆⸻ •

Toes spread, tail out for balance, Arniz sank into a deep squat and rested her head on her folded arms. The pre-destruction road was as clear as it was going to get. She listened to the Polint, Yurrisk, and Qurn tromping through the underbrush with Salitwisi and his ancillaries, their grief over Magyr's death muted by a chance to examine ruins they had no legal access to. Her opinion hadn't been sought. She didn't do buildings and she'd already commented on the destruction of vegetative evidence. Forcefully. In an extended duet with Tilzon. They were ignored. Fine. She'd take this chance to rest and fantasize about what she could do if she'd only had the rudimentary poison sacs a small fraction of Niln were born with.

Sareer crouched beside her, weapon pointed across the road where Lows held his weeping bondmate, their remaining ancillaries at work clearing the ruins. Arniz assumed Yurrisk had put the geophysicists under guard in case their grief turned to violence. Violence was the Younger Races' default, after all. Ganes, their own violent default, had been escorted back to the anchor by Gayun and Mirish so that he could help bug proof the visiting Humans. With any luck, he'd get the repellent and the acid wash confused.

"He wasn't always like this."

Had she said that out loud? "Ganes?"

"What? No. Commander Yurrisk." Sareer spoke softly enough Arniz doubted anyone else could hear her. "The commander saved our lives when the *Paylent* was destroyed back in '08. Mine, Beyvek's, Gayun's and Prius'. And ten more."

"And where are they?" She peered around, the movement as silently sarcastic as she could make it. "The other ten?"

"Some are still serving. Some went home and actually fit in."

Arniz snorted at the qualifier. "There's counseling available, you know. It's generally considered the civilized choice over kidnapping and murder."

"I went to counseling." Sareer didn't pretend to misunderstand. Arniz appreciated the self-awareness, if nothing else. "Then the commander called."

"And you answered."

"And we answered. Counseling . . ."

The air tasted bitter.

"The Navy couldn't figure out how to fix what they'd broken. Saving our lives destroyed his."

"That doesn't make you responsible for him."

Arniz felt as much as saw Sareer shrug. "Maybe not. But we weren't doing all that well without him. Now we have purpose again."

"Because you're killing again."

"No!" She spat to one side and lowered her voice. "We're keeping the *DeCaal* flying. The rest, that's all Martin. If that *serley chrika* wasn't here, no one would have died. We didn't want him and his crew—not because some of them are Primacy, I mean, Qurn's Primacy, or she was—but they don't get it. They don't understand."

"Then why are they here?"

"The buyer insisted we bring muscle. They're paying a lot for that weapon; they wanted their investment protected. But the commander made it clear when they came on board, they're working for him."

"Which makes him responsible for the killing."

"No."

Arniz sighed, suddenly as emotionally weary as she was physically exhausted. "Yurrisk all but gave the order for Martin to kill Dzar."

"The commander's not himself sometimes." The words tumbled out as though they'd been said over and over and over again. "Martin plays on that."

"You don't like him."

"Martin?" Her lip curled. "I don't have to like him. I have to follow orders. That's what keeps us safe."

Arniz wished she knew more about the Krai as a species. Did they all require the safety of hierarchy? Or had the war twisted them—some of them—enough that they needed to bury personal responsibility in the structures that supported it? "Sareer . . ."

A muscle jumped in Sareer's jaw. Surprise that Arniz had known her name?

". . . why are you telling me all this?"

"You watch us. You question. You're grieving the death of your student and you think you're too old to worry about living or dying, so you say what you think. You're wrong." When Arniz finally turned toward her, Sareer moved her shoulders up and down in an awkward copy of a Human gesture. "I told you, I went to counseling. You want to know why, and I'm telling you so that you don't ask the wrong question and have Martin answer. He's here because he's a professional bully. His people are here because he's paying them. But the rest of us? We're here because the commander is ours."

"Loyalty."

"Yes."

The air tasted of sadness. "I don't think you understand what loyalty means."

Sareer's mouth twisted into a painful looking curve. "Fuk you. We saw people we'd shared our lives with blown apart. Pieces of them, pieces of the ship, all tumbled together. Blood spinning by in perfect spheres. The living sucked into vacuum too fast to scream. *You* don't understand what that shit does to the survivors."

"Some of the survivors," Arniz pointed out tersely. "There's ten who didn't answer your commander's call."

Nostril ridges closed, Sareer straightened and stared down at Arniz, lips lifting off her teeth. "And four who did."

• ⸻◆⸻ •

"First of the DLs went active, Boss."

Torin and Vertic moved together to the front of the VTA, Vertic waving Torin on ahead. "I can see over you, Gunny. It doesn't work the other way."

Unless they were advancing toward open fire, officers went first. And old habits died hard. Torin put a hand on Alamber's shoulder and leaned in. "What've you got?"

"The Krai we got the best look at on Ganes' image . . ."

"Lieutenant Beyvek, Engineering."

His hair flipped out, stopping just short of Torin's chin. "Yeah, well, Beyvek's hanging out in a bunch of ruins all by himself, eating snakes. Even Werst could've made the shot from where they set up the look and listen."

"Except that we're not shooting."

"Yet," Vertic added.

She wasn't wrong. It always came to shooting in the end.

"Ah, that are being the first sight of the enemy." Torin held her ground, so rather than push her out of the way, Presit squirmed in beside her, peering over Alamber's other shoulder. She'd been resting, overwhelmed by the fall from the shuttle and the trek through the jungle. Temporarily overwhelmed as it turned out. "I are going to be needing to access the feed."

"You may have access to a copy of the feed," Torin told her. "Not to the live signal."

"I are . . ."

"You are not going to endanger lives," Vertic growled.

"I . . ."

"No."

Torin kept her eyes on the monitor. And didn't smile.

Beyvek ate two more snakes. His weapon rested against a carved block of stone a meter, a meter five away from his perch and, if he'd been ex-Corps, Torin would've been pissed on principle. Justice wanted criminals captured, and she bet her pension that after finding one of the enemy alone, Werst had considered a "dragged off by wild animals" scenario. Two on one, Krai to Krai, they might've been able to manage it without giving the alarm. Unfortunately, *might have* crossed the preventable risk to the hostages line that had been the reason behind the no contact order.

The second and third DL went active seventeen and twenty-four minutes later.

"Are they clearing a road?"

"Looks like a road, Boss. They're down to stone here, and here." Alamber aligned the two screens. "Also, looks like the road leads straight to those ruins where Beyvek's waiting."

"They are thinking the weapon are having been hidden in the ruins!"

"Possibly."

"Possibly?" Presit's response was shrill. "Then why else are they going there, Gunnery Sergeant?"

"For the snakes?" Alamber nodded at the screen. "I mean, why eat packaged when you can eat fresh, right?"

"Are your name being Gunnery Sergeant?"

"Alamber, tag both hostages and mercenaries as we see them. I want a full head count as soon as possible." The Niln squatting at the side of the road looked old, the two Katrien distressed, and all three of them exhausted. The Krai squatting beside the Niln, probably Sareer, looked annoyed. Better than angry, Torin allowed. "D . . . Vertic, any thoughts on why Netrovooens is standing in the underbrush, weapon ready?"

"I assume he's standing guard? Other than that, no."

Why would he be standing guard so far from the road? What could they be guarding while standing withers-deep in foliage?

Two Niln, a Katrien, and Commander Yurrisk rose up out of a tangle of green and brown until they stood only ankle-deep. Netrovooens adjusted until all three were in his line of sight. One of the Niln began talking, hands and tail waving.

"I think they're looking at pieces of buildings, Boss. They're standing on a corner of wall. See?"

Alamber sketched a bright pink highlight across the screen. "Runs across here and back that way."

Now he'd pointed it out, she could see the underbrush grew over stone differently than it did over clear ground. Commander Yurrisk was being admirably thorough, heading for the intact ruins but not discounting the bits along the road.

"Is that Niln helping the mercs?"

The heavier set of the two Niln, his scales a soft charcoal gray, was speaking directly to Commander Yurrisk, both hands sketching what Torin assumed was emphasis on the air. There could have been Niln still in the VTA, a Niln on the mercenaries' side however unlikely, a Niln not caught in Ganes' image. "Can you clear up the sound?"

"Can I clear up the sound." Alamber snorted. "Already on it, Boss. Dropping animal life and tree noises to background hum on three, two . . ."

"Although erosion has smoothed out the marks, I doubt very much these blocks were machine dressed, but that's neither here nor there. Many primitive societies built in stone and some created elaborate underground chambers. I'll have a better idea when we finally reach those ruins your companion returned to inform us . . . you . . . of as actual intact walls that would allow me to build a hypothesis. In the meantime, I really think we should take the opportunity this illegal entry under the canopy gives us to note the foliage before tearing it away from any remaining flooring. Harveer *Tilzonicazic, who, if you'll recall is our xenobotanist, would like to record her findings and is more than willing to have you remove any communication ability from her slate. And* Harveer *Arniz, I'm sure, would be willing to assist as most botanical growth offers insight into the soil and . . ."*

"I don't care about any of that," Yurrisk responded.

The Niln spread his hands. "Well, why would you? I understand, I'm all about the architecture myself, but . . ."

"Clear the plants away. Continue searching for underground chambers."

"But . . ."

Yurrisk's nostril ridges slammed shut. "Now."

"I'm feeling a certain sympathy for Yurrisk."

Torin tightened her grip on Alamber's shoulder. "Yeah."

"There were four Humans in the original image and none in this one. Or in the ruins." Vertic leaned close enough Torin could feel her body heat raise the temperature of the air between them. "Adjust until we can see the other side of the road. There . . ." She pointed a thick, clawed finger at the screen. ". . . lock on the Druin." The DL rotated until Alamber locked on, red clothing more visible against the green than even Netrovooens' fur. The Druin directed the Niln and Katrien tearing vines aside, while Camaderiz, the black Polint, stood guard. "Where are the Humans?"

Presit combed her claws through her ruff. "Perhaps they are already having been eaten by insects."

"Reducing our opposition by four ex-Marines?" The Primacy officer sounded amused. "It never works that way."

"Gunny! We found ruins!" Torin turned as Firiv'vrak launched herself into the VTA and rose up, arms unfolded and waving, the remainder of her legs still propelling her toward the front of the vessel. "Intact buildings . . ." Her eyes shifted toward the screen where Beyvek now stood on the carved capstone, staring up at a small bird like he was staring up at dessert. "Buildings like those!"

Stepping into the jungle on the far side of the clearing was like stepping into deep water. Deep, dirty water. There were no clear lines of sight and the late afternoon sunlight through the leaves had tinted the uncomfortably humid air green. Heavy artillery could have been preparing to fire a hundred meters away and Torin wouldn't have been able to see it. Granted, *had* the mercs put heavy artillery into play, her helmet scanner would've registered the energy signatures, but equipment could fail, had failed, and in this kind of environment, eyes and ears—Human eyes and ears—were at a disadvantage.

The dense growth wasn't notably different than the tangle she'd faced leading Presit back to the shuttle, but the reporter and the adrenaline burn of the jump sizzling under her skin had been a distraction. Here and now, all she had to concentrate on was the jungle.

The amount of insect life nearly overwhelmed her helmet scanner, so she filtered them out and maintained only the mammal and reptile scans. The paler fronds of a bracken clump waved from behind the trees to her far left, but she could see no indication of ruins beyond the crumbling sections of broken wall that drew humped lines under the tangled mats of vines and organic debris. If they were standing on the route the two Artek had taken, Torin could see no sign of it. "Did you PIN the ruins?"

"No need, Gunny. We'll follow the scent trail." Firiv'vrak folded her arms back in under the edge of her carapace and disappeared, reappearing a moment later in a break between two tufts of fern, sliding through the foliage like a missile through vacuum. Between random glimpses and the out-of-place scent of cherry candy, Torin managed to follow until they moved into even denser foliage. Stopped at the edge of a clump of chest-high bracken, she took a moment to consider the path to the right versus the path to the left. Firiv'vrak had to have gone around, but damned if Torin could tell which side. And while the air at ground level was still and damp, the leaves up above moved constantly, their pervasive whisper drowning out any noise the Artek made as she moved.

Only officers considered themselves so omnipotent they never had to ask for directions. "Firiv . . ."

A hiss from her right dropped Torin to one knee and an insect about as large her palm dove through the space her head had just occupied. Flat enough that only its spread wings gave it volume, the big bug looked disturbingly like an Artek. Her finger outside the trigger guard . . .

"Because shooting at shit gives your position away, Corporal. What part of recon do you not understand?"

. . . Torin watched the insect hit the ground and run for the nearest tree, wings a blur as it climbed. About a meter up, the bark bulged and a pale appendage whipped out of a crevasse. An instant later, the insect had been dragged, fighting and hissing, in under the bark and out of sight.

"Gunny?"

She stood, uncertain if the hunter had been another insect or the tree itself, and reinstated the scanner's defaults.

"Looks like a symbiotic parasite," Firiv'vrak observed.

"I hate jungles," Torin replied.

She could see patches of pale blue sky above the ruins, through the same tall, slender trees she'd seen around the image of Lieutenant Beyvek. Huge heavy roots, out of proportion to the trees they supported, grew through and around buildings made of dark gray stone. The same stone, pitched and uneven, covered a courtyard twenty meters by thirty. Along the far end of the courtyard, a building two stories high with ornate stonework around the arched windows and doors and the edge of the roof, anchored the narrow end. Adjusting the zoom on her scanner, Torin could see carvings on larger blocks set into the lower levels. Language or abstract art—she neither knew nor cared. The buildings on the shorter arms of the courtyard were a single story high and significantly less decorated.

No, not buildings. Building. The single-story structures connected to either end of the larger building, and it looked like all existing divisions between them had been made by time.

At the remains of the wall that had once enclosed the area, Torin kicked at a broad-leafed plant exposing paler stone with lines of brown through the gray. She followed the lines of the crumbled wall first to the left then to the right past the dark gray arch that had probably once topped the gate but now lay on its side partially buried in the ground.

"Gunny?" Firiv'vrak paused in turn, halfway to the two familiar figures perched on the curve of a root. Their positioning suggested Keeleeki'ka had been telling Craig yet another story of her people—her oratory posture had become familiar over the last couple of days. Before Torin left the VTA this second, less spontaneous time, Vertic had explained that she'd sent the two Artek out exploring to force them to learn to cooperate.

"And because they're the next thing to indestructible," she'd added too quietly for Firiv'vrak to overhear.

"Both valid points," Torin had acknowledged.

As neither Vertic nor Firiv'vrak had mentioned they'd be meeting up with Craig, best guess said Keeleeki'ka had gone looking for him while Firiv'vrak went after her.

Torin wasn't happy about Craig being away from the relative safety of the clearing. He was their pilot. How the hell was he supposed to pilot half a klick away from the VTA? Not to mention how much the jungle resembled a deadly accident waiting to happen.

Her grip tightened on her weapon. If Craig was killed, the peace would definitely be sabotaged.

Without enough light to see into the buildings, even at full zoom, every room could be hiding multiple threats. Pity she hadn't thought to bring Dalan—his camera had a higher resolution. On the other hand, Presit wouldn't have allowed him to go alone and Torin had done Presit-in-the-jungle once already today. The dark stone walls were too thick for her scanner to pick up heat sources, but both the not-a-spider and the web it had built across one of the upper triangular windows registered high thermal signatures. High enough that Torin spent a moment wondering what the web had been built of.

Had they planned to take the ruins, she'd send Binti around to a secure location on the far side of the buildings where they could map out the interiors by bouncing a signal from scanner to scanner.

Stations, ships, and anchors were constructed to prevent the enemy doing the same, but, as far as she knew, no one had applied the tech dirtside pre-Confederation.

Lines of data scrolled past as an amphibian—ninety-seven point four probability—waddled out onto a cracked threshold and sat blinking in the dim light. Torin recorded the possibility of interior water and finished her scan back where she'd begun. The not-a-spider now read thirty-seven point two degrees; the web two point four degrees cooler with a line of weak joins flaring cooler still in the upper right quadrant—the best access to the window if breaking the web became the only option.

She took another look at the information she'd been able to gather: depth of rooms directly back of openings, thickness of walls at the openings, surface temperatures that allowed for the mapping of weaknesses, position of every exterior life-form. Data representing four-point seven percent of the scanner's potential. Threat due to lack of information: high. In the environments they were designed for . . .

Torin frowned.

In the environments they were designed for, combat scanners could expedite illegal activities. In the wrong hands, a scanner could bypass security systems and indicate how to break into buildings where cases of weapons were stored waiting legitimate shipment. The Wardens investigating the MI robbery had never used a combat scanner, and the Elder Races who considered war to be socially primitive would have little trouble considering the equipment used to fight it equally crude. Evidence suggested the Corps was having a hell of a time keeping track of billions of pieces of decommissioned equipment, and the gunrunners had known they were coming.

"Gunny?"

Experience insisted the courtyard was a kill zone.

"New experiences," she reminded herself, and stepped onto the first angled slab. When the hostages were safe, she'd suggest Justice have someone run an MI robbery simulation in full combat gear. Craig opened his mouth; she raised a hand and kept it raised until she was close enough he wouldn't feel he had to shout and attract unwanted attention.

He looked unimpressed. "I'm here because Keelee came and got me."

"I figured."

Firiv'vrak snapped her mandibles over the end of Torin's response. "I told you to wait here."

"You don't tell me what to do." Keeleeki'ka flicked her antennae back. "Warden Ryder hears the story. Besides, you were gone for so long, I thought you weren't coming back."

"I thought you *were*," Craig added, "so we waited. Right here. No exploring without an armed escort."

"Thank you."

"I haven't got a death wish, luv." The corners of his eyes crinkled. "Nor did I want to get jumped by something nasty in the ruins."

She returned his smile, amazed by how well he understood her. "How did you get here before us?"

"I didn't have to wait for orders to leave." He spread his hands. "Told Bertecnic where I was going and left." He waved a small cloud of insects away from his face and turned the movement into a gesture that took in the entirety of the ruins. "It's something, isn't it?" Born on a station, he'd spent most of his life in space.

Not the time nor place to challenge his belief about her relationship with Vertic. "It is. I've never seen anything this old." For the most part, the war had taken her to OutSector colony planets and stations. "Except for the H'san tombs," she amended.

"Doesn't count. They were tidy. Sterile. This isn't. This looks . . ."

"Old."

"Yeah."

Shoulder to shoulder, the Artek squabbling quietly in front of them, they walked around the courtyard, Torin explaining how the mercenaries seemed to be heading for a similar set of buildings.

"So you're here doing recon?"

"Always."

"That's a terrifying work ethic, you that know that, right? You've got a light," he continued grabbing her arm before she could sink an elbow into his ribs. "Should we eyeball the interiors?"

They needed to gather any information that could help take out the mercenaries without harming the hostages. "I'll look . . . first," she added at his expression.

He shrugged. "You've got the gun."

Torin checked her scanner was still recording and stepped through the open door. And stopped. Roots nearly filled the room. From hair-thin to as big around as her thigh, they burst through the rear wall and down past the shattered slabs of the floor like tentacles, leaving no more than a square meter of empty space. Space Torin shared with a pile of beetle shells.

Firiv'vrak thought she might be able to get through the tangle. Keeleeki'ka called her an idiot. Craig stepped between them.

In the next room, the roots emerged from the rear wall at the opposite angle. Torin stepped back into the courtyard to try and spot the tree they supported.

"What if the weapon's buried in that mess?"

"The weapon's not our job. Our job is to free the hostages and arrest the . . ."

"But if it's a weapon that could destroy the plastic aliens?" Firiv'vrak insisted. "Don't we owe it to the casualties of the war to find it? And use it?"

"Doesn't matter if we dig it up. Unless they change the Class 2 Designate, we can't take it off world."

"Commander Yurrisk will."

Craig shrugged. "Yurrisk is broken. Functional, but broken. We don't have that excuse."

The next room was empty: no roots, no beetle shells, only a four-by-five-meter rectangle with an interior door near the far end of the opposite wall and a slab floor that looked as though something had broken through from beneath. Where the stone had been broken, or was missing entirely, shadows hinted at a lower level.

"Good place to hide a weapon."

"Not our problem." Torin stepped off the threshold. The floor seemed solid.

"If we found the weapon, we could trade it to Yurrisk for the hostages and then take him down with no innocents in the line of fire."

"We're not here to look for the weapon."

"Then where are we going?"

"We . . ." She turned, but Craig was already in the room. He shrugged as both Artek skittered past. Curiosity—the commonality of sentience. "It seems we're mapping the room."

"I'm pretty sure there's something alive down there, Gunny." Firiv'vrak's back end clung to a nearly vertical slab, her head half engulfed by a triangular hole—one antennae in, one out.

"Is it a constructed hole?"

"No. Looks like the dirt collapsed into a tunnel." She shifted back, moving easily in spite of the angle. No surprise. As long as they moved fast enough, Artek could defy gravity for short periods.

Keeping her boots in contact with the stone, Torin shuffled away from the wall. If she could get a clear ping . . .

"Hold on." Craig grabbed her webbing. "I don't like the look of those fractures."

"There's less damage back here." Keeleeki'ka waved from the other side of the vertical slab. "Stay close to the wall until the patch of lichen that looks like a *seeter* . . ."

"Ceremonial building," Firiv'vrak explained when the translation program failed to find the reference. "Round. Pointed top. Most boring place in the universe."

"Ritual holds our history!"

"Blah blah blah. There, Gunny." Firiv'vrak unfolded an arm and pointed. "Then cross on the diagonal, aim for the interior door, and make a hard right when you pass me. The fractures predominantly run toward the interior wall."

Predominantly. If she only had credit for every time that word had been used in a briefing only to come back and bite them on the ass.

"We're heavier than we look."

"Most mammals are."

The patch of lichen didn't look like a round building with a pointed top. When she reached it, Torin stepped away from the wall. This time, Craig let her go. She moved quickly, her helmet light throwing a broad band of illumination that sent the Arteks' shadows dancing. Past the slab, she dropped to one knee and leaned forward.

"Well?" Craig demanded. He'd followed as far as the lichen. "Cellar or hole?"

"Can't tell. Floor's three point five seven meters down, a little deeper than the ceilings are high, and it's flat . . ." The ping, even at its broadest setting, had shown only minor deviations from the one eighty. ". . . but I can't be sure from here if it's natural or worked."

"From here? So you're considering a drop, then?"

"No." She stood. Disappointment smelled like fried shrimp; it wafted up from both Artek, and she suspected Craig would have projected it also had he been able. "We don't have time. Once all the DLs are active, we need to make plans. It's enough to know that there's a possibility of lower levels in these kinds of ruins. However . . ." She cut off Firiv'vrak's protest. ". . . we should take a look through the interior door since we're here."

Keeleeki'ka reached it first, Firiv'vrak close behind. "I believe it's nothing more than a cor-ee-oh-door, like on your stations."

The Artek didn't build with corridors. Good to know. In case. Torin had ended one war; she wasn't vain enough to assume she'd ended all of them.

"Instead of corridor, try hall," Craig offered, still standing by the lichen. "Shorter word. And come on, even the H'san like halls."

"The H'san story is not our story."

"Join the club," Torin muttered, shoved Keeleeki'ka back with her lower leg, and leaned out over the threshold. Her helmet light illuminated a meter-wide hall—whoever built it wasn't broad—and, to the left, one door on both sides, although there'd been two rooms off the courtyard. To the right . . .

"Do you smell bitter . . ."

The translator lost the last word in the scrabble of claws on stone and a sound that was both a roar and a scream, both low and high . . .

Because there were two of them.

Larger and slightly smaller.

They had four legs, the front noticeably longer than the back. Narrow shoulders, pointy faces, short muzzles, a flat nose with a single horizontal nostril, lots of teeth. And scales. Niln scales, not fish. Two eyes closed to dark slits in the helmet light as they charged toward the door.

Torin caught the blow from the larger on her KC. The three claws on each front foot were short although the toes themselves were long. A fourth claw came out of the side of the foot at the point where the toes separated, the toe it was attached to extending back toward the . . . Was it called a wrist on a foreleg?

The teeth—and Torin was getting a good look into the animal's mouth—were made for tearing at meat. She heaved her KC up until it fought for balance on two feet, then tipped it over to expose multiple swollen nipples.

Crap!

Mammal.

She spun on her heel.

One running stride back into the outside room.

Firiv'vrak had almost reached the exterior door. Craig was by the vertical slab, Keeleeki'ka beside him.

Two.

"Run!"

Three.

"Torin! Down!"

The cracked stone shifted when she hit it. It didn't break, so she counted it a win as the smaller of the two animals leapt over her and sank its claws into the rear curve of Keeleeki'ka's carapace.

Torin grabbed a back leg and yanked, rolling onto her back to bring her legs into play. As she braced her boots in its stomach, it pissed all over her.

Which made her feel better about slamming it into the corner.

"Shoot it!"

"Not if I don't have to." It had hit the wall hard enough to stun it. With limited choices, Torin turned her back on it again, braced one hand on the top of the slab, and jumped over the hole. "They're protecting young!"

Muscles bulging across his back, Craig had covered half the space between the attack and the courtyard with the wounded Artek in his arms. "Firiv! When we reach the courtyard, we'll need to barricade the door!"

"On it!"

The part of the floor she stood on seemed solid underfoot, so she spun around in time to see the larger animal charge through the door, up onto the slab, and launch itself into the air.

Torin ran forward, dropped, rolled, and sprang back onto her feet as it landed where she'd been standing.

It roared.

Torin roared back.

She had to keep it focused on her.

Fortunately, it seemed more than willing to stalk slowly forward, eyes locked on Torin's face.

"One eighty!" On her verbal command the right side of her scanner showed her the view from behind. The smaller animal remained in the corner, blinking rapidly, short ears flat against its head. Still stunned.

Good.

That gave them all a chance to get out of the encounter alive.

Matching the advancing animal growl for growl, Torin shifted to the right. From the top of the slab, she'd command the high grou . . .

The floor tipped. She slid backward, the smooth surface of the stone and gravity combining to defeat the grip of her boots. Letting her KC hang, she grabbed for the edge with both hands as she went over, the new break creating a large enough hole to drop her into the lower level.

Her right hand caught the slab. It had held her weight once and it did again. Her left fingers found purchase in a diagonal crack. It widened. Stopped. Widened again. Stopped.

Her left elbow was still on the floor, her right arm fully extended, her lower body hanging over a three-and-a-half-meter drop.

The animal roared again and charged.

"Fuk it." She'd be out of the fight at the bottom of the hole.

Maybe she could taunt it enough it would follow her down.

Hopefully, she wasn't dropping into a pack of the things.

She snatched her fingers back out of the crack just before teeth scraped against the stone. Swung to the right and . . .

The animal collapsed, draped over the top of the slab.

Torin anchored her left hand again and stared at the tranquilizer dart in its shoulder. "R&D came up with a tranquilizer for a species they didn't know exists?"

"No." Craig leaned over the body and checked it was breathing. "It's the 'if it needs oxygen and weighs less than eighty kilos, it should work' load."

"And you were testing it for them?"

"Some of us don't feel the need to test drugs on ourselves." He wrapped both hands around her right wrist. "On three. Two. One . . ."

When she got a boot on the floor, she used it to push them both away from the edge. Then, shoving Craig behind her, she turned, weapon raised, as the right side of her screen showed the second, smaller animal leap to the top of the slab.

Then lie down beside its mate.

Torin backed toward the door.

Eyes locked on her face, it didn't follow.

Once over the threshold, Torin exhaled and stepped aside so

Firiv'vrak could block the lower half with a stone slab. There was a rectangle of fresh dirt in the courtyard.

"We're stronger than we look. And I used a lever."

"Basic mechanics. Well done." Craig wiped sweat off his brow. "You think they'll follow."

She thought of the expression on the smaller animal's face. "No. But that's no reason to linger."

Keeleeki'ka smelled strongly of lemon furniture polish, but the claws had only cracked the upper layers of her carapace and taken a triangular chunk out of the rear curve. Torin quickly sprayed the crack with sealant to keep out dirt and spores and whatever else the jungle was flinging around, then Firiv'vrak pushed her toward the outer wall of the courtyard and Torin spent a moment leaning against Craig's side.

"Seriously? Tranquilizer gun?"

"I don't want to kill things, but it's a strange planet and I'm not entirely stupid."

"How many shots?"

"Just the one. They haven't exactly started mass production and were only willing to hand it over if I called this a field test."

"What difference does it make what you call . . . never mind. Let's move. Back to the VTA. We've got the pertinent details about this type of ruins."

"That they're dangerous?"

"Got it in one. We know. The mercenaries don't. We can use that."

They caught up to the Artek in time to see Kleeleeki'ka reach back to stroke the raised patterns on her carapace and wail, "But it'll leave a scar."

Firiv'vrak cracked her mandibles together. "Then you'll finally wear a story you're a part of!"

Profanity smelled of acetone although Torin could barely smell it over the urine dripping off her uniform.

SEVEN

"THIS IS WHAT I WAS LOOKING FOR! This is how you begin a new day!" Standing where the road ended in a series of broad steps covered in lichens and mosses, Yurrisk threw his arms wide as though he were introducing the ruins in front of him to known space.

Which he was, Arniz acknowledged silently. It had been millennia since the destruction, since eyes belonging to a sentient species had looked on the dark stone buildings and seen the shadows of the builders.

The use of the rarer, harder stone on a large, impressive building indicated a social hierarchy, an unfortunate requirement for sentience among every species who'd made their way to the stars. As Arniz understood it, fully cooperative societies missed opportunities for technological advances, but since she was no anthropologist, there was always a chance she'd gotten that wrong.

Both doorways were tall enough for most Humans, but narrow. While a short species could pass through a tall, narrow door, the rise of the steps required legs longer than that of the Niln, the Katrien, or the Krai. The width could have been a defensive decision, that theory supported by the difficulty in accessing the ground-floor windows which were horizontal openings a meter long by half a meter wide almost two meters from the ground. The windows on the second floor were isosceles triangles.

Architecture. Arniz's tail rose. It wasn't difficult.

Weighed against the implications of a defensive design was a clear love of nonrepresentational decoration; abstract swoops and curls visible on cornices and pilasters in spite of wear.

"We still have to find the weapon."

Arniz blinked as Qurn's voice brought her back to the road and exhaustion, and the presence of creatures with guns. She reminded herself that she hadn't been clearing the way to the ruins but to what the ruins were thought to contain.

"It'll be in there." Yurrisk sounded like a true believer. Arniz could hear desperation under his words. "Just look at this place. It's important. It looks like a government building, and government buildings hold armories."

It might look like a Krai government building, although Arniz couldn't see how given the whole arboreal living. It could have been a pre-destruction bakery.

Qurn pulled off a glove and laid the backs of pale fingers against Yurrisk's cheek. "We'll keep her flying."

"Yes." He pressed his face into her touch for a moment, then straightened, twitched his cuffs into place, and swept a slow, considering gaze around those watching. "You."

Arniz considered glancing behind her. Decided she didn't have enough energy to be the gravel in the tilth. "Me?"

"Yes, you. Find a latrine. Scan it. Find another. Scan that. The largest concentration of plastic residue will mark the definitive battle."

That almost made sense. It had nothing whatsoever to do with where the alleged weapon might have been placed afterward, but, on its own, recognizing that it was speculation and not science, it wasn't entirely without merit. She swayed, shifted her stance to stay upright, decided not to bother, and sank into a squat. Dreams of finding plastic bodies with familiar faces had woken her six or seven times during the night. When they'd been hustled out of the nest just after dawn, every muscle in her body had ached and her joints felt as though they'd been filled with grit. "I'll need assistance," she said. The air tasted bitter. "You murdered my ancillary."

"I didn't . . ."

"Fine. You allowed my ancillary to be murdered."

"She threatened the lives of my crew."

"Not the second murder." Arniz slapped her tail against the ground. "The first."

Leaning in, Qurn murmured by his ear.

"Of course," he said when she pulled back, "Martin's lesson."

A quick glance showed Martin standing with one hand on his weapon, waving away insects with the other.

"That death expedited cooperation and not only kept us safe, but kept more of you safe." His nostril ridges closed. "Although it didn't stop the other from being an idiot. From endangering everyone. I'm sure she wouldn't have considered being a hero if she knew what that entailed. Heroes are . . . Heroes don't . . ." He swayed to the right, then right again. Arniz tasted the air. She'd seen him do that before. He swallowed and made a second sweep of the crowd. "You!"

Salitwisi had been staring at the structure in silent admiration, hand opening and closing around the place his confiscated slate should be. He jerked at the sound. "No, not me. This is actually architecture, not buried lines of rock and educated guessing. I'm . . ." Arniz could almost see him searching the most convincing way to make his point. "My specialty is architecture." He clutched at the fabric over his chest. "No one on this planet is better suited to helping you right now."

"And you agree with me that this is an important building?"

"Of course I do. Harder stone means it's harder to work and a pretech society doesn't spend that kind of effort for an unimportant building." He twitched his tail in a way that clearly said how he felt about having to provide such rudimentary information.

Qurn cocked her head. "You want to explore the ruin for your own benefit."

"Of course I do." Tail lashing, he spread his arms. If Yurrisk had been introducing the ruins, Salitwisi was embracing them. "Our permit doesn't extend under the canopy, and yet here we are. After this debacle, the Ministry will never allow me to return, so this is my only chance. Which doesn't mean I'm not still *your* best chance of finding the hiding place of that weapon," he added. "Secret passages, hidden doors, *netan* holes; you'll never find them without my help."

"Fine." Yurrisk shifted and glared over Salitwisi's head. "You . . ."

Hyrinzatil's tail dropped, and he took a step back.

". . . go looking for latrines. You can do the old lizard's heavy lifting."

"Oh, no," Salitwisi protested, "I'm his prime. He's mine."

"Wrong. He's mine. Until we find that weapon, you're *all* mine." Yurrisk's nostril ridges closed. "All of you. Martin, send one of your people with them. I don't want them sneaking off and sending a message."

"I don't think there's much sneak left in the old lizard," Zhang snickered.

"Is your name Martin?" Qurn asked quietly.

"You can't . . ."

"Corporal."

At Martin's interjection, Zhang huffed out her displeasure and settled back onto a mossy log, shoulder to shoulder with Malinowski. Martin motioned Trembley forward. Without speaking, the younger Human grabbed Hyrinzatil's shoulder and dragged him first to the stack of equipment and then, once he'd picked up a scanner, over to Arniz. She leaned slightly sideways into the warm curve of the ancillary's leg. Perceived affection was less embarrassing than toppling over. Trembley took up position behind them.

"If they try anything, Trembley, shoot the old lizard." Martin laughed. Arniz wasn't seeing the humor. "That'll keep the young one in line. Kids don't care about their own safety." Arniz could feel Hyrinzatil tense as Martin gestured rudely at Salitwisi. "And if *this* old lizard gets us lost in that building . . ."

"How could I possibly get you lost?" Salitwisi protested. "At the most, it's six rooms long, two high, and two deep."

"Yeah, well, he's Navy," Trembley muttered. "Vacuum jockeys can't find their ass with both hands and a homing beacon," he added when Arniz glanced up at him. When she frowned, he rolled his eyes. "Hey, I'm not the first to say it."

"The building's the front of a 'U,'" Beyvek called from the roof. He drew the gaze and stiffened the postures of every one of the ex-military personnel. Arniz heard Trembley snap out of his slouch, heard the slap of his hands on his weapon. She wondered what it would be like to live like that, always assuming danger. Wondered if it ever wore off. Beyvek pointed toward the tall skinny trees rising behind and, in one corner, through the building. "There's two wings off the back, Commander. Single story, longer than this front section. The courtyard between them has been chewed up by tree roots, but there's hardly any other growth."

"Something in this particular stone must be suppressing it. Feeding the skinny trees." Arniz reached up and tugged at the scanner in Hyrinzatil's hands. "Give it here." Across the road, Tyven ran her hands over a thick root that flowed down the steps and dove into the ground, the slow movement of growth captured as living sculpture.

Hyrinzatil hung on—she remembered she didn't like him much; he'd learned bad habits from his prime.

"It seems as though we have significantly more real estate to explore." Yurrisk squared his shoulders, hands in fists by his sides. "I want eyes on every square centimeter of it. We will find that weapon."

"Bet your ass," Martin muttered. "So, Trembley . . ."

Trembley shifted his weight from foot to foot.

". . . if this old lizard emerges from these buildings without us, for any reason, shoot your old lizard."

Salitwisi actually shifted his attention from the building. "I'm not old!"

"Of course, that's what you take away from that," Arniz sighed.

"Pyrus!"

"Commander?" The di'Taykan's hair moved listlessly, weighed down by the heat and humidity. He hadn't done any of the heavy lifting, but he looked as if he had. In fact, both di'Taykan present looked frayed. The Taykan's homeworld had nothing that could be considered tropical by right-thinking people and rather a lot of snow. Niln, like most reptilian species didn't do well in snow.

"Go back to the anchor, take . . ." Yurrisk pointed at a pair of ancillaries. Arniz thought they were from the cartography department. ". . . them. I want every light you can find."

"What about Ganes, sir?"

Having spent the morning forcing herself to keep clearing the road—to put one foot in front of the other, to lift another armload of debris—Arniz hadn't missed him, but he wasn't in sight. She tugged on Hyrinzatil's overalls. "What *about* Dr. Ganes?"

"I heard he tried something last night."

"Got into the anchor's office. Got his ass kicked," Trembley added.

"Ganes stays where he is, safely contained," Yurrisk told the di'Taykan. "You stay with him, take advantage of the climate control in the anchor, and have Gayun bring the lights back."

"Sir."

"Mirish, you good for a little longer?"

She nodded, deep blue hair moving in counterpoint. "Yes, sir. I'm good for a while."

"Glad to hear it. Pyrus, go."

Even worn down by the climate, Pyrus moved like a dancer. Arniz took a moment to admire the lithe grace. The di'Taykan all moved like dancers. Dzar loved dance. She danced with an amateur Sand-and-Moonlight troop. Had danced. The ancillaries followed less gracefully, their notable lack of enthusiasm overlooked in the shadow of the ruins.

"Well?"

Tongue tasting the air, she tilted her head back and met Trembley's gaze. "Well, what?"

"You were *told* to search for latrines."

"I'm resting. Because I'm an old lizard and spent yesterday doing the kind of physical labor that reminded me of how far I am from the egg."

"Get up."

She felt Hyrinzatil twitch, and she glared at the young Human until he held out a hand.

"Do you need *help*?"

"Do I need help?" She wrapped her fingers around two of his and levered her legs straight, absently noting that her weight had no visible effect as she hung off the end of his arm. Humans were not only larger and stronger than many of the Elder Races—the Dornagain, the Ciptran, and the H'san being the obvious exceptions—but they augmented that strength with weapons. No wonder so few species trusted them. Of course, those same species had brought the Younger Races into the Confederation specifically to use those weapons, so the mistrust was disingenuous at best and blatantly hypocritical at worst. And that was without factoring the differences between Humans like Dr. Ganes who was almost civilized—the qualifier in place until she discovered just what, precisely, he'd tried last night—and Martin who wasn't civilized at all. Trembley was young enough; he might still be . . .

"*Harveer*?"

She blinked both eyes, inner lids dragging, and realized she was still hanging off Trembley's arm.

Hyrinzatil tasted the air. "Are you having a problem, *Harveer*?"

"I'm thinking. I know Salitwisi doesn't do it much, so I don't blame you for not recognizing it."

"That's not . . ." He blinked, inner lids flicking back and forth as indecision over how polite he needed to be to someone not his prime but still a full *harveer* flickered over his face.

"Let's go." Trembley shook her free and sighed. "Shit pits aren't going to find themselves."

Arniz rolled her shoulders and flicked the stiffness from her tail. "If they did, I'd be out of a job, wouldn't I? Come on, then. We need to make our weary way around back."

"Why?"

"Because no one shits outside their front door," Hyrinzatil said, stepping back to give her room. He bumped up against Trembley's legs in a stupid display of bravado and was kicked away.

"Careful." Arniz shifted just far enough his flailing arm didn't take her down with him. "If you're too much trouble, he'll shoot you. He is, after all, willing to shoot me."

"I'm not going to *shoot* you," Trembley muttered.

"Are you confused about who the old lizard is, then? Because Martin said . . ."

"I heard what the sergeant said." He placed a hand in the center of her back and shoved. "Now move!"

"The latrines that go with these buildings, they'll most likely be at or just beyond the open end of the courtyard." Arniz led the way off the road. She paused, went around a hummock covered in pale moss, and continued her lecture. She'd been too long a *harveer* to waste what might be her last audience. "Sentient beings have relatively few ways of disposing of their own waste and, as we already know, the predestruction people used pits, that narrows the possibilities even further."

"I heard some members of the Methane Alliance reabsorb theirs," Trembley offered, stepping over a fallen branch she'd had to climb.

"Essentially the same thing you do on a ship or station," Arniz pointed out.

Both Hyrinzatil and the young Human made exaggerated gagging noises.

How nice, she thought. They're bonding.

• —◆— •

"Two dead," Torin growled, ducking under a branch, left arm up to keep the dangling snake from dropping onto the back of her neck.

Stalks snapped behind her under the sudden rapid movement of Binti's boots. "Fukking snakes," she muttered, then added, "It could be worse, Gunny."

"And it could be better."

"It doesn't count as losing them if we didn't have boots on dirt when they were killed."

Since Binti knew how she felt about that, Torin let the comment stand. The primary mission directive had been to free the hostages, not free those hostages who were still alive. She *knew* the second part of that statement was the only part that applied. No one expected her to bring the dead back to life, but if politics hadn't delayed them at the station, they might have arrived in time. Two more Confederation dead because of the Primacy. With the ease of long practice, she pushed those thoughts deep and locked them down.

Boss? Zero implant use by the bad guys. The only thing I can read is a carrier wave from that science scanner Firiv'vrak mentioned on the DeCaal. We're a go for conversation, and I can route the DL feeds to the slates.

Binti snorted. "Little hard to watch a slate while humping through the . . . Goddamned, fukking snakes!"

I will, of course, continue to paraphrase the action, so you don't need to have eyes on. And if you're bored, I do amazing aural.

Up ahead, Bertecnic tripped over a hummock, propelling dozens of tiny red lizards into the air. Torin took it as confirmation the Primacy implants had fully integrated with the shuttle's communication system. And that the translation program was having trouble with homonyms.

"Not while you're working, you don't."

"Your word, Boss, my command."

"That's how it's supposed to work."

"Why wouldn't they be scanning?" Binti wondered. "Commander Yurrisk's Navy, but Martin and two of his people are ex-Corps. You'd think they'd have a better grasp of covering their asses."

"They think they're in the clear until the next supply ship. They've no reason to scan. Given the state of the ship, they may not have the equipment."

Yeah, they're not in helmets, but—come on—I could write something that'd work off a slate, bounce off their ship, and get kickass ground coverage.

"You're wasted in fieldwork."

Not getting rid of me that easily, Boss.

She could hear him preening.

"Why aren't they using their implants?" Binti asked. "All but one of the commander's crew were high enough rank to have mustered out with them, and Martin made sergeant, so he's got one for sure."

"The Commander had a brain injury," Ressk called down from above. With Werst taking his turn out front, he was pacing the group. "Whatever caused the injury could've damaged his implant and left him too neurologically scrambled for them to put it back."

Torin nodded and carefully moved a red lizard off her sleeve. "That's possible."

"You don't monitor your injured veterans, Gunny?" Merinim asked from behind Binti.

"We do—there's a Ministry of Veterans affairs, both the Corps and the Navy do rehab, and there's at least a dozen private programs—but the Confederation covers a lot of space, and some people slip through the cracks."

"Some people slip through deliberately," Binti muttered.

Merinim's considering hum blended into the hum of small red wings. "And thus your Strike Teams were formed."

"The Wardens don't bring in injured veterans."

"Not the physically injured, perhaps."

Torin remembered the feeling of stability the Durlan's mere presence caused. "Fair enough. And in the Primacy?"

"The engineered war with the Confederation forced an internal peace. There's already signs of fractures, and most of our governments are concentrating on not going back to war with each other."

"Sounds like you'd be better off still fighting us," Binti said thoughtfully.

After a long moment, Merinim said, "Yes."

The sounds of birds and insects filled the next half kilometer of awkward silence.

Although her shoulder continued to ache and the band of bruises across her hips occasionally pulled, they were on pace to cover the three klicks back to the drop clearing in a quarter of the time Torin and Presit had taken. First, because Torin, Werst, and Ressk had been over the ground before and had begun to learn the particularities of the jungle. Second, because in the Primacy military, Polint males were used to break trails. With the Krai in the trees scouting the fastest route, Bertecnic and Dutavar, wearing the Primacy equivalent of combat fabric over their lower chests and front legs—an apron/leggings combo secured over their withers—used size and strength and broad blades that were almost swords to take down underbrush, vines, and small trees. They switched out of the lead position every half kilometer. Watching them, Torin was grateful that the one time she'd met the Polint in combat, they'd used more conventional weapons.

"Trust me, Gunny, they've been cooped up on ships and stations for almost a tenday now. You want them to work off some energy."

"I want them in shape to fight when we arrive."

"That won't be a problem. With the edge taken off, we've raised the odds Bertecnic will wait for orders and not charge in bellowing a challenge."

"And Dutavar?"

"He made Santav Teffer; as I said, an unusual achievement for our males. I believe we can count on him to resist his instinctive urge to challenge the enemy. I take it this is a problem Confederation forces don't have?"

Torin thought of the Silsviss. The big lizards gained rank by challenge and although their entry into the Confederation had been slowed by the end of the war, the offer hadn't been withdrawn. "Not yet."

Conditions by the Ministry for the Preservation of Pre-Confederation Civilizations—agreed to on Strike Team Alpha's behalf by the Justice Department—had included a ban on clearing paths through the jungle by mechanical means. They'd never considered the damage one hundred and forty kilograms of sentient quadruped under orders could do.

The third reason for their speed was the lack of Presit.

Presit had declared her intention to accompany them, to witness and record.

"And gather material for your program."

"That are being my job, Gunnery Sergeant. You are being required to be giving me full access." Presit's muzzle had wrinkled in a self-satisfied smirk. "Just like it are being the old days."

The argument had lasted until she'd seen the size of the insect Bertecnic had twisted out of his foreleg and then she'd retreated to the shuttle, demanding Dalan section and brush every square centimeter of her fur. The insect's moist, bloated body pulsed a deep purple as it humped toward the rotting debris on the jungle floor and sprayed blood a meter out when crushed between two rocks.

"I'd rather be shot at," Freenim said quietly under the background rubble of translated profanity.

Torin agreed and instructed the Polint to stay out of the bracken as it was evident their blood had the same local food value as hers.

Presit had agreed to help monitor the DL feeds when Craig had informed her that they'd appreciate her insight on the Katrien among the hostages. Torin wouldn't have been able to deal with the artillery barrage of Presit's opinions, but if Craig could, more power to him.

Torin followed the two Polint, Mashona followed her, and, after Mashona, the Druin, who lacked both the ability to take the high road and the size to make it easier for those behind. On their six, Vertic kept her blade holstered along her side, tucked through the strapping the Ner used riding into combat. She carried her RKah diagonally across her chest in the same easy access position Torin and Binti carried their KC-7s. Like the sevens, the RKah used explosive charges to propel metal rounds out of a rifled barrel at high speeds. Anything more complex could be—and had been—taken out at distance by the enemy. With computer guidance off the table, infantry had been a skilled trade on both sides of the war. The RKah fired a higher caliber than the sevens; essentially, given differences in systems of measurement, as high as the KC-12s the heavy gunners carried. The Polint, with four clawed feet on the ground and ropes of muscle over shoulders, arms, and torsos, didn't need augmentation. In the Primacy's ranks, they *were* the heavy gunners. Torin could work with that.

The Artek flanked the path. Firiv'vrak had pointed out that she was a better shot with a ship around her, but carried two slender rods strapped to the sides of her abdomen and a double line of chargers along her thorax. Energy weapons were useless in infantry battles where both sides used EMPs, but for Justice work, it was only familiarity with the weapon that kept the KCs—and on this mission the RKahs—in play.

Keeleeki'ka, duct tape stabilizing the rear curve of her carapace, had refused to be left behind.

"I go," she repeated, eyestalks drawn in close to her body, "where the story is."

Torin abandoned diplomacy. "You go where I tell you to. You're staying in the VTA. You're not trained for this."

"Neither is Firiv'vrak."

"But she has had training."

"You can't keep me from the story. It's why the council agreed to my presence."

Presit applauded. "I are liking her persistence. I are being right beside her if not for the discovery of murderously large insect life. Which she, of course, can be ignoring being murderously large insect life herself."

"Presit . . ."

"I, of course, am not requiring story. I are having plenty of story to go around. I are requiring facts and I are able to get them from the helmet scanners without needing to be risking my life."

Keeleeki'ka rose onto her rear legs. "To risk a life in the pursuit of story creates shadows that highlight meaning."

"I are having no idea what you are talking about."

Torin ignored them both and turned to Vertic. "Could my keeping her safe in restraints derail the Confederation/Primacy cooperation attempt?"

Vertic spread her hands. "She's a representative of a powerful lobby."

No one was saying "more important than she appears" in either Artek's hearing if only to keep Firiv'vrak from overreacting.

"Look at the bright side, Gunny, they're hard to kill."

"Gunny!"

"Don't do that!" Binti stumbled, grabbed a vine to stop herself from tripping over Firiv'vrak, and snatched her hand back as soon as she had her balance. "Son of a . . ."

Torin ducked, and the partially crushed, yellow slug Binti flicked off her fingers flew past her.

"Gunny!" Firiv'vrak tapped Torin's knee. "A herd of mammals are about to cross our trail."

"Bertecnic, Dutavar; hold!" In the sudden silence that marked the interruption in Polint trail making, Torin heard nothing she hadn't already begun to internalize as *normal*. "About to?"

"They're moving fast, almost as fast as we do," Keeleeki'ka said, from Torin's other side.

"Wear a fukking bell," Binti muttered.

"They look like the animals we fought in the ruins. Same hide. Same coloring. Same shape." She tucked herself in behind Torin's legs. "Smaller, though."

"How many in the herd?"

"Fifteen." Firiv'vrak's eyes swiveled toward the left side of the path. "Maybe sixteen. They've excellent . . ."

Bertecnic dropped low and reached for his RKah as the first crossed no more than a meter in front of him.

". . . camouflage."

The scales that had appeared silver gray inside the ruins were greenish gray surrounded by foliage, light shifting the shades as they moved. Most of them were half a meter at the shoulder, but a few in the center of the herd were smaller. They all had the slightly out-of-proportion look of juveniles and none of them did more than glance at Bertecnic as they raced across the path. Now she wasn't fighting an enraged parent, Torin could see they had curled their long front toes under during the upper extension of their stride, extending them again as their heels hit the ground.

"Locals passing." Torin pitched her voice to carry to both ends of the march, both through implants and air. "We won't be stopped long."

"Any danger?" Vertic called.

"No, I think they're juveniles." The foliage closed unmarked behind the last, and Torin had to use the zoom on her scanner to see footprints, speed combined with the broad splay of their feet keeping them from sinking into the jungle floor. "It must take them a few years to get to breeding size." She frowned at the uneven claw indents around the faint imprint of a front paw and tried to remember why they looked familiar. "Alamber."

On it, Boss. Sending images from your helmet scan down the line.

"Those are big babies." Binti leaned around Torin and stared at the place they'd crossed. "You think there's something in here big enough to eat them?"

"If there is, high odds we neither smell nor look like it. Which is interesting as the insect life finds both Polint and Humans edible."

"Nice of the Ministry to mention the possibility of large predators. Of course," Binti added, "the Ministry didn't mention the things large predators eat either."

"Incomplete surveys aren't that unusual. And we heard a group go by when were out before." Ressk dropped down to a lower branch, Werst crouched on a branch over his head, knees raised, KC held across his shins. "They can't be seen from up here unless they cross an open area and the little fukkers are fast. I think they're livestock gone wild."

"Livestock?" Merinim pushed her helmet back to stare up at Ressk. "Seriously?"

"There was a city here, right? That kind of population density eats a lot of food. Pre-tech means they probably had livestock inside the walls. The population's gone. The walls are gone. That doesn't mean the livestock's gone."

She blinked as a bug fried in the helmet's adaptive shielding. "You've studied ancient civilizations?"

Ressk grinned. "I know food."

"They have teeth and claws," Keeleeki'ka protested.

He shrugged. "Maybe they preferred food that fought back."

"Lots of food does," Torin pointed out. "When my brother was eight, he had his arm broken by a goose."

"I don't even know what a goose is," Binti admitted. "And I'm the same species you are."

Torin grew up on a farm. Craig sounded amused. **In the country. Not yeast vats and hydroponics, but animals and seeds grown in dirt. She can identify multiple types of shit.**

"Multiple kinds of shit?" Binti folded her hands on her KC and rocked back on her heels. "Why did I never know this about you, Gunny?"

She's never taken you home to meet her parents.

"True. She hasn't. Why is that, Gunny? Are you ashamed of me?"

"Yes, that would be why. Listen up." Torin raised her voice. "The parade's past, let's move. Best speed. Those hostages aren't freeing themselves."

"Humans are weird," Firiv'vrak said, and, for a moment, the smell of cherry candy overwhelmed the smell of jungle.

•—◆—•

With the occasional exception of the large roots, very little foliage grew around the outside edge of the building. Even mosses and lichens stayed nearly a meter away. Arniz tottered a little just because she felt like allowing Hyrinzatil to carry part of her weight—whether he wanted to or not. She missed sand and dry air and the way the dig had started, varying disciplines taking their time mapping the plateau, primes turning everything up to and including unpacking the supplies in the anchor into a lesson for their ancillaries, no one dying . . .

Trembley turned to walk backward in front of them. "Are you all right?"

"I'm exhausted and annoyed, and I don't like jungles. The soil contains far too much botany. Why?"

"You were hissing."

"Your superior has killed two ancillaries; I'm entitled."

"Magyr picked up a gun," Hyrinzatil began.

Arniz smacked him on the back of the thighs with her tail and tightened her grip when he tried to pull his arm away. "Which she didn't know how to use."

"That actually makes her *more* dangerous," Trembley pointed out.

"In what universe? Because it's somehow worse to be shot by accident than on purpose? Are you saying that Martin, twice her size and trained in violence, couldn't have taken the weapon away from her?"

"No . . ."

"Good." She cut him off before he added a *but* and she lost what little control her exhaustion was allowing her to maintain. "And don't tell me Dzar's death taught us not to fight you. We're scientists and are all perfectly capable of applying the concept of consequences. Except for Dr. Ganes, none of us had seen a weapon fired; Martin could've shot and destroyed a . . . a chair."

"I *know.*"

She opened her mouth and closed it again.

Trembley shrugged, stumbled over the edge of a canted paving stone, and said, "Lieutenant Commander Ganes said sort of the *same* thing last . . ." He jumped backward, fumbling to aim his weapon. "What the flying fuk is that? It's got like a *billion* legs!"

"Don't be ridiculous."

"Well, how *many,* then?"

"How would I know? I'm not an entomologist." It was about half a meter long, shiny bronze, and moving very quickly between the stone Trembley had tripped over and another triangular crack. "Don't!" she snapped when he moved toward it. "They're on the toxic list, I very much doubt you lot took precautions before you arrived, and we've already determined the native insect life likes how you taste."

"Then why not let it taste him?" Hyrinzatil muttered.

"Did you miss the part where Martin killed two people for no reason?" Arniz took her hand off his arm and pushed her face so close to his, he had no choice but to recognize the scent of her anger. "Do you want to give him a reason to kill more?"

"No, but . . ."

"No buts. What has Salitwisi been teaching you? Honestly. And you . . ." Her tone jerked Trembley's attention off the last few pairs of legs disappearing into the hole. ". . . stay out of the bracken, don't stick bare body parts into dark corners, and rock any stones before you sit on them. And, since I would like to get out of the looming shadow of a depressing building built by long-dead enigmas, I suggest we get moving."

"So, to *you*, me staying alive is only a means to keeping the sergeant from taking revenge."

She sighed. "Your life, Emile Trembley, is as important to me as every other life. Empathy is one of the building blocks of sentience, and I've long considered myself a sentient being." Sweeping her gaze over the two of them, she snorted and started toward the distant end of the building at the best speed she was still capable of. "Opinion, however, is still out on the two of you."

The latrines were right where she expected them to be. Once a species began living in cities, quite a number of functions began to follow form.

"Look at the differences in the vegetation." Arniz gestured at an area just inside the remains of the wall that defined the open end of the courtyard. "In the same way that the dark stone clearly only feeds the slender trees, look at the pattern, at the oval created by these plants with the tufted stems."

"Doesn't look like an oval to me, *Harveer*."

Trembley nodded. "Or me."

"Fine." She sighed at the rigidity of youth. "Extended oval."

"And there's another bunch of those plants there, on the other side of the wall."

"And there," Trembley added.

She sighed again. "Yes, but they're thickest here. Just the same way they were thickest over the latrines out on the plateau. Fortunately, because of the openings in the canopy, they're getting enough sunlight and making identification easy for us. Dr. Tilzonicazic took samples, and I expect we'll find that these particular plants prefer highly acidic soil given the amount of urea found in the one latrine I was actually allowed to do a comprehensive scan on, but only because it was done before . . ."

"*Harveer*?" Hyrinzatil dropped to his knees and ran his hands in under the edge of the patch of plants in question. "There's no stone under these plants and the last slab has a finished edge. Wouldn't that be a better determinant than plants. Plants die."

"Yes, but . . ."

"Then can we drop the probes?" He stood and stretched. "I now know all I need to know about these particular shit pits—they haven't any architecture."

"That," she said over Hyrinzatil's head to Trembley, "is the problem with ancillaries these days. Narrow fields of study."

The ancillary in question blinked, inner eyelids flicking back and forth at speed. "You study soil, *Harveer*. When my prime asked you to identify a spur of surface rock, you asked him if you looked like a geologist."

"Which I don't." She brushed at a sap stain on her overalls. "Get started, then." Perched on a curve of root, she watched him adjust the scanner and log the position before he sent the first electronic probe. To give Salitwisi credit where it was due, his ancillary appeared to have been well trained on this particular piece of equipment. She supposed that meant there was a way to use them architecturally, but she didn't care and was content to give over control. With the scanner calibrated exclusively to find plastic residue, she had little interest in the results.

As Hyrinzatil swept the probe slowly back and forth, Trembley approached, weapon hanging across his back, holding two of the tufted, stemmed plants. "What are they called?"

"They haven't a name yet. The Ministry will take the specifics Dr. Tilzonicazic sent them under advisement and eventually agree on a scientific designation. Don't hold your breath."

He pressed the plants together. The tufts interlaced and held, even in the face of a vigorous flailing. "I'm going to call them sticky stems."

"Seems apt."

"Do you think the Ministry will use the name?"

"Only botanists care what the Ministry calls things. Sticky stems is a name that the rest of us would use."

"Really?" He beamed down at her with so much innocent pleasure on his face that she revisited the idea of poison sacs and a moment spent with her teeth in Martin's forearm, wondered if she could prove corruption of youth to the Wardens, then realized she didn't have to. They'd be after him for two murders.

Her tail twitched.

Martin had to know that.

Was he arrogant enough to assume there was enough empty space to hide his . . .

Head cocked toward the two-story part of the complex, she tasted the air. Tasted only jungle. Didn't hear a repeat of the shout that had attracted her attention.

"*Harveer*! There's no plastic residue in here."

"That you found," Trembley pointed out smugly.

Arniz felt it fortunate Trembley couldn't read the complete disdain of Hyrinzatil's expression. "If I haven't found it, I can't speak to its existence, can I?"

"If you can't find it, Martin's going to be pissed." Trembley rocked back on his boot heels. "Likely order me to shoot you both." He rocked back a little further and murmured, "But I'm not going to shoot you."

"Good." Arniz rubbed her temples. In spite of the constant humidity, her scales felt dry.

"All right. Fine." Hyrinzatil set the scanner carefully on the ground and crossed his arms. "If you don't consider the scanner and my analysis of its data sufficient, we should go get the digger."

The digger had been returned to its charging station in the anchor. Ganes had been unable or unwilling to reprogram it to clear the road, claiming that as the only part of the site they had legal access to was essentially grassland, the digger wasn't designed for rough terrain.

"The digger has destroyed one set of historical data, let's not compound the folly." Moving slowly, sore muscles having stiffened during her rest, she got down off the root and shuffled over to Hyrinzatil's side. "This is still a Class 2 Designate. If we dig in here, if we disturb the site to that extent, the Ministry . . ."

Trembley's raised hand cut her off. "We're not worried about the . . ."

A high-pitched shriek cut him off. Arniz couldn't tell if the sound was pain or anger, but it hadn't come from the throat of a Niln or a Katrien, that much she could guarantee.

She turned toward the sound in time to see a large quadruped flung out of a first-floor window at the far end of the courtyard. It had claws and teeth and the gleam of scales although the air was scented with angry mammal.

And blood.

It got to its feet. Shook itself. Deep-red drops sprayed from a wound in its side, splashing black against the stone.

It took a deep breath, mouth open, tongue out.

Stared directly at her.

Started to run.

Hyrinzatil screamed, and Arniz could hear the snapping of stalks and branches as he ran.

The quadruped ran silently, although the motion of its front paws suggested it should be slapping at the ground.

She felt as though roots had poured from her legs and anchored her.

She would die here. Like Dzar had. And for no better reason.

Then she heard a crack. A clean sound. Like a pickax splitting stone.

The quadruped stumbled. Kept coming.

And Trembley was in front of her.

Instinct forced her legs to carry her aside when he fell, the quadruped on his chest, teeth snapping in his face. He brought his legs up, threw the quadruped off him, rolled onto his side, pressed the muzzle of his weapon up under its ribs.

She heard the crack again.

Oh.

Heard him fire his weapon again.

Of course.

The quadruped jerked. Stilled.

Trembley flopped onto his back, sucked in loud lungfuls of air, and bled.

Arniz dropped to her knees by his side, fingers of one hand sliding through blood to press the edges of flesh torn from his stomach together while the other hand searched her pockets for the tube of sealant. Sprayed it on his throat. Pressed and sprayed again.

His eyes closed.

"Don't you dare die!" she snapped, over the near deafening pound of her heart. "I won't have it!"

His tongue came out to wet his lips and on his next breath, his eyes opened and he said, "Because Sarge will take revenge?"

"Well, yes, for that as well, you stupid boy." As the sound of boots against stone grew louder, as the sound of shouting separated into words, she touched her tongue to his cheek.

He smiled.

•—◆—•

No, the animal's really most sincerely dead, Boss, but the kid's still alive. They've made a stretcher out of branches and overalls and are carrying him back to the anchor. Anchor's registered as a 277, so not full colonial version, but they have an autodoc in the infirmary that's up to tissue repair.

"Glad to hear it."

Alamber had pieced together most of what had happened from the conversations the DLs had picked up as they brought Emile Trembley out from the rear of the ruins. One shot and then hand to claw? Whoever had been in charge of that boy's training had a lot to answer for.

Martin's declared he's going to stay in the anchor with Trembley. Not sure if it's because he feels responsible or because he thinks he's the best person to run the autodoc. He's not, by the way. Not unless he's taken a lot of medical training lately. Dr. Ganes is on the expedition roster as emergency medic.

Vertic snapped a branch off a tree with enough force the crack sounded all the way up the line. "If these animals have a set breeding season, that means they'll have also killed its mate and young."

"The fukkers," Binti muttered.

"There's three Krai, Vertic. They won't waste the meat." It was a point, if only a minor one, on the positive side of the ledger.

"We'd all eat the meat, Gunny," Freenim's gesture took in his bonded, the Polint, and the unseen Artek. "Why of your people would it be only the Krai?"

"If there was time to test the meat for incompatibilities and then to cook it, I might eat it rather than allow the animal's life to be wasted. Craig and Binti and Alamber only eat vat-grown proteins." She grinned at the muffled sound of Craig keeping Alamber from broadcasting his eating preferences. "The Katrien . . ."

This Katrien are not eating anything that are having been running about, and are having mates and young and a face! It seems Dalan are having eaten such in his misspent youth, but he are not eating such anymore.

Not if Presit had anything to do with it.

"We have vat proteins on ships and stations of course, but planetside most species eat living protein sources. Previously living," he amended, when Binti turned far enough to shoot him a disgusted expression.

"Not taking a life in order to live is one of the tenets of the Confederation. With a few terabytes of codicils."

"There's a terabyte on the Krai alone."

Werst dropped to a foot grip, swung over the path, and up onto a broad branch. "What can I say? We're efficient eaters."

Torin raised the volume on the DL feed and listened to the accusations and counteraccusations. Their blood up, Martin had to threaten bodily harm to prevent Tehaven and Netrovooens from charging off on the hunt.

They'd covered five klicks, were still three out from the ruins and were almost close enough they'd have to trade speed for stealth. "Gather in, people." When they all had a line of sight, she pulled out her slate and flicked up the hard light map. "Bottom line, we don't want to have to attack the anchor. Even with the codes for the air lock, there's still the mechanical security system. We couldn't get past it without sending up flares and endangering the hostages."

"Simple metal bars are surprisingly difficult to defeat. I believe your people called the mining colony Puhgit," Dutavar added, speaking directly to Torin.

She remembered Puhgit. "Primacy attempts at entry were stopped by the metal bars, so you flattened the anchor."

He didn't deny the pronoun. "We'd have preferred to keep the anchor intact so that it could be used." The shrug rippled down his torso and along his withers. "Time ran out. The Primacy wanted a new supply of the metal as much as the Confederation did."

"Eighty-two people died. We took it back."

"When seventy-eight people died."

"And in the end," Freenim interjected, "it was the plastic aliens that wanted us all to want the metal."

Torin frowned. "No, I'm fairly certain the Confederation actually wanted the metal."

Dutavar nodded. "Same."

They stared at each other for a long moment, then Torin nodded and said, "Anyone else with experience attacking an anchor? Anyone else from the Primacy," she clarified as Binti and both Krai raised their hands.

"My ground assaults were against more established colonies." Vertic folded her arms over her RKah. "And, of course, Ter-deevan."

"We have always fought together." Freenim leaned into Merinim's shoulder. "Ter-deevan, numerous other large infantry battles, two stations."

Merinim nodded.

"I did a moon," Bertecnic offered, crest up.

They ignored Alamber's reaction as Firiv'vrak dipped her antennae and said, "I was vacuum all the way."

"All but Ter-deevan," Vertic reminded her.

"Almost all the way," she amended. "I needed to requalify in atmosphere, so Command put me on strafing runs."

Binti folded her arms. "You were shooting at us?"

"We were *all* shooting at you. You were shooting back."

"Fair enough."

"Back on topic, people. If our targets are in the anchor with the hostages, we won't be able to get them out. They searched for that weapon until sunset yesterday, and since they're not going to find anything, odds are they'll stay out as long today. They're spread out, separated in and by the ruins, out of sight from each other. We take them down, quick and quiet." Torin swept her gaze around the team. "Nonfatally, if possible."

"So they can be rehabilitated." Freenim shook his head. "So strange."

"A story can change," Keeleeki'ka said, rising up and pointing at Bertecnic. "You have a large winged insect on your back. It appears to have a stinger."

The insect achieved an impressive air speed propelled by a flicked claw.

"Even if they send the stretcher bearers back," Binti said thoughtfully as Bertecnic ran his hands over every bit of fur he could reach, tail flicking jerkily, "Martin and Trembley will still be in the anchor."

"Good."

She frowned. "Did you mean to say *good*, Gunny?"

"Martin will kill an innocent without hesitating. We know that. If he sends the stretcher bearers back, the only hostage in the anchor is Lieutenant Commander Ganes."

"I know he's Navy, Gunny," Ressk began.

"By definition, no innocent," Werst interjected.

". . . but we should save him, too."

"Agreed. But given an opportunity, I expect he'll be able to take care of himself."

Dutavar swept another of the stinging insects away. "So, how would we get into the anchor, Warden?"

"We wouldn't," Torin told him. "Werst would. We take down all targets at the ruins, dress Werst in Yurrisk's clothes, and send him to knock on the door. Martin will see the commander and we'll have access."

"*I'll* have access," Werst pointed out, settling into a squat on the branch beside Ressk. "Then what?"

"Then you take Martin out and we wrap things up."

He nodded slowly. "That should work." And showed teeth. "But only because we all look alike to you."

"If *you* means Humans . . ." Binti spread her hands. ". . . then, yeah."

"Your noses suck."

"And our eyes take time to distinguish subtle variation." Torin zoomed in on the ruins. "Alamber, add current positions of both targets and hostages."

Targets in blue, hostages in orange, Boss. As well as the individual lights, eight arrows, three blue and five orange pointed at the ruins. **Those eight, I know they're inside. Targets include Yurrisk, the Druin, and Corporal Zhang. Hostages are* Harveer *Salitwisi, Dr. Lyon, and three ancillaries. You'll be the first on the list if I get visuals.**

All three Polint and one of the Human mercenaries were outside. Malinowski stood, feet on the road, at the stairs leading into the building. Standing guard. "I'm sending Werst and the Artek on ahead. Get them in place on the far side of the ruins—between the ruins and the anchor in case anyone makes a run for it. We still need you up high," she said before Ressk could protest, "to lay out the fastest route for the rest of us."

Vertic crushed another stinging insect and flicked the body away. "Why the Artek?"

"We don't send one Warden to cut off a retreat." Torin ignored Werst's muttered protest. "And the Artek can keep up with him at speed."

"The pacifist will be of no use in a fight."

Torin caught a whiff of cherry candy. Firiv'vrak was amused. "If a fight overruns her position . . ." She captured as many of Keeleeki'ka's eyes with her gaze as she could. ". . . then she will hide in the underbrush so that she will not put other lives at risk because I will go back to war with the Primacy before I allow that to happen. Do you understand me?"

Keeleeki'ka flattened closer to the ground. "Yes, Warden."

"Sounds like the best options for all concerned," Vertic said. "And I'd prefer to send out the Artek together."

"I'm glad you agree." Torin had planned to, regardless, but Vertic knew more about working with the Artek than she did and she'd have listened to an objection.

What died in your ass?

"Problem, Alamber?"

Not me, Boss. Craig's twitchy.

"Craig?"

It's nothing.

She doubted that, but whatever it was, she couldn't deal with it now. "Keep it off the coms. Werst, position Firiv'vrak before you position yourself."

"Not my first *herlakir*, Gunny."

"Didn't need to know that."

"Standard operating procedure if there's a chance of subvocalizing being overheard?"

"That's why they're standard." She had no idea how the Artek's implants worked, given the lack of ears, but then, she didn't need to. "Discretionary contact only."

"Discretionary contact; got it." He stood and stretched.

Ressk stood beside him. "Discretionary?"

"Gunny trusts my ability to know when things are going to shit."

"Gunny?"

"He's not wrong." She met his eyes. "Still not wet or cold."

"Still likely to be miserable."

"Recon go."

He nodded and defied gravity, leaning out far enough to see both Artek. "Can you two follow me from down there?"

"Well, you are very slow," Keeleeki'ka said, tucking her arms in and streamlining her body.

Firiv'vrak flattened her antennae along her back. "True."

At least they were getting along.

Werst flipped them off, touched his forehead to Ressk's, and jumped for the next tree. The rustle of his passage faded in seconds. The Artek made no sound at all as they disappeared.

"All right, people, let's move." Torin settled her pack. "Best speed for the next kilometer, then we regroup."

Bertecnic slapped a palm against his chest, smearing the body of a stinging insect against his uniform.

"Fukking snakes," Binti muttered, leaping sideways.

• ——◆—— •

When Trembley had been carried off, Martin at his side and two di'Taykan holding the ends of the makeshift stretcher, when Beyvek had taken the body away and everyone had silently agreed not to ask what he was going to do with it, Arniz realized she hadn't seen Hyrinzatil since he'd screamed and run off. When she opened her mouth to ask if anyone had seen him, she realized she was alone.

She could hear Yurrisk yelling inside one of the buildings and Salitwisi yelling back at him—or perhaps the other way around. She could taste blood on the air, knew Camaderiz guarded Lows and two ancillaries as they cleared the remains of a less solid structure just out of sight, but she could see no one. This would be the time to attempt an escape, to slip unseen back into the anchor, to make her way to the anchor's office and the satellite communications without being heard upstairs in the infirmary, to send for help—or it would be the time were she significantly younger and considerably stupider.

Young enough to cover the distance quickly and quietly. Stupid enough to believe she wouldn't be caught.

Smarter to find Hyrinzatil before he got himself into trouble, being as how he was both young and not very bright.

The remains of the wall that had once enclosed the courtyard had been crushed under heavy boots and claws, so the undergrowth next to it had had little chance of surviving intact. Having spent her entire working life learning to make as faint an imprint as possible on ancient sites, Arniz wasn't happy about the damage, yet—honestly—after watching the digger excavate a latrine, destroying any scientific value, it was hard to care. She couldn't locate Hyrinzatil's path within the destroyed foliage and boot-sized patches of green pulp, so she moved farther away from the building, until the damage lessened to the point where she could see a trail of broken plants leading off into the jungle.

And if she could see his trail, Hyrinzatil had been flailing about like a *nok* in a *sebitle*.

Within the privacy of her own thoughts, she acknowledged that had terror not locked her in place, she'd have done the same thing.

She pushed aside a tangle of vines, ignoring the scattering of the insects that had been sheltering below it, and climbed a low wall, half expecting to find Hyrinzatil crouched on the other side. But no. It seemed panic had kept him moving.

"Of course, it had," Arniz muttered. He wasn't behind the next wall, or the next, and she paused on the top of the wall after that, breathing heavily and wondering just how much farther he could've run, a little surprised that even panic had motivated him to cover this much ground. Looking back, her trail to this point cut an obvious, if unexpectedly sinuous path. Looking ahead . . .

Looking ahead, the underbrush was undisturbed.

Either Hyrinzatil had tucked himself up at the bottom of a wall and she'd missed him, or she'd lost his trail.

"Well, if that's just not the perfect end to the day. Next time," she added, climbing wearily back to the ground, "he's on his own. It's not like I'm responsible for Salitwisi's ancillaries," she muttered as she stumbled against a hummock and the air filled with tiny, red, flying lizards. They should be responsible for her. She was old. "And tired, little cousin." Scarlet wings whirred and a yellow tongue tasted the air. When it landed on her shoulder, she smiled. "And, also, glad of the company. We will, of course, have to part ways before I rejoin the others. I don't trust Yurrisk not to make a snack of you."

She couldn't see the ruins; hadn't been able to from her vantage point on the wall. Clearly overgrown lines of sight were to blame because she knew she couldn't have gone that far.

She should have reached the second last wall by now. Be almost back to the ruins. Given the differences in growth that marked the differences in soil composition, the space between the walls, between these particular two walls, was most likely an interior space, and she enjoyed a lovely little daydream about actually being able to test and record and theorize.

She was too old for adventures.

She missed Dzar.

She should've reached the second last wall . . .

There was a flash of scarlet at the edge of her vision as the ground gave way beneath her.

Falling.

Screaming.

Pain.

Her eyes snapped open to see a blur of red and a green-gray square above that. She blinked, once, twice, and the flying lizard came into focus about ten centimeters from the end of her nose.

"I'm alive," she told it. At the moment, that was all she was willing to commit to.

Everything hurt, but nothing hurt specifically. A constant throb, pain pulsing in time with her heartbeat, but no bright shards of agony. Arms, legs, tail; careful movement proved they all continued to function. Slowly, very slowly, she sat up and did nothing but breathe for a moment or two.

The air tasted stale and a little like ozone.

"I'm okay," she told her companion as it settled back on her shoulder, tiny claws dug into the fabric of her overalls. "Not good, but okay. I landed flat, probably why I didn't break anything." A tiny tongue touched the side of her head. "And also why my tail feels like it's going to become one big bruise."

Enough light spilled in from above for her to see she was in a . . .

Room.

Not a rough hole, or a cellar built of stone blocks, but a room with smooth walls and—extrapolating from the one she could see—perfectly squared corners. The floor, where it wasn't covered in a pile of organic debris, was as smooth as the walls, warmer than it should have been, and had a slight give under the pressure of her hand.

"Apparently, it wasn't only distribution of mass but final impact on a surface with a generalized elasticity. That," she added as the tiny lizard settled inside her collar against the side of her neck, a cool patch of comfort that rapidly matched her body temperature, "is science for fell and bounced."

Still moving carefully, she stood, staggered as her tail adjusted to the new position with a cascade of pain, put out a hand to keep herself from falling, and touched the wall.

The lights came on. She instinctively looked up to see at least half a meter of organic debris above the level of the opening she'd fallen through, one side sloped, clearly delineating her passage. The layer of the debris closest to the opening looked to be humus, densely packed enough that visible roots passed above it. Given her equipment and enough time, she could use it to date approximately how long the room had been buried.

Educated guess—pre-destruction.

Then she glanced around the room, and snorted. "Unless the builders threaded indestructible solar gathering filaments through the canopy, I call bullshit. The satellite surveys would have picked up an active power source . . . and everlasting passive sources are fictional. And bad fiction at that."

Illumination followed her as she followed the wall to the corner, brightening as she moved, dimming behind her. Her stride was approximately a third of a meter, and she took seven, eight . . . or was that ten? Did she miss two? Not important. It wasn't a large room and, except for her, her little red companion, and the debris that had fallen with her, it was empty.

The fourth wall lit up when she touched it, and she stared at the orange rectangle mounted in the center, sudden shadows throwing raised patterns into sharp relief.

The team hadn't included a linguist. They hadn't expected to need one while mapping the plateau. Any symbols discovered were to have their location precisely recorded and high-resolution images acquired for further study.

Arniz recognized nothing on the wall she could call language although some of the patterns had a familiarity that spoke to the commonalities of science. She touched three symbols she saw repeated multiple times and a unique symbol in the upper right corner expanded a full centimeter up from the background, moved, and became a part of the symbol below it, now also raised although not as high.

Would a plastic life-form use plastic as a building component?

Or . . .

She stepped back.

Was this particular building component made up of plastic life-forms?

Was Yurrisk not so much delusional as right?

Had the plastic aliens left a weapon behind?

"Oh, get a grip," Arniz growled, provoking an answering hiss from the lizard tucked in her collar. "There are a lot of plastic-using species in the universe, and there's nothing to say that any species with the technology to get here while the pre-destruction society was still pre-destruction wouldn't have been one of them."

Another unique symbol slid sideways through one of the repeated symbols, both of them morphing into new shapes when it emerged out the other side.

"Stop it!" she snapped.

Whether it had finished a pre-set program or whether it understood her better than she did it, it stopped.

"Well, that's mildly disturbing."

EIGHT

WERST SQUINTED DOWN into the underbrush beside the cleared road and couldn't spot either of the Artek. Credit where credit was due: if a Krai, evolved to spot prey through shifting foliage, couldn't see them, no one could. He flicked his helmet scanner back on, noted their position under a pile of debris, absently ate an egg sac that had been webbed to the tree beside him, and settled at ninety to the road. Minimal movement gave him a clear view both toward the anchor and toward the ruins. Thanks to the DLs, they'd have plenty of warning before they had company, but he preferred eyes on.

He plucked a catkin dangling in his line of sight and ate it to cut the bitter taste of the egg sac, resting his KC behind the angle of his leg to take advantage of his uniform's camouflage. There were times when he couldn't tell the difference between being a Marine and being a Warden; ass down, waiting for the shooting to start was one of those times.

Once the shooting started, even a H'san with their head up their ass could tell the difference. Wardens didn't face enemy combatants, the battle field divided conveniently into *us and them*. Wardens faced *us and those of us who need rehabilitation.* Or possibly, *us and those of us who think they're* serley *hot shit and really aren't*.

He grinned.

Us and those of us who need to grow the fuk up and realize it's not all about them.

Us and those of us who are mistaken about where the center of the serley *universe is.*

Us and . . .

Werst. Tech just powered up.

Angling his face into the trunk to block sound waves, he ducked his chin and replied, "In the anchor?"

Not as I understand your anchors, Firiv'vrak said thoughtfully. *_The ground is vi . . ._*

Singing.

It's not singing!

He switched to the group channel, tongue probing the protrusions along the inner left side of his jaw. "Gunny, the Artek report tech powering on. Ground is vibrating." Neither of them had said vibrating, but given the rising hints of spice and mint, the semantic argument was still going on.

DLs are picking up SFA. Targets and hostages are moving deeper into the jungle. Emphasis made it sound as though the jungle had gotten under Gunny's calm. Werst grinned. *_Martin, Trembley, and Lieutenant Commander Ganes remain in the anchor. Hold your position. We're picking up the pace._*

"If the Artek are reacting to a perimeter defense, I should go take a look."

She could tell him once again to hold his position, but they both knew neither Firiv'vrak nor Keeleeki'ka would recognize a Confederation perimeter pin if a H'san shoved one up under their collective carapaces.

Keep our noncombatant on a tight leash.

"Roger, Gunny. Out." His tongue tip found a missed bit of catkin as he switched off group. He listened to the continuing argument as he descended. Off implant, he could hear a few clacks, smell a little stink. Best they got it out of their system.

Flattening, he crawled under the debris pile on elbows and knees, tucked into the space between the hard edges of their bodies, and pressed both hands and feet into the ground. "I don't feel anything."

He could smell damp, rot, and cinnamon, though.

Antennae touched his cheek. "We feel it."

"I believe you." On the prison planet, all three Artek had been the only species able to feel vibrations that had led them to the control room. "Can you find the source?"

Firiv'vrak shifted, Werst's uniform stiffening under the pressure of a wayward leg. "It's stronger that way." Her antennae pointed to the forty-five.

"Not stronger, louder," Keeleeki'ka clacked.

"For the last time; it's tech, not a song!"

"Quiet."

"You . . ."

Her outer mandibles were far enough apart, he could barely get his hand around them. "When I say quiet, you—you both—shut up. Is that clear?"

"Yes, Warden."

Keeleeki'ka huffed a cool breath down his left side and, when he took his hand away, muttered, "Yes, Warden."

"Good." The debris had been stacked loosely enough he could scan through the spaces. When the area read clear of both asshole and held-hostage-by-asshole life signs, he crawled out into the open and stood. *Open* being relative. The piles of debris and the undisturbed underbrush were taller than he was. And the same color. He was nearly as well camouflaged on the ground as he'd been in the tree. He'd rather be in the tree. Of course, he would; he was Krai. A sudden wave of sympathy for Commander Yurrisk held him in place for a moment, then he shook it off. No time. Not right now. "All right. DLs will give plenty of warning before we have company, so we find the tech, we deal with the tech, and we return to watching the road. Let's go."

After a little jostling for position, Firiv'vrak moved to his left, Keeleeki'ka moved to his right, and they both moved out front, antennae held nearly parallel to the ground. Not that any of the assholes with the guns would notice, but professional pride put his feet onto patches of ground already destroyed.

At the crumbling remains of a wall he couldn't quite see over, the Artek pulled the vegetation away and pressed into the angle between the worn stones and the ground. A blue beetle tumbled off a discarded vine, landing on Keeleeki'ka's carapace. Her back end rose, the beetle slid forward, she twisted her head and snapped her mandibles together, the movement smooth and practiced.

"Tell me you didn't eat that," Firiv'vrak said, eyestalks swiveling toward Keeleeki'ka.

The scent of roast potatoes momentarily overwhelmed the smell of jungle. "I'm familiar with the concept of alien species."

"Either of you familiar with shutting the fuk up?" Werst growled. "If you can't find it . . ."

"On the other side of the wall," Keeleeki'ka began.

"Vibrations are stronger at the base of the wall," Firiv'vrak interrupted. "We can't know what's on the other side."

"Yeah, well, there's one way to find out." Two quick strides, his toes found a hold, and he was up and over, Firiv'vrak following close enough behind him that a waft of cherry made him want to sneeze.

He slammed his nostril ridges shut, landed on yet more crushed vegetation, and wasn't sure who was more surprised, Tehaven, the variegated Polint who had fukking awesome natural camouflage, or him.

As he ducked the first swing, the sudden, overpowering smell of lemon furniture polish nearly took him out.

"And *most* surprised goes to the Artek." Werst grabbed a vine, climbed up out of the Polint's reach, switched once, twice, three times as the vines were yanked out from under him. He went down with the fourth, back into range of six sets of claws.

Tehaven roared a challenge. His translator ignored it.

"Yeah, yeah . . ." The challenge gave him time to roll clear. Challenges were fukking stupid. ". . . yours is bigger."

Truth.

Alamber, off com!

Sorry, Boss.

Up on one knee, Werst couldn't take the shot without hitting Keeleeki'ka. "Get clear, you *serley* bug!"

She ignored him, flowing up and over Tehaven's haunches as quickly as if she were on flat ground. Damp patches that might've been blood darkened the variegated fur.

Werst, multiple targets returning. Get out, now.

"Negative Gunny; Keeleeki'ka has engaged."

Say again!

"Keeleeki'ka has engaged. And she's kicking Dutavar's brother's ass."

That's not possible! Dutavar snarled.

"Hey, I'm here, you're not. Suck it up."

Claws caught the edge of the duct tape covering the cracked edge of Keeleeki'ka's carapace. Muscles bulged as Tehaven used the torque of his twisted torso to fling the Artek off his back. She tumbled twice when she hit the ground, got her legs under her, and rushed back in.

A blue energy bolt took out the tree to Werst's right. "The fuk,

Firiv!" he snarled as his uniform kept him from being shredded by shards of wood.

I'm better in a ship, Firiv'vrak muttered from his implant. Another tree shattered six meters out.

"What the serley fuk are you shooting at now?"

There's a Polint and a di'Taykan incoming. I've slowed them down.

The trunk of a third tree shattered. The crown dropped to hang up in a fourth, much larger tree.

"Stop defoliating the jungle!" Werst raced for cover as a line of KC rounds chewed up the ground where he'd been standing. The approaching di'Taykan, blue hair so either Mirish or Gayun, stood on Netrovooens' back, holding the strapping that crossed his shoulders with one hand while the other continued to spray the area.

Vine to branch to vine to the tree Firiv'vrak's ray gun had hung up; Werst ran up the trunk, dove through the canopy, and launched himself at the di'Taykan as Firiv'vrak sped past the deep red Polint's front legs and slammed him in the knees. The timing was so perfect, they couldn't have planned it better. Mostly because everyone knew plans went to shit when the shooting started.

Netrovooens stumbled and fell as Werst took Gayun—Gayun light blue, Mirish darker—to the ground. The Polint roared a challenge as he scrambled back onto his feet and took off after Firiv'vrak who led him away. If it came to number of feet on the ground, he'd never catch her. Had he gone for the stationary target, Werst knew he'd have been fukked. He knew how strong and fast the Polint were. A slash to rip off his helmet, a slash to rip off his face—game over.

Tehaven's vocabulary had the translation program substituting *bug* for a dozen other words Werst bet were less neutral. Seemed that fight was still going on.

That left him with Gayun, who'd gotten to his feet and pulled a knife. He was Navy, not Corps, and he didn't look all that familiar with knife fighting. In Werst's neighborhood, a knife fight meant it was Foursday.

"33X73 is a Class 2 Designate," he snarled as he raced in, braced a foot against Gayun's thigh, climbed the side away from his knife hand, and slammed an elbow into the side of his head. "You're under arrest for . . . for messing shit up."

He twisted in the air, landed on his feet, and snarled as Gayun whirled around trying to keep his balance, staggering close enough he could see the light receptors in the di'Taykan's eyes opening and closing. Opening and closing. "Go down and stay . . ."

The ground dropped away beneath them.

They exchanged a momentary, mutual recognition of shit hitting the fan and fell.

As the canopy retreated, Werst bellowed, "Go to ground!"

Werst? The weight of Firiv'vrak's pause told him she'd turned in time to see him disappear. *_Warden!_*

It was a long way down.

"Now!"

Impact.

Pain exploded across his back, through his head, under his chin.

Then darkness.

The darkness didn't last long.

The pain seemed to be hanging around.

He blinked, spit out a mouthful of light blue hair, and pressed his fingers into something soft—no idea of what body part—in search of a pulse. Alive and unconscious. Well, good. He'd still have the chance to kick their . . .

The darkness returned.

Turned out it was too much to ask for accurate intelligence from the Ministry. A muscle jumping in her jaw, Torin ducked through a door, lintel intact, unable to tell which of the surrounding pieces of buildings it belonged to. Length of day, ambient temperature, necessary supplements, radiation levels; all that was useful. Random pits in the jungle; that would've been more useful.

"Gunny . . ."

Torin looked at the piece of stone in her hand, had no memory of breaking it off the decorative carving on the side of the door, and without breaking stride, threw it so that it smashed against a fallen pillar. Half a dozen multihued insects scattered at impact. If the mercenaries thought they now had leverage, they were right. If they thought making it personal would strengthen their position, they were idiots. She checked her cuff again. "His life signs are strong."

"We evolved falling out of trees," Ressk growled, nostril ridges opening and closing as he ran to the end of a branch and swung into the next tree, his landing sending a black bird with purple highlights screaming up into the sky. "Of course, he's fine. Bertecnic, path goes right of the dead tree. Your military right, for fuksake!"

Data streaming to Torin's cuff indicated deep bruising in multiple sites both front and back and blunt force trauma to the back of Werst's head. Only Krai bone could remain intact when a combat helmet shattered on impact. His pulse and respiration were labored enough for concern, but good enough not to turn concern to worry. The dropping blood pressure, however, that was cause for worry. The shattered helmet had clearly caused lacerations outside the area his uniform covered. Head. Throat. Either could be very bad.

The Justice Department had disapproved of the Strike Team's uniforms using military medical tech, protesting that sending comprehensive medical data to the team leader was an invasion of privacy with the potential for bio-terrorism should it fall into the wrong hands. Although she'd made her opinion of that clear, Torin was aware that Captain Kaur's more diplomatic report to committee had carried the day.

Ressk's medical data noted elevated heart and respiration as well as increased muscle tension. All within an acceptable range after having listened to his bonded plummet into a pit.

Torin ducked a branch without losing speed, her boots slamming down on the crushed vegetation that marked the Polint's path. They'd traded Werst for the ability to move at full speed.

Ex-Marine Lance-corporal Brenda Zhang and ex-Navy Gunner Jana Malinowski have joined Tehaven at the pit. The words tumbled over each other so quickly Firiv'vrak's translator had trouble separating them. *_They have rope . . ._*

So does Camaderiz, Boss. All three of them ran past the DLs on the road; he kept going when Zhang and Malinowski peeled off. He's past the ruins now and out of visual.

No point in asking if Alamber had forced his way into the mercenaries' slates yet. He'd tell her when he had.

Zhang and Malinowski appear to be arguing over who will descend into the pit. Firiv'vrak clicked a pattern that didn't translate. *_I have a clear shot on Zhang._*

Werst needed medical attention. Torin weighed the odds of him receiving it against the odds of him being shot in the head. If they intended to shoot him, it would be a hell of a lot safer to make sure he was dead before going into the pit for the di'Taykan.

Zhang has informed Malinowski that the Warden is a scout. That were the rest of the team in range, she'd already be dead.

Zhang recognized the uniform. Which was the point of uniforms; couldn't have a battle if no one knew who to shoot. The Justice Department had insisted that the Strike Teams be easily identifiable. While the other guys get to wear civvies and hide in plain sight, Torin grumbled silently, jumping a low wall. If the ground on the other side could hold the Polint, it could hold her.

In some ways, the job had been easier back when they'd been contract players.

She's tied off the rope, and is about to descend. Do I take the shot, Gunny?

With Zhang down, Malinowski and Tehaven would go on the defensive. There'd be a standoff while Werst bled out.

Gunny?

"Negative on the shot." Torin shouldered the weight of Werst's life and lengthened her stride. "Bertecnic, pick up the pace!"

• ——◆—— •

Blinked at the sound of shouting. Human voice. Female.

"Gayun is broken, but breathing." She snickered. "B . . . b . . . broken but b . . . b . . . bleeding. The Warden cushioned his fall! Tehaven, go for a stretcher!"

Blinked at a rumble from above.

"Fukking hell. Language. Malinowski, tell him to go for a stretcher."

"How?"

"I don't care. Use interpretive dance!"

"Regular stretcher or AG?"

"What do you think? Gayun's in the bottom of a hole!"

"So we leave him there. Why make the effort? He's not one of us."

"He's on the same side we are as long as we're on this shithole of a planet. Marines don't leave people behind! Stretcher! Now!"

Blinked as a light shone into his eyes.

"So, you're alive, eh?"

Was he? Good.

"Sarge, the Warden's alive."

Blinked at the voice. From a slate. Not using implants, must be using slates. Communication important.

"Bring him in, Zhang. I'll send Tehaven back with both stretchers."

A hand closed around his jaw. “Fuk me, that’s a lot of blood.”

He couldn’t remember how to snap. Remembered what sealant felt like, though. Warm. Then cool.

“You got enough juice left inside, you should be fine. You don’t, well, not my problem, is it?”

Boss, I’m in Werst’s implant, switched it to group, and boosted the gain on his mic. As long as his mouth is open, we’ll hear what he hears.

“But Martin has your military’s implant,” Freenim called out. “Wouldn’t he know that’s possible and have the implant physically turned off?”

Torin’s lips pulled back off her teeth. “Martin’s served with Krai. He knows better than to put anything organic in their mouths.”

Blinked at the light back in his eyes.

“Come on, Warden, I know you’re awake. Since we’re stuck in the bottom of a pit clearly not dug by the people from upstairs—unless they enjoyed primitive living and kept their ability to create seamless walls made of who the fuk knows what for the important things like, well, pits that look a lot like big empty, creepy cylinders—and because Gayun entirely sucks as a conversationalist, right now it’s up to you. Because this is the pits.”

Blinked.

“Oh, come on, that deserved better. It was funny, right? Look, if you’re going to die on me, I should at least know your name.”

Werst!

Blinked. “Ressk . . .”

Blinked again.

Thought he heard someone say, “Fuk me. You hear that, Sarge?”

Blinked . . .

The darkness settled in to stay.

“I’m confused,” Binti called out from behind her. “Why is it a good thing those assholes think they’ve scooped Ressk instead of Werst?”

“You heard them.” Zhang’s conversation with Martin had been enlightening. Martin’s crew knew Ressk was a member of Strike Team Alpha, they were familiar with his skill set, and were aware he had a better chance of understanding alien tech than anyone currently dirtside—regardless of how vehemently Alamber had protested that opinion. He'd continued protesting, his discontent a background rumble in everyone's jaw, until Craig threatened to take the VTA up a kilometer and drop him on the anchor if he wanted to be there so badly. "They need Ressk," Torin reiterated. "That need will keep Werst safe. He can fake it until we arrive."

"They know we're coming now," Freenim said, a bland statement of fact. "They'll retreat to the anchor."

"So we change the plan." Torin matched his bland. "We needed someone inside to open the door. Werst is inside."

"Yes, fine, you'll both soldier on regardless because senior NCOs don't worry. They adapt." Vertic snorted. "I, personally, am bothered about the depth of their knowledge, particularly concerning Strike Team Alpha. They know your strengths, they know your weaknesses."

"They know individual strengths and weaknesses." Torin stepped over one of the winged snakes, grounded by Bertecnic's passing, and ignored Binti's expletive. "They don't know how we integrate them."

Sergeant Martin and Corporal Zhang are having indicated their knowledge are being several layers beyond what are being readily available to the public.* Presit's claws tapped out a background rhythm behind her words. **It are obvious, to me at least, that they are being fed information. There are being no other possibility. I are willing to sift data on Strike Team Alpha—all of which I are having on my slate—and attempt to be finding the connection between the Justice Department and these mercenaries.

Torin felt her brows rise. "All of which?"

Besides my own sources, Commander Ng are having been most forthcoming.

"Wonderful." Torin was aware that Presit considered Strike Team Alpha to be hers; she wasn't thrilled to find Commander Ng agreed.

And, as our presence are no longer being a secret, I are able to bounce a signal through the* Promise *and into the net.

"To cross reference."

Aw, Gunnery Sergeant Kerr are having expanded her vocabulary. That's . . . She broke off with a huff of air, and Torin sent a silent thank you to Craig. **Yes, I are able to cross reference.**

"Then see what you can find. I'm less concerned about the extent of their knowledge of the Strike Teams," she continued, when Presit remained silent, "as I am about them knowing we have Primacy members with us."

"But they don't. Unless the Artek were seen . . ."

We weren't.

". . . my brother wouldn't have told Corporal Zhang of the attack. Zhang is female," Dutavar explained. "Tehaven lost the fight."

He didn't lose it, Firiv'vrak began.

Keeleeki'ka cut her off. *We went to ground as Warden Werst instructed before I could defeat him.*

Not for the first time, Torin wished the fight had been within the range of the nearest DL. Artek against Polint. That would have been something to see. "You have unexpected skills."

Hardly unexpected. I told Warden Ryder I carry the story of Tyar Who Defeated the Warlord. Stories worth carrying have substance.

When Torin turned, Freenim shook his head. Seemed this was new information to him as well. If a durlave who'd served with Artek in the war didn't know, the Primacy as a whole—Artek excluded—didn't know. Facing forward again, she ducked a low branch and said, "Craig, go through that list of stories. Check for any other with potentially useful substance."

On it.

"It doesn't matter if the fight ended without a clear winner." Without pausing his forward momentum, Dutavar snapped a slender tree off at the base and tossed it over a low wall. ". . . my brother will be shamed he didn't win and won't want to lose honor in front of a female."

Bertecnic barked out a laugh, and smacked Dutavar on the rump. "Honor?" He sidestepped away from the return blow. "Like it's ever that simple."

"Simple?" Binti scoffed. "Zhang's a female of a different species."

"In a combat situation, it doesn't matter."

Torin tuned out Alamber's enthusiastic agreement. "That must make for interesting battle plans."

"Tell me about it," Freenim muttered.

No one suggested Werst might surrender the information.

"Can Werst convince the hostiles he's analyzing alien technology?" Merinim called out loudly enough Torin heard her external to the implant. The Druin, for all their shorter stride, had easily maintained the pace set by the Polint. "He punched the coffee maker."

"He's smarter than he looks," Ressk growled above Binti's laughter.

Torin glanced up to see Ressk directly above her. "Hold!" *Infantry*, as they said, *could stop on a dime and give you nine cents change.* She had no idea what a dime or cents were, but both the Confederation and the Primacy had planted mines—stopping on command beat taking one more step and losing body parts. Resettling her pack, she reached up and squeezed his foot. "Sitrep."

His answering expression insisted he was holding it together, but his toes tightened around her fingers. "We're twenty meters out from the bog."

During their run to plant the DLs, he and Werst had mapped the bog at five hundred and seventy meters across. They'd been unable to ping either end. Combined experience in humping gear over various unfriendly landscapes had identified the area as an old river, silted up and spread out beyond its original banks.

"We can't go around." Vertic gouged a trench with one front foot as they gathered for another look at the map rising off Torin's slate. "And it's barely over a meter at the deepest point."

Torin, they know we're incoming. Craig sounded impatient. **Why walk?**

"At this point it's just as fast. And considerably more subtle."

Yeah, yeah, they have hostages, no backing the bodgie gits into a corner. His eyeroll was nearly audible. **They've got one of ours now.**

"I know." He hated feeling useless as much as she did. "I'll let you know when you can drop the shuttle on their heads."

Are she meaning that literally?

"Until then, keep Presit off the coms."

Because I can do the impossible.

She could hear the reluctant smile in Craig's voice and answered it. "Yes, you can."

"A meter deep is nothing," Binti pointed out.

Merinim raised her hand. "Uh, Freenim may be a meter five, but just over a meter tall here."

"Just over," Binti repeated. "Over's good. And you have boats. A kind of variable definition of *boat,* sure, but I want to see them in action."

The Druin equipment had included personal flotation platforms. Freenim had explained the inclusion with the observable fact they were, as a species, short.

"We're wasting daylight." Torin clipped the slate back on her belt. "Ressk, find a path."

"On it, Gunny. There's a ruined wall that'll take us a third of the way, water barely over your ankles. Bertie . . ."

"Don't call me that," Bertecnic sighed.

". . . forty-five degrees to your left until your feet get wet. Two, maybe three meters. Move!"

"He's motivated," Vertic said approvingly as Ressk swung out in front again. "Danger to a bonded will inspire extraordinary effort."

"Yes, sir. That's how I explained it to Commander Ng."

She heard Binti snicker behind her. "You figure Werst'll hold off on kicking ass until we get there, Gunny?"

Torin felt her forward boot sink into the loam. "He doesn't usually."

• ——◆—— •

"What were you doing out here?" Yurrisk shoved his face toward hers, teeth bared, eyes gone crystalline. "You need to stay where I put you. You need to hold the line. I didn't clear you to leave the ruins."

Arniz slid her gaze past his—he wasn't seeing her, so what did it matter—lowered herself onto a piece of carved and broken stone, and waved a hand until she could catch her breath, trying not to inhale a cloud of tiny silver insects with each gasp. Who knew being dragged out of a pit would be so exhausting? All she'd done was dangle. "I was looking," she managed at last, "for Hyrinzatil."

Yurrisk whipped his head around to glare at the ancillary. "You were trying to escape, were you? Planned to circle around and use the shuttle's com system to send for help? Thought I wouldn't notice one lizard missing? I know how many enemies I have. No matter how many come, I'll protect what's mine." He'd taken a single step toward Hyrinzatil, fists clenched, when Qurn's gloved hand on his wrist brought him to a sudden stop. He swayed right and right again, took a deep breath, then said in a less terrifying tone, "Answer me."

"I ran because an animal was attacking." Hyrinzatil folded his arms, unfolded them, folded them again, tail tip lashing. Had he stopped there . . . but then he wouldn't be Hyrinzatil if he stopped on the prudent side of the line. "It was the *sensible* thing to do."

Arniz ached with how much she missed Dzar and the emphasis of youth, then she pushed the ache aside because Yurrisk had drawn himself up to his full height, his spine a rigid line, and Hyrinzatil's primary was, as usual, useless, more concerned with a hole in the ground than the life he'd been given responsibility for. "Sensible only because Trembley was willing to fight," she snapped. "Am I wrong?" she demanded as Hyrinzatil leaned around Yurrisk to glare at her. "The greater majority of predators chase running prey. If Trembley hadn't been there, the animal would have caught and eaten you."

"I don't have to run faster than the animal, *Harveer*, I only have to run faster than you."

"I should have let you stay lost in the jungle."

"I wasn't lost! And let's not forget, I'm the one who found you and went for help!"

"Enough." Yurrisk pointed at Hyrinzatil, the weary officer back at the surface. "Get in the pit. Assist in the removal of the plastic."

"Removal?" Salitwisi, kneeling between the two ropes dangling over the most stable point of the crumbling edge, twisted around and glared. "We need to examine it *in situ*!"

Yurrisk took three fast steps, grasped Salitwisi's tail with his right foot, and swung him out over the edge. "Then look at it *in situ*," he said calmly over the shriek as he let go. "And you . . ."

Hyrinzatil opened and closed his mouth, tongue darting out to taste the air and disappearing again.

". . . go in after him. Under your power or mine, I'm not fussy."

"Go on," Arniz said when the ancillary glanced over at her, willing to yield to her authority now, when his other options had run out. "It's not deep, and there's a pile of debris built up from all the in and out to soften the landing."

Beyvek and Mirish di'Yaunah—both of whom turned out to be engineers although Arniz thought that might mean something different in the Navy than in academia—had been gratifyingly enthusiastic about her discovery and had brought quite a bit of fresh debris in with them. She'd kicked some in herself on the way out.

"Commander?" Beyvek's voice rose up out of the pit. "We seem to have acquired a civilian."

"He's there to observe the plastic *in situ*."

Arniz was too tired to prevent her smile at Yurrisk's tone.

"That a Niln word for hole in the ground?"

"I assume so. Don't let him delay you, Lieutenant, but don't allow him to assist either. He wants to observe, let him observe. His ancillary is on his way."

"Good. We could use another set of . . ."

"Great egg! Don't touch it with your bare hands!" Seemed Salitwisi had survived the fall. "You need the prop . . . Awk!"

"You heard the commander, tail against the wall and observe."

A meter from the edge, Hyrinzatil stood frozen in place. Yurrisk sighed pointedly. "Did you need me to assist?"

"Go on," Arniz repeated. "I'm a lot further from the egg than you, and that hole wasn't open to the sky when I discovered it. I went through the roof."

Hyrinzatil shuffled forward, tail tip twitching. He balanced on the edge for a moment and, when Yurrisk stepped toward him, jumped.

Niln bones weren't as light as the Rakva, but they were light enough to keep the mass part of mass times acceleration low. From the near instantaneous sound of both Beyvek and Salitwisi ordering him around, Hyrinzatil had landed safely.

"Did you have an observation to make?"

When she realized Yurrisk was speaking to her, Arniz snorted. "I'm surprised Salitwisi didn't expect the figurative kick in the ass, but, other than that, no." His response, given that Salitwisi still hadn't internalized the current power structure, had been reasonable. No one had been hurt. Martin wouldn't have been so restrained.

He waved at the black Polint waiting by the tree where the ropes had been secured, pointed at his eyes, then pointed at Arniz. "Camaderiz. Watch her."

Camaderiz made a derisive sound and, while Arniz had no idea what she was likely to do except sit and feel old, she didn't think the derision had been aimed at her. Nor did she think the Polint were as unaware as they pretended. It was highly unlikely the bright green item she'd glimpsed through the fur in Camaderiz's ear was purely decorative.

"The Niln has a point," Qurn said quietly. "Not her," she added when Yurrisk frowned at Arniz. "The other, Salitwisi. To remove unfamiliar tech and expect it to keep working . . ."

"You don't understand." He took hold of her arms, gently, barely indenting the fabric, and met her gaze, nostrils open. "The Krai from the other pit, Ressk, he's an advance scout . . ."

"For Strike Team Alpha. So you said."

He shook his head. "Strike Team Alpha is a designation, nothing more. Ressk is one of *her* people. We have to get back to the safety of the anchor if we want to survive this. We're dead if we stay here, and I can't leave the plastic behind."

Qurn stared at him for a long moment. "Who is she?"

Yurrisk frowned as Arniz silently applauded the Druin. Exactly the question she'd have asked.

"You said Ressk is one of *her* people," Qurn prodded. "Who is she?"

"Gunnery Sergeant Torin Kerr."

Pale lids flicked across both eyes. "The gunnery sergeant who exposed the plastic aliens?"

"The same. The same gunnery sergeant who saved a training platoon on Crucible, who got her people off Big Yellow even with the handicap of Commander Carveg. The same gunnery sergeant who brought the Silsviss into the Confederation. Who got a Marine armory out of the hands of pirates and who added a strong arm to the bureaucracy of justice. Do you know what they say about her?"

"That she's trying too hard?"

"That she was only doing her job. And now, we're her job."

"So we leave. Take the VTA back to the ship and go."

Arniz was fully in favor of that plan.

"We can't. Without that weapon to sell, there's nowhere we can go. We've barely enough fuel to get us to the meet with the buyer and, without the buyer, no way to get more. The buyer will pay us enough for a converter, and that will free us. We can't leave, but if we stay—if we get to the anchor—we're safe." Yurrisk stared through the trees as though he could see the anchor in the distance. "Kerr is smart enough to understand the principle behind hostages. As long as we have them, she can't attack. As long as we have them, she has to negotiate, and that's not one of her strengths."

"The strengths she does have seem to be working for her," Qurn pointed out. "And there's six or seven on a Strike Team. One of the others can do the negotiating."

"The Strike Teams are weapons aimed by the Justice Department. Ex-military who can't leave institutionalized violence behind."

Hello, irony. Arniz moved a stick bug to safety on a bit of undisturbed greenery.

"They have no subtlety. They can't charge in, guns blazing, and put civilians in danger. The civilians are a shield. So when we leave with the weapon, we take one with us." He glanced over Qurn's shoulder at Arniz. "Not you."

She waved it off. "I'm crushed. Truly."

"Ganes would be best. He's Navy. He knows how this works."

"He doesn't seem to," Qurn muttered. "Martin's people barely stopped him from getting to the communications equipment."

Again, exactly what Arniz would have said.

"That's how it works," Yurrisk insisted. "Ganes has been captured, it's his job to make it difficult for the enemy. He doesn't surrender. He keeps fighting. And fighting." His nostril ridges slammed closed. "Stop fighting and everyone dies. Stop fighting and it's over and lives were spent for nothing. Stop fighting . . ."

"Come back." She pressed her hand to his cheek. After a moment, his nostril ridges slowly opened.

"Commander!" Beyvek's voice pulled him around toward the pit. "The plastic wasn't connected to the wall; it was hanging. There's six holes and six hooks, hard to see because they're the same material. The plastic itself is like a semirigid plastic curtain. I have no idea what's powering it, but it shifted another symbol when we were taking it down."

Arniz could see Yurrisk's chest rising and falling, too quickly to be normal, but he had control of his voice when he asked, "Is it still functioning?"

"Can't tell, sir. The moment the bottom edge hit the floor, it rolled into a tight tube."

"Well, you couldn't transport a semirigid curtain, now could you?" Salitwisi's voice drifted up out of the pit. "Those who hung it here had to first bring it in. It clearly wasn't built on site. If you insist on moving it, the odds of damaging such a priceless artiFACK! Uncalled for," he muttered a moment later, barely loud enough to be heard outside the pit. "Totally uncalled for."

"He's annoying, but he's not wrong, sir. Rolled, this thing is going to be significantly easier to transport. We're securing the lines."

"Well done. Camaderiz!" Both hands curled around nothing, Yurrisk mimed pulling motions. "Ready on the ropes."

Camaderiz pointed at his eyes and then at Arniz.

"Not now!" Yurrisk made the motion again. "Now, you pull the ropes!"

Highlights rippling through ebony fur, Camaderiz shrugged.

"It's not a weapon," Arniz said, before the pissing contest could escalate. While she was all in favor of them fighting among themselves, even an elderly *harveer* with no experience in violence could see who'd win. With Yurrisk gone, Martin would have no restrictions, so best distract him before Camaderiz took him apart.

Yurrisk turned back to the pit, lips still off his teeth. "It will lead us to the weapon."

Arniz shrugged and stayed silent. If there *was* a weapon, it might.

"Camaderiz."

Other than the Polint's name, Arniz understood nothing Qurn said. Camaderiz, however, clearly knew exactly what she was saying if his wide eyes, flattened mane, and sudden interest in the ropes were any indication. Qurn was Primacy. Arniz had forgotten that.

• —◆— •

Boss, Werst's signal just dropped out.

"He's unconscious," Ressk snarled before Torin could respond.

I didn't say he stopped chatting me up. The words were flippant, but Alamber's tone was kind. **I'm not reading the signal from his implant. I could hear muffled and unoriginal speculation about Strike Team Alpha from the people around the stretchers as they reached the anchor, then blip. No signal.**

Military grade implants kept functioning eighty-one hours after death. If Martin had stuffed his fingers into a Krai mouth to destroy the implant, well, he'd be a lot easier to beat with bloody stumps where his hands should be. And Werst's implant would still be sending. They needed "Ressk," so Martin wouldn't have blown Werst's jaw off. Only one option left. "The anchor has an implant blocker."

"Why?" Vertic sounded confused.

"Best guess . . ." Without slowing, Torin ducked a branch too thick for the Polint's machetes. ". . . one or more of the other scientists didn't like the thought of Lieutenant Commander Ganes having sole access to certain tech."

"He's the only one with an implant. Who do they think he'll be talking to?"

“They’re not thinking. They’re afraid he’ll use it to record them.”

“Your implants record?” The durlin hadn’t been an officer long enough for her attempt at mild curiosity to sound like anything but the suspicion it was.

“No, it’s against the Confederation’s privacy laws. But that doesn’t stop civilians from making assumptions and spreading rumors.”

Binti snickered. “I heard there’s a subliminal that’ll turn anyone with an implant into a mindless drone.”

“That’s not true here?” Freenim, on the other hand, had been an NCO long enough to make his mild curiosity entirely believable.

The only sounds were the whistle/chunk of Dutavar’s machete, birds, and insects. Even the surrounding foliage had stopped rustling.

Bertecnic broke first, his sputter turning to a deep belly laugh. A moment later, the other members of the Primacy team joined in.

“Your face,” Merinim giggled. She must have been referring to Binti’s face because Torin hadn’t turned, had locked her gaze on Bertecnic’s haunches, and started working on a way to deal with the worst case scenario.

They couldn’t make a subliminal we couldn’t hear, Firiv’vrak pointed out, still chuckling.

Oh, that are being very funny. I are including it in the final edit for sure.

Neither Presit nor Dalan had an implant; they couldn’t know how it felt the first time a tech cracked a jaw. Before going under, even the most badass sergeant fixated on the rumors. Torin had no intention of adding a new rumor and would have a word with Presit later. Or, she’d ask Craig to have a word. Although, now she thought of it, Presit had used illegal tech in the past and implants could definitely be built to record if someone ethically flexible got into the right position, so it was possible the Katrien would know if . . .

No. Not her problem. Not her job. Not even very likely.

She shortened her stride to avoid a low black mat of fungus, then tilted her head up toward the canopy when she caught sight of pale fronds between hanging vines and underbrush. “Ressk! We go around the bracken! Left or right, I don’t care, but we don’t go through.”

“Sorry, Gunny. I was . . .”

Taking the fastest route to Werst, Torin finished silently when his voice trailed off.

•—◆—•

"So it looks like Strike Team Alpha's here to rescue you."

"What?"

Werst didn't recognize either voice although he did recognize the familiar feel of antigravity shutting down. The Corps never used the AG stretchers in a war zone, too susceptible to EMP, but he'd caught a ride between the VTA and Medop a couple of times.

"Don't get your hopes up," the first voice sneered. "We're not going to make it easy for them. Keep the Krai alive. There's an artifact coming in we need him to take a look at."

Blink.

Human. Male. Look way up. Short pale hair. Red cheeks. Robert Martin.

"What's wrong with him?"

The second voice sounded overwhelmed. Angry.

"He fell down."

Sounded like Martin hadn't changed. Still a *serley chrika*. Mashing his tongue against the inside of his jaw, Werst pressed three times against the implant controls. One short, two long. Conscious, but injured. They had no code for captured; the Primacy didn't take prisoners.

Which made Martin a bigger ass than an enemy they'd been at war with for centuries.

"Once again, I'm an engineer, not a medic!"

"Don't care. You've already proven you can operate the autodoc; operate it again."

"Trembley is Human. The autodoc was preloaded with Human parameters. It isn't set up for Krai. Do you see a Krai on the science team?"

Blink.

When the light returned, Werst gritted his teeth and flopped his head to the right. Voice two belonged to a Human male. Dark skin. Narrow gaze. Warm hand.

"Yeah, and you were Navy." Martin. Still sneering. "Fuk of a lot of Krai took the easy way and went Navy. Use your superior Human brain and figure it out."

Blink.

Navy. The second voice and the warm hand belonged to Lieutenant

Commander Harris Ganes. Good. He'd made contact with a potential asset. Go him.

"Warden's name is Ressk, if you want to sweet talk him. Your choice if you keep the di'Taykan alive. I don't care either way and Yurrisk will blame you. That could be fun."

Fukking Martin.

Blink.

• ——◆—— •

Boss, targets and hostages are back in range of the DLs by the ruins. They're carrying . . . a large roll of . . . It looks like plastic. Hang on. Yep. Pinged it. It's plastic.

"Briefing packet said this was a pre-plastic planet," Binti argued. "Wiped out before they figured out how to stabilize hydrocarbons. You need to ping it again."

It's plastic, Bin. Also bright orange. Danger orange. Do not advance orange.

"Why would they warn the di'Taykan off?"

Everybody loves us.

"You wish."

"DLs recorded a lot of rope moving past the ruins." Torin cut off the banter. "Past the ruins, away from the pit Werst opened up. What would keep this lot, looking for a weapon to use against the plastic, away from a sudden pit? Another pit. A second pit that held a sizable piece of plastic."

The natives didn't have plastic, Craig continued, **so the plastic had to have come out of that second pit.**

"Then the plastic in the latrine that started this mess could be discarded scrap," Vertic said thoughtfully. "Or, the plastic dug the pits as blinds in order to observe another social experiment."

The plastic's social experiments tended to happen on the scale of intergalactic war—but they had to have started somewhere. Torin definitely understood why the possibility of a weapon had everyone's hands in their pants. Confederation. Primacy. They all wanted to get some of their own back.

Running blind in ankle-deep water so brown it was entirely opaque, she led the way along the top of a ruined wall, trusting her boots as much as her scanner. They were making good time.

"So the dumbass story about a weapon that can destroy the plastic is maybe not so dumbass," Binti observed.

"And the destruction of this civilization?" Vertic wondered. "Populations failing worldwide at approximately the same time?"

"You were there when plastic spoke. Seemed to me it was . . ."

"A molecular, hive mind of shapeshifting manipulative, murdering, shitheads?" Binti offered.

". . . pragmatic. Although Mashona's not wrong. If the plastic became aware of a weapon here . . ." Torin let her voice trail off into the obvious implication. The edges of the stone underfoot crumbled. She shifted her stride back toward the center of the block.

Torin had taken point back at the water's edge; followed by the three Polint, Binti on their six. The two Druin had sped across to the other side, ready to provide covering fire if needed. Their floaters had looked like a wad of thick paper when pulled from their packs, but had expanded out to two slick surfaces with honeycombed folds between them. For all their apparent fragility, they held the weight of a Druin and accompanying gear. Propulsion came from a single-use charge about a centimeter square. Torin wanted one. A larger one. Wanted one reverse engineered and made part of standard equipment. She also wanted aftercare for veterans without the cracks too many fell through, and for the violently antisocial to get some fukking therapy and take up flower arranging, putting the Strike Teams out of work. It didn't look like she was going to get those anytime soon either.

"And if there is a weapon?" The undertones in Vertic's voice, sharp through both translation and implant, reminded Torin how short a time the war had been over.

"If a weapon exists, and they've found it, we'll confiscate it."

And have Anthony Marteau make a million of them, Craig drawled. **That ought to polish his nuts.**

"Why does it go to your people?" Vertic asked coldly. "The plastic are responsible for as many of my people dead as yours."

Makes no difference; that fukker Marteau will sell to both sides.

"I understood we were on the same side, Warden Ryder."

Yeah? Then forget I said anything.

Fortunately for Torin's decision to stay clear of Craig's problem with Vertic, whatever it turned out to be, Alamber stepped in.

Unclench before you turn your shit to diamonds, he muttered, presumably to Craig, then added, *_The roll's about two and a half meters long and if it's consistently the same thickness as it is at the edge, it's around three meters unrolled. And I suspect it's unrolled by now inside the anchor and being stared at by all the hostages and all the bad guys. Who are also all in the anchor._*

"They've secured their position." Secured after a quick walk down a cleared road and a sunny stroll across an empty plateau. Torin squinted across the murky water, at the floating debris, at the slender trees growing up through water and debris both, at the visually impenetrable mat of vegetation waiting behind the Druin on the other side. She really hated jungles.

I could drop the VTA on the plateau.

"They'll tell you to leave or they'll kill a hostage."

"They can't kill all of them," Binti pointed out, "or they have no bargaining position."

This job, unlike her last job, came with no justifiable body count. "Doesn't matter. We're the good guys, we're doing it the hard way."

"We're attacking the anchor?"

"They know we're here; we'll give them the option to surrender first."

Bertecnic snorted. "Does that ever work, Gunny?"

"Hasn't yet. Might someday." Her scanner pinged. They were out of wall. She looked up as Ressk climbed to the top of a slender tree and rode it down across the four treeless meters that ran down the center of the bog—the path of the original river. As the tree reached the widest point of the arc, he jumped, catching the closest tree on the other side. The first tree snapped back, sending leaves and small branches showering down into the water and a small flock of birds shrieking up into the sky. The birds had been a good ten trees away; Torin hoped that wasn't an indication of a predator large enough to use Ressk's maneuver on the hunt.

Did the scaly mammals climb?

She dropped off the end of the wall, the water now over her knees and increasingly murky from the debris her boots stirred off the bottom.

Scanners showed the water either empty of living creatures or so filled with life-forms that individual readings were impossible. Torin didn't like either option.

Bertecnic dropped into the water with a splash that sent ripples a good five centimeters up her legs. "Feels good!"

Yes, you do.

"Alamber."

Credit where credit is due, Boss.

• —◆— •

"Bruising. Minor lacerations across the back of your head. A piece of helmet went into your throat, just missed major blood vessels, and probably hit a few minor ones. Concussion. The helmet broke, but your head didn't. If you weren't Krai, you'd be dead."

Werst watched Ganes finish inputting the data from his cuff into the anchor's medical unit. Nothing much had hurt upon regaining consciousness this last time, so he assumed he'd been shot full of the painkillers in his pack.

"The piece of helmet is still in your throat. When Zang sprayed the wound, she sealed it in. My last field first aid course was a long time ago, but I believe that if I pull it out, the sealant will close behind it. Any chance you've got field experience to support that belief?"

He did. He mimicked a Human thumbs-up.

"Good. If I ignore species parameters and concentrate on repairing the blood vessels, make it structural rather than medical, there's an outside chance I can use the autodoc to repair the damage to your throat."

How far outside? Outside the room? Outside the anchor? Lieutenant Commander Ganes had been an engineer in Naval R&D. Dr. Ganes had gotten himself attached to an archaeological expedition as tech support—Werst was sure Ganes had a reason for the lateral move, he just didn't care what it was. Neither career suggested extensive medical experience.

"The theory's sound, and it should give you your voice back."

Werst made a sound somewhere between a growl and a gurgle. His bonded's name might be the last word he ever said. Well, fuk that.

"As long as you weren't hoping for a second career in di'Taykan opera, I can get you operational. I can't, however, replace the blood volume. You'll have to be careful."

Vague memories, recent memories of making a run for it, of throwing himself off the stretcher into a spinning room followed by the slow collapse of consciousness, supported the commander's concern. But he gave himself bonus points for the attempt.

"I put Trembley on saline, but there's nothing labeled Krai compatible and without Krai specs in the autodoc, I don't know enough to adjust the content. We'll have to replace blood volume the old-fashioned way with supplements and liquids taken orally." Picking up another tube of sealant, Ganes leaned forward until his breath lapped against Werst's ear. "You want my advice? When you can talk again, when Martin starts pumping you for information, keep pretending to be Ressk."

Werst's nostril ridges slammed shut.

Ganes lightly touched Werst's bare shoulder. "It's the name you gave Zhang. It's not the name in the medical information I pulled off your cuff, but it is the name of your first contact. Martin seems to think he knows what Ressk can do. Werst will take him by surprise."

Fuk taking him by surprise, Werst would take him by the throat. Cut off his air. Leave him barely alive enough to arrest.

"All right." The commander took a deep breath and let it out slowly. "Let's do this."

•—◆—•

When she stepped off the soft/hard flat/buckled bottom that had, back in the day, been pavement along the river bank, the water rose up above Torin's waist. She held her KC in her right hand, finger flat against the trigger guard, bent arm at shoulder height, well aware that should anything attack, she'd have to depend on Freenim, Merinim, and Ressk on one side of the four meters of open water and Binti on the other. In other words, she was in no great danger.

Her uniform seals were holding and she could cover four meters of waist-deep water in . . .

"Sorry, Gunny."

She'd lifted her weapon above her head before the chest-high wave of Bertecnic's entry hit and decided not to waste her breath telling him not to do it again.

Her boot sank through silt and onto . . . The new surface had give. It had bounce. Rubber?

Natural rubber wasn't high tech. It was possible the builders of the ruins had also built the water course, laying rubber to keep it confined where they . . .

The rubber moved. Rose. Twisted.

"Something touched my leg!" Water sloshed higher as Bertecnic reared.

Big enough to be under her foot and by his leg. Or there was more than one. "Prepare to fire!"

"I can't see it!" Binti called.

Torin's scanner continued to read everything and nothing.

The rubber jerked left. Torin shifted her weight to her other leg and stayed standing.

Twisted. Looked back . . .

. . . as a loop of something wrapped around Bertecnic's lower body and dragged him down.

Bertecnic's reddish brown fur was a little darker than the color of the water. The thing wrapped around him—thrashing body parts surfaced and disappeared again—matched exactly.

Feet swept out from under her, Torin curled her legs up, reaching for the knife in her boot sheath.

The Corps had underwater weapons; they shot a 120-millimeter-long, 5.65-millimeter-caliber steel bolt out of an unrifled barrel, had shit aim on dry land, and Torin wasn't carrying one. Her KC could handle the dunking, but it wouldn't fire, and an edge would do more against rubbery flesh than blunt force. She let her KC hang. When they got back to the station, she'd revisit the bayonet argument.

Her scanner pinged a proximity alert. No shit.

Arm around an undulating oval tube as wide as her torso, she drove her knife in, took a breath as they broke the surface, felt the tube jerk as a shot hit, and went under again, losing her grip on the knife.

Currents put Bertecnic's fight to her left. She reached out, touched fur, touched a strap, touched the end of his machete—filled in Alamber's response—drove her hand under the rubbery whatever-the-fuk was wrapped around his withers, felt something scrape against her arm, and grabbed the machete's hilt.

Something rubbery and detached bounced off her chest. Bertecnic had taken a piece out with his claws.

She clamped her thighs tight enough to dimple the sides and used both hands to drive the machete through center mass. Jerked the blade to the right.

The thrashing turned her upside down. Sideways. Dragged her

through the mud on the bottom. She grabbed for her dropped knife as the familiar hilt bounced off her cheek.

Slammed into solid muscle.

A big hand shoved her away.

She kept cutting.

When the heavy blade reached the edge, she hacked back through the other way.

The pieces separated.

The top piece jerked away. The piece her legs were around thrashed harder.

Sliding backward, her legs were gripped in turn by solid pincers.

She broke the surface again, sucked air through her teeth, twisted around as she went under, and thrust the machete through the segment behind her.

It stiffened. Either she'd hit something vital or it had finally realized it was in pieces. She kicked free as it sank. Her boots found the bottom and she straightened, blinking the water away from her eyes in time to see Bertecnic surge up into the air, swinging a clawed hand at a red-brown loop already missing triangular pieces.

He spat out a mouthful of water and gasped, "Did I hit it?"

"You did." Her scanner continued to read nothing or everything. "Mashona, Bertecnic, Dutavar, Vertic—other side, single file, top speed!"

Blood, or whatever, in the water was never good.

"Ressk, Freenim, Merinim, lay down covering fire along both sides of them!"

She let the Polint's bow wave help push her to shore, machete in one hand, knife in the other, her legs fighting the water with every step, the familiar sound of weapons fire a comfort.

She kept them moving until the ground only squelched.

Dragging her helmet off, Torin could hear Bertecnic still hacking up water and the rumble of Vertic's voice. "Injuries?" she asked.

"Nicks in his legs—already sealed," Vertic reported. "A strap cut nearly all the way through . . ."

"Guts," he coughed, "under claws." More coughing. "Hate it."

". . . and he'll be taking an antibiotic as soon as he can swallow. You?"

Her vision was blurry, but clearing. Her mouth tasted like she'd been chewing raw liver, but she'd managed to keep from drawing any liquid into her lungs. Torin took a drink from her canteen, and spat as she checked her cuff, fully aware adrenaline could hide any number of injuries. "I'm good."

She heard a sound that might've been Craig exhaling. He had access to the medical feeds, but never entirely believed them.

"Uh, Gunny?" Binti pointed at the backs of her legs. "What about them?"

Them?

Definitely Craig that time.

The pincers attached to her calves, one set on her right leg, two on her left, had curved arms approximately five centimeters long, were a paler red brown than the body of the beast, and looked like horn more than bone or teeth. Over the course of her career in the Corps, Torin had been both gored and bitten and seen more bone than she'd ever needed to. The triangle of rubbery tissue that held the curved arms together had pulled free of flesh with edges so intact, its separation had to be part of the design. Fine serrations along the inner edge of the curves made it impossible to pull them off without damage.

Polint hide was tough if Bertecnic had *nicks*.

Torin?

"Pincers. They're clamped to my calves, they haven't penetrated the uniform."

They look like mouth parts.

She had no idea which helmet feed he was using, but he was right. Her legs hadn't been clamped near either end of the creature—the places where mouths usually ended up—but Torin had been around enough to know how variable life could be. She rose up on her toes and dropped down again. "They're barely affecting my movement." More bruising, but that was all.

Take a minute and get rid of them anyway.

"Planned to." Had Craig been with them, he'd have ripped the pincers away by now, and not only because he was their field medic. She tossed Bertecnic his machete and passed her KC to Ressk.

"Field strip it, Gunny?"

Hands on her weapon that weren't hers weighed against Ressk's need for distraction while they paused the run to Werst. She trusted Ressk. The water had been foul. "Do it." She slid the point of her knife into the small triangle under the pincer's hinge. Unable to lever it off, she worked the flat of the blade up tight against her uniform and cut through the serrations moving the blade around toward her shin. Horn, or bone, or teeth—none of them up to Marine Corps steel.

Merinim dropped to a squat beside her, held up her own knife for Torin's approval, then mirrored Torin's cut under the pincer's other arm.

Torin took a moment to consider an ex-enemy combatant wielding a sharp object against her body and said, "It's trying to reestablish contact behind the blade."

"I find it slightly concerning how a detached pincer with an agenda no longer surprises me."

"Likewise. Cut quickly."

When they'd cut all the way to the ends of the arms, they pressed new bruises into Torin's leg prying the pincers off.

The first triangular piece of flesh hit the damp ground, the pincers snapped shut, and a clear liquid oozed from the severed serrated surfaces.

"Tox screen on Bertecnic, now!"

Freenim had the kit out before Torin finished speaking and an instant later jabbed the prongs into the heavy muscle of the Polint's front leg.

"Insects snack on both of us," she snapped, cutting off Bertecnic's profane protest. "That raises the odds Humans and Polint are more susceptible to this world's poisons."

The tox screen finished before they'd pried the other two pincers off.

"Slight irregularities," Freenim announced.

"Could just be Bertecnic," Merinim muttered, forehead against Torin's thigh as she worked her knife around.

"Administer a general antitoxin," Vertic ordered. "Unless you had another plan, Gunny?"

"No, sir." Torin recognized an adrenaline-fueled need to be in control—if only minimally—when she heard one. "Mashona; anything?"

Binti had remained a few meters behind, weapon pointed toward the watercourse. "Nothing, Gunny. No visible friends, no visible dinner companions. Don't know what's happening under the surface, but that's not our problem."

"Come in, then."

"On my way. You know," she continued, jogging toward them, the impact of her boots marked by small fountains of dark water, "I hit that thing half a dozen times during the fight. I put at least one round into every part that surfaced."

"Segmented nervous system." Torin swallowed both antibiotic and antitoxin. "Wouldn't be the strangest thing we've seen."

I are once having seen a threesome that involved erotic cannibalism.

Pictures, or it didn't happen.

"Alamber . . ."

What?

After scanning her hair for life signs, Torin resettled her helmet, took back her KC, and ran a quick, involuntary check knowing Ressk would understand. "Dutavar, take point. Vertic, if you would, remain on our six. Bertecnic, thank you for the use of your blade. Middle of the line until we're sure you're not going to turn blue and fall over."

"Your concern touches me, Gunny."

"As it should." She rolled her shoulders, settling her pack. "Let's move; we're racing the heat death of the universe here. Ressk, find us a path. Everyone else, watch out for those pincers as you pass."

"Fukking snakes," Binti muttered, falling in behind Torin's left shoulder.

The creature had been more like a flatworm, if anything, but Torin let Binti have the last word. The profanity, at least, was accurate.

• ——◆—— •

Werst curled his toes and rolled his shoulders, nostril ridges closing as his shoulder blades shifted and sent jagged lines of pain down his back. Then he swallowed. Took a deep breath. Let it out slowly . . .

Commander Ganes had expressive eyebrows.

He felt he should say something profound. Or not, if that's how it went.

Then his stomach growled.

"Fuk, I'm hungry."

The commander laughed—high-pitched and nervous, but Werst figured he couldn't blame him for nerves all things considered—and crossed the infirmary, returning with a protein drink in each hand. "Now we know your voice works, we might as well find out if you leak."

He didn't. "Kill them to give these things a flavor?"

"I once watched a Krai eat the centerpiece at a diplomatic dinner. She was making a point, but . . ." The commander spread his hands.

"Doesn't change the fact these taste like H'san sweat."

"Fair enough. Any pain."

"Not in my throat." His voice sounded like he'd been chewing mortar rounds. The immediate area of the wound ached slightly. Nothing more. "Back hurts. So does my left heel. And left elbow." His left side had probably hit bottom just before his right. "My right thigh keeps twitching."

"You've got a bruise there as big as my palm." Ganes flipped up the pale gray thermal blanket. "Another one overlapping your abdomen and the bottom of your right ribs. I expect it's where Gayun made initial contact. You're lucky you were close enough together the impact didn't do more damage and that di'Taykan are light enough you weren't crushed beneath him."

Both bruises were an ugly purple against the mottled green skin. His back had to look worse. "Did Gayun . . . ?"

"He's in stasis."

Werst lifted his head far enough to stare between his feet at the pods. Two were occupied. "The other?"

"Dzar. *Harveer* Arniz's ancillary. She's dead. Martin shot her to make a point."

"Asshole."

"Yurrisk won't allow us to return her to the sun."

"Commander Yurrisk is . . ." Werst discarded a few descriptions and returned to the vague, but accurate, ". . . broken."

"That's not an excuse to . . ."

Werst cut him off. "It's a reason."

"Not a good one." Ganes leaned forward, his hand braced against the edge of the stretcher. Snatched it back when the padding adjusted for his weight and Werst hissed. "I don't know how much contact Martin will allow between us once you're mobile. You saw combat, I didn't; I'll back your play."

"My play, sir?"

"None of that, I'm a civilian now." He sounded defensive. "The mercenaries are inside the anchor with the hostages," he continued, speaking quickly, quietly. "Your Strike Team needs to get inside. You're their ace in the hole."

Fuk him; he was right. That was the plan. Scratching the tight skin at the edge of a bruise, Werst checked his implant. "Is your implant working . . . ?" He bit off the *sir*, matched volume and speed to the commander. "I have power, but no signal, and I guarantee the VTA's scanning for me."

"Before the mercenaries landed, there was no one else dirtside with an implant, and I haven't . . ." Ganes ducked his head.

Embarrassed, if Werst had to guess. "Hey, you've been busy. Try a two seven three." Two seven three was a distress call; too tight to be stopped by an anchor and eight klicks of foliage and designed to be picked up by any Confederation CP. They'd change it at mission end; no way Parliament would give the Primacy a free pass to their communication system.

"It's on. I have power." Ganes tipped his head as he pushed his tongue against the inside of his jaw, visible movement a tell that he hadn't used his implant much. "No signal and I should, at least, be able to pick up the carrier wave from the Ministry satellite."

Not a lot of reasons Martin would disable tech potentially useful to him. Easier to disable Commander Ganes if he didn't want the other man listening in. Werst rolled his shoulders, the pain enough to sharpen his focus. Ressk was so much better at figuring out this kind of shit. "They've put a block on the anchor." Yeah, that sounded right. "Can't block the whole planet, so they blocked where you are. Explains why Martin locked you in."

Ganes stared down at Werst. Over at the window where an insect flared against the force field. Down at Werst again. "They've blocked the anchor to keep me from contacting a theoretical rescue party?"

"Why not? You got a message out. That's why we're here."

"Yes, but through my slate. Before they confiscated it."

Werst shrugged and regretted it. "Or it's to keep you from creating a weaponized signal you could send from your implant to theirs."

Ganes frowned. "That's not . . ."

"Not theoretical. We used a weaponized signal on some gunrunners. Okay, we sent it from the VTA, but same principle."

"Not quite." He shook his head. "And it wouldn't work; they're not using implants."

"Then the block *has* to be there to stop you. Bad luck the fukker's also blocking me."

"All right. Why not. You can't get word out, so we'll assume Martin blocked the anchor." The commander squared his shoulders, confidence rising as the engineer emerged. "The emitter needs to be central, but it doesn't need to be very large. A slate with a decent processor could do it. You need to find out which slate, and disable it."

"I need to? No. Ressk does the tech stuff, not me."

"They think you're Ressk."

Werst sat up, carefully. "I can fake it. I can't actually do it. It's got to be you."

Ganes held up his right hand. "I can't leave the infirmary."

The band around his wrist looked to be about two centimeters wide, half that thick, the surface matte black against the rich brown of Ganes' skin. Could've been decorative. Humans wore those kinds of things.

"It's an adaptation of a precision mining tool. Quite a clever bit of work, really." His fingers curled into a fist. "I input the wrong code, I try to cut it off, I tug on it too vigorously, I go out the infirmary door, and the cutters activate."

"Cutters?" Werst frowned. Put the pieces together and snarled, "You lose your hand?"

"He's the only science type who might be dangerous. Navy, and an officer, but still." Martin leaned on the side of the open door. "Can't let him wander around, he gets into trouble. But he's useful. And he might remember where his loyalties lie."

How much had Martin overheard? They'd kept their voices low until that last reaction. Werst readied himself to jump. No chance of bleeding out, thanks to the commander. He could take Martin down.

"Is he fixed? Hard to tell with his lot." Hands on his weapon, his chest pushed aggressively out, Martin crossed the room and sneered down at Werst. "We have a use for you, Warden Ressk."

"Fuk you." From his angle on the stretcher, Werst could see Ganes removing his identification from the data on the screen.

"That's fuk you, *Sergeant*." He gripped Werst's elbow, digging his fingers into the joint. "You do what I tell you, and the hostages live. You don't, they die."

"Gre ta ejough geyko," Werst growled, hand spasming. If he killed Martin, would Martin's people kill the hostages? He couldn't risk it.

Without breaking eye contact, Martin reached out and closed the fingers of his other hand around Ganes' shoulder, dragging the commander around to face him. "Where's the di'Taykan?"

"In the stasis unit."

Werst was impressed by the way Ganes made a statement of fact sound like, *fuk you.* More impressive, given he'd been a staff officer.

"Why?"

"He's too badly hurt for me to risk an amateur medical intervention."

"I said you could let him die."

"I didn't care to."

"You didn't care to," Martin mocked. "Did you fuk him before you fridged him? He'd have apprecia . . ."

The crack of Werst's teeth coming together rang out like a gun shot.

Martin jumped back, KC in his hands. No one who'd served with the Krai ignored that sound. The butt of the weapon swung around and hit Ganes in the forehead. He dropped.

The *serley chrika* was fast, Werst reluctantly gave him that, the weapon now aimed at Werst's chest before Werst's feet were on the floor.

"Aren't you supposed to be the smart one on Strike Team Alpha?" Martin sneered. "What part about your behavior having a direct effect on the hostages didn't you understand?"

"The part where you . . ."

Commander Ganes grunted as Martin's foot came down on his ankle.

Werst closed his mouth.

• ⸻◆⸻ •

One of the Krai who isn't Yurrisk has left the anchor and is heading into the shuttle.

"Which Krai?" Torin asked. "The engineer or the bosun?"

Firiv'vrak paused for a moment and Torin knew her antennae would be sweeping the air for clues. **Sorry, Gunny. No idea.**

Alamber sighed loudly, clearly intending to be overheard. **Up to me to magnify the image and match the mottling, then.**

"Why does it matter?" Bertecnic wondered, hacking a vine in half.

"Different skill sets, different possibilities," Freenim told him.

And it are important to be getting the names right in the credits.

•—◆—•

"Why would we have a linguist?" Arniz asked, staring at Yurrisk. "We were here to do preliminary studies, to map the plateau."

Yurrisk waved at the plastic sheet, hanging from the ceiling of the common room. "To find the weapon, we need to decipher this language."

"It's an alien technology!" Salitwisi pushed between Yurrisk and Arniz. "It might not be language. Which is what linguists deal in. What we need, is to go back to those ru . . ."

When she saw Yurrisk's lips pull back off his teeth, Arniz poked Salitwisi hard at the base of his tail.

He whirled around to face her, fists clenched. "What was that for?"

She sighed. "Don't provoke the people with the guns."

"I wasn't provoking anyone." He turned back to Yurrisk, and Arniz threw up her hands. "Was I provoking you?"

"Yes."

"Yes?"

"You're very provoking." But he seemed almost amused, so Arniz started breathing again. "Sit down. Be quiet. It's safer when you're quiet."

Salitwisi's tongue tasted the air. "I'm not feeling very safe."

"You're not being very quiet. Sit!"

Arniz had begun to reach out for Salitwisi's tail when he snorted and headed for the designated hostage area. She hadn't been told to leave, so she stayed.

"We need Beyvek here for this. I saw Beyvek by the door." He twisted around, gaze searching the common room. "Where is he now?"

"Martin sent Beyvek to the shuttle." Qurn leaned closer to the plastic, the highlights on her eyes, orange. It was the first time Arniz could remember that the Druin's focus wasn't locked on Yurrisk.

"Sergeant Martin doesn't give orders to my crew." Yurrisk's nostril ridges closed. Arniz stepped back. Suddenly, sitting with Salitwisi seemed like the smarter option. "You!" He pointed at Malinowski. "Where is the sergeant now?"

For a moment, Arniz wasn't certain Malinowski would answer the question. She wasn't familiar with Human body language, but a Niln holding Malinowski's position could only be called insolent.

"He's up with the Warden. In the infirmary."

"Go and get him."

Malinowski shrugged. "He told me to stay put."

Yurrisk's chest rose and fell. His fingers twitched below Qurn's loose grip on his wrist. "Staying put is safer."

"Yeah." Malinowski's pale lip curled. "That's what he said."

Taking another step back—because this was one of those times when staying put was definitely not safer—Salitwisi's muttered complaints wrapping her in a familiar background buzz—Arniz realized that Martin wasn't working for Yurrisk, regardless of what Yurrisk thought.

NINE

"WARDEN KERR, you'll want to see this." Dutavar stepped aside as Torin came forward. He pointed to three pale lines on an exposed arc of root, deep enough that under the layer of feeding insects, sap still seeped from the wood.

Torin's first thought was predator. Her second that Vertic had gouged similar lines in the dirt at every stop. "Polint." Her first thought hadn't been wrong. She moved ahead, noting the broken foliage. "They had a rope around the tree here." Bark had splintered off the far side of the tree just over a meter up. Insects fed on that wound as well. Following the line the rope had taken, she stepped up on a fallen block, over a half wall, and froze.

"Dutavar, hold." She swung her KC around to the front of her body. "I've already lost one team member to gravity. Let's not make it two." The hole was centered in one of the areas with no sizable trees, or she'd have sent Ressk above it. Her helmet scanner read solid ground right up to the edge of the hole although that seemed unlikely. "Merinim, you're lightest; get up here and take a look."

Merinim had already hooked a carabiner to her webbing, and she handed the other end of the rope to Dutavar as she passed. "Only the Ner are lighter," she said to Torin's raised brow. "Not my first crumbling edge."

Torin could feel the others gather behind her as Merinim carefully approached the dark oval in the earth. When the rope stretched taut, she raised a hand and leaned out. "No life signs. I have a ping on the floor at two point four five meters, adjusting for angle. There's a debris pile point six two meters deep. Walls and floor are . . . an artificial substance."

Looks like plastic.

Merinim's shoulders rose and fell. "It could be. I can make the jump, Gunny," she added over Alamber's background sputtering about being doubted. "If you want me to take a closer look."

"Gunny . . ."

Torin glanced up at Ressk. "We can't advance on the anchor until dark. They think he's you, and they need you to examine the artifact. We can't leave the unknown at our backs." His toes clenched around the branch. Torin gave him a moment and, when he didn't reply, said, "Go, Merinim."

The hole *was* lined with plastic. A kind of plastic.

Torin followed Merinim down and ran her hands over as much of the walls and floor as she could reach.

No reaction.

Walls, floor, what remained of the ceiling—all seamless. Both Confederation and Primacy scanners showed solid matter behind the walls. A scan of the ceiling pulled the same result, even though they could see jungle and a small patch of sky through the hole.

"If we're not getting an accurate reading through the walls . . ." Merinim tapped her helmet, and the blue glow across her face disappeared. ". . . there could be anything back there."

True enough, and the odds were high that *anything* would be dangerous. "We won't linger."

"This must be where they found the data sheet." Sweeping a light across the extruded hooks high on one wall, Merinim frowned. "Why would the plastic build with plastic? Wouldn't it be like us building with meat?"

"*If* the plastic is responsible."

"Molecular remains in the latrine, Gunny."

"This . . ." Torin waved a hand around the hole, ". . . makes better odds that those remains are building debris. We don't, we can't know who built this, and it's not our job to know. There are neither hostages nor hostage takers in this hole. That's all that's relevant to the matter at hand." Torin nodded toward the debris pile and the rope. "Let's go."

"Any chance the mercs found the weapon down there?" Binti asked as Torin emerged. "Maybe behind Alamber's data sheet?"

It's not* my *data sheet.

"If they did, it was too small to ping when they passed the DLs at the ruins." Torin drew the rope up, coiled it, and reached for her pack.

"You saying size matters, Gunny?"

"In so many things, Mashona." The rope tucked away, she shrugged into the pack. "Get ready to move out."

"Why would the plastic have left a data sheet down there?" Vertic asked, stepping back as Bertecnic flexed to test his repaired strap, sending the small lizard on his withers scuttling for safety.

"It's blah blah blah plastic aliens, Durlan." Binti shrugged. "They kept an intergalactic war going for centuries as a social experiment. Who knows how they think?"

Freenim held a triangular piece of stone against a broken block, tossing it away when it didn't fit. "If they think enough like us—like *any* of us—then that's a station designed to collect data from above. Abandoned when the people of this planet were destroyed."

"Or abandoned after the plastic destroyed the people of this planet," Torin said.

We've all been thinking it, Craig muttered. **Trust you to say it.**

"Or abandoned when a species not the plastic destroyed or didn't destroy the people of this planet. Just some of the many possibilities we don't need to care about." She glanced up to find Ressk as far from the hole as he could get without explicitly disobeying orders. "All right, people, let's go. We'll follow the mercenaries' path from here on. Ressk, join the Artek at the edge of the plateau."

He'd started moving on *Artek* and was out of sight by *plateau*.

If we are speaking of the Artek and the plateau, I are having identified the Krai in the shuttle as being Lieutenant Beyvek, engineering.

And I've identified Naval search pattern 01022. Lieutenant Beyvek is using the shuttle to scan implant frequencies. I'm reading their wave and shifting our frequencies ahead of it. That's me off breaking the slates, Boss, and all in keeping the implants clear for the duration.

"Good work, both of you. Craig?"

Yeah, I'll keep running the cracker on the slates.

Warden Kerr? Keeleeki'ka sounded like she was having a wonderful time, hiding at the edge of a jungle, watching the outside walls of a distant spaceworthy cube. **Should we meet you at the hole where Werst fell.**

"No. You two keep watch on the plateau. Immediate word if anyone moves."

But don't the DLs . . .

"Eyes on, Keeleeki'ka."

But . . .

Firiv'vrak cut her off. **Eyes on, Gunny.**

"You know," Binti said thoughtfully, falling into line, "if there's two underground rooms, who's to say there isn't more? We could be running over a plastic city."

Torin thought about the tunnel under the ruins. She hadn't scanned it; the smooth surface might have been plastic.

• —◆— •

Rummaging through clothing in the Niln nest room, Werst kept the greater part of his attention on the Human who leaned against the wall by the door. Jana Malinowski, ex-Navy gunner, had the default medium-dark beige species coloring, a right arm with skin visibly younger than the left, short brown hair slicked back, and creases at the corners of both eyes and mouth. Unlike most mercs, she wore civilian clothing rather than a repurposed combat uniform, although she'd kept her boots. No one with two brain cells to rub together gave up their boots. He'd kept his even though he hated wearing them.

The biggest difference between the Malinowski in the Navy's records and the flesh-and-blood Malinowski here and now was the anger that simmered under the surface. Anger held close. Held tight. Anger that was as much a part of her as the scars on her younger, regrown arm.

Scars were a choice. No one had to keep them.

Another casualty of the war who'd been failed by the programs intended to turn soldiers and sailors into civilians. About seven percent of the people they arrested fell into that category. Ninety-three percent were power-tripping assholes laboring under the mistaken belief that an ability to use a weapon made them the center of the universe, but seven percent was too fukking many living broken and lost.

Her eyes narrowed and her lip curled when she realized he was watching her.

Werst winked. It was a Human thing.

Her anger didn't scare him. He worked with Gunnery Sergeant Torin Kerr. Malinowski's anger was a pale imitation.

"You're not suiting up for a Star Cluster ceremony," she snarled. "Find a pair of overalls, cover your ass, and let's go."

"Easier to give me my uniform back."

"Easier to shoot you, but it's not my call. What's wrong with that pair?"

He held them up. "Too short." He tossed them aside and picked up a blue pair.

"Put. Them. On."

Too long beat too short, given that anything beat being crotched twenty-seven/ten. The shoulders were tight and the ass was breezy where tail cuff gaped open.

"Oh, for fuksake." She beat her head against the wall. "No one wants to look at that."

"Then don't." Shrugging the overalls down around his waist, Werst picked up a strip of faux leather—no idea what it was for, didn't want to know—and tied the tail cuff closed. "All right, let's . . ."

Blinked.

He tipped forward into the nest as the room tipped and spun.

"Get up!"

Blinked.

Nest was warm.

"I said, get up, you asshole!"

Was green-gray like the skin behind Ressk's . . . Ressk's . . . Ressk's . . .

Blinked.

They'd put Trembley in Dr. Ganes' room, the only one with a large enough bed. Arniz didn't believe Humans needed specific Human furniture, but that didn't change the fact that the nest in the Niln sleeping quarters wouldn't have worked any better, given his injuries, than the Katrien beds half his size. The bed took up most of the space in Ganes' small, but private quarters. As the expedition's only Human, he hadn't had to share space. Of course, the bed was so small it wouldn't have fit even Ganes and a di'Taykan, had there been a di'Taykan with them, making sharing moot.

And why was she standing in the doorway, listening to her brain babble at her about beds?

No one had ever risked their life for hers before. Particularly no one who was also part of a group keeping her captive and killing children. It had been the latest in a series of unique experiences. She didn't know how to respond, confusion rising to redirect the anger.

Trembley was barely more than a child himself.

"You coming in or what?"

"Possibly what." She moved closer to the bed. "Hard to say." He sounded weak, the bed only just low enough for her to get a good look at him.

Four purple lines cut through the thin hair on his chest, the skin around them green and yellow. Ganes' inclusion had meant Human specifications in the autodoc, but it had never been intended for anything more than basic repairs. The plan had been to stasis serious injuries and message the university for a retrieval. Martin had refused to put Trembley in stasis. Trembley was alive, thanks to Dr. Ganes, but he still had healing to do.

The young Human had half circles a darker purple than the scars under both eyes. She didn't know their function. His lips were so pale, they nearly disappeared into the surrounding skin. Humans had such obvious lips. Well, Humans and Taykan and . . .

"Stop *staring* at me." Bed fabric rustled as he shifted under it.

Arniz blinked.

"Why?"

It took her a moment to realize it had been her voice asking the question. A moment, and Trembley's frown.

"Because it's creepy as *fuk*!"

It took her another moment to realize he'd thought she'd been responding to his demand when the word had instead been a spontaneous bleed-off of pressure. "Why would you risk yourself for me?"

"I didn't." He released a long, shaky breath. "Okay, obviously, I *did*, but I wasn't thinking about you, not *specifically*. I mean other than you being small and helpless. The attack needed to be *stopped*, so I stopped it. Sarge was pissed because you aren't even Human, but . . ." She saw another purple line on the inside of his arm when he lifted his hand to rub at his jaw. "It's a Marine thing. I don't expect *you* to understand."

"Ignoring for the moment that Sergeant Martin should understand a Marine thing, you're no longer a Marine. The military is a profession. You're doing a job." She spread her hands. "I believe you're now what the news calls a mercenary."

"Yeah, well . . ." He plucked at the blanket. ". . . that doesn't mean I'm one of the *bad* guys!"

"Historically, I believe it does."

"No!" A vigorous shake of his head had him paling further and sucking air through his teeth. When he caught his breath, he added, "We're here to protect the weapon for the buyer. That's *all*. Protection."

"Protection?" She hissed. "What were you protecting when you allowed Martin to murder Dzar?"

"Why do you keep coming back to that? I didn't *allow* . . ."

"Did you stop him?"

"How could I stop him?" His eyes were wide. "And besides, you know he did it so *more* of you wouldn't be shot!"

"How effective. After all, Martin didn't shoot Mygar when he murdered her."

"Stop saying murdered! She had a *weapon*!"

"And Martin didn't have the skill to disarm her."

"I *told* you he . . ."

Arniz waited while Trembley licked his lips. Plucked at the blanket. Shook his head. Winced. Licked his lips again. She looked at the glass of water on the chest beside him and continued to wait.

Finally, he sighed. "You make hard decisions in the heat of battle." It sounded as though he were quoting. "Everyone *knows* that."

There'd been no heat; people on both sides of the equation had frozen in place when Mygar picked up the weapon. And no battle, only an ancillary holding a killing tool she had no idea of how to use. But Arniz would allow Trembley's statement in a specific instance. "The decision *you* made in the heat of battle was to save lives. If you believe that's a Marine thing, then there must still be a Marine inside the mercenary." Buried a few strata down under whatever Martin had been shoveling, but it was there. Up on her toes, her tail extended, she held the glass of water to his lips. When he finished drinking, she returned it to the chest and stepped back. "Thank you for saving my life."

Trembley muttered a string of words she couldn't make out and pointedly closed his eyes.

Arniz paused at the door, turned to face the bed again, tongue flicking out. The room tasted of pain. "Do you know the difference between a profession and a job?" she asked softly. "You can walk away from a job. Realize it's not right for you and stop doing it."

She hadn't expected a response. She didn't get one.

The upper level of the anchor maintained a constant temperature of twenty-three degrees, a little cooler than the Niln preferred, a little warmer than the Katrien liked. She doubted anyone had asked Dr. Ganes his preference, but the Younger Races were supposed to be adaptable. Shoving her hands into the pockets of her overall and tucking her tail in close against her body, she headed for the stairs.

"Hey, lizard! What the fuk are you doing up here?" Malinowski stood sideways in the door of the infirmary, angled to keep part of her attention in the room.

Trembley's curiosity meant Arniz had had little contact with the female Humans. They scented the air differently than the males, but otherwise seemed remarkably similar. Malinowski's arms were bare, skin darker than Martin's, a little lighter than Trembley and Zhang. Bare skin was sensible given the heat and humidity outside. The air around her tasted of salt. "I came to thank Trembley for saving my life."

"Sarge clear you?"

"I didn't ask his permission."

"He's in charge, lizard. You ask him if you can blink."

Arniz blinked. Without permission. Because Martin and his people were there to protect the weapon for the buyer. They hadn't been hired by Yurrisk, they'd been assigned to him. Why would Martin bother pretending Yurrisk was in charge? Layers under the surface . . .

"Move, lizard. Get downstairs with the rest of the degenerates." She'd thought she'd controlled her reaction, but Malinowski sneered. "Yeah, degenerates. The so-called Elder Races."

"The H'san were old as a species before Humans discovered fire. Your qualifier is objectively inaccurate."

Malinowski's pale eyes narrowed, and her lip curled up off her teeth. "Can you fly, lizard?"

Strange question. "There are Niln with residual gliding wings, but . . ."

"Get down the stairs before I put my boot in your ass."

Ah. Not so strange a question after all, rather a somewhat lateral threat. As she passed the infirmary, Arniz saw a Krai foot, toes curled in, and part of a leg. She assumed both remained attached to the Krai.

If this was the Warden here to rescue them, they weren't going to be much help.

• —◆— •

"Why Polint?"

Vertic cocked her head as Torin dropped back to run beside her, the underbrush sparser around the ruins, boots ringing against stone as often as soil. "Why Polint what?"

"Why hire Polint mercenaries? To fight a few scientists and one ex-Marine who'd never seen combat? That's like bringing a gun to a knife fight. Overkill given Commander Yurrisk's situation. He's not going to waste resources he could put into his ship."

"We're very fast. Perhaps he intended them to chase down escapees?"

"Through this?" They ducked a branch together, and Torin ignored both the slender loop that fell to brush her shoulder and Binti up ahead muttering, *fukking snakes*. "He's Krai. He can't take the high road, but he has two Krai in his crew."

They cut through a roofless ruin as Vertic thought about it, humming low in her throat. After a moment, she said, "He brought them to fight you."

"To fight the Strike Team sure to be sent should the Wardens discover he's here. Insurance, then." Made sense. If he knew 33X73 was a Class Two Designate, he knew it was under satellite surveillance. Fairly useless surveillance given the distance the Ministry had their collective heads up their asses, but still. Three Polint would change the odds in a fight out in the open. Out on the plateau. They were faster, stronger, and if a way existed for a biped to take them out in close combat, she hadn't discovered it. Yet. After a retreat to the anchor, three extra sharpshooters would make more sense, but young male Polint, eager to prove themselves, probably came cheaper.

"No, Gunny. *You*." Vertic hit the emphasis as though it should be obvious. "You're the logical choice to send after a mixed group of mercenaries. You haven't only fought against us, you've fought with us and, thanks to Presit . . ."

You are being welcome.

". . . everyone knows that. Commander Yurrisk had to have known that should a Strike Team be sent in, it would be Strike Team Alpha."

Torin considered that as they moved away from the ruins and back into the perpetual twilight of the closed canopy. The commander wasn't a mercenary regardless of his current situation. He wanted to keep his ship flying. So he hired Martin. It was in Martin's best interest overall, not merely for this job, to take out the Strike Team with the best arrest record. So Martin brought Polint. Made sense.

Mercenaries were more flexible than governments; he couldn't have expected Strike Team Alpha to bring Polint of their own.

Torin pulled ahead as they reached a partial wall, braced one hand on the top stone, and went over. Her boots sank six centimeters into the soft soil, but the ground held under her weight, the added weight of her pack, and gravity. The ground had broken twice; all bets were off.

Vertic all but floated over the same wall, landing lightly, legs folding enough to absorb the impact so that her belly fur just brushed the ground. "A weapon to destroy the plastic could bring us all together."

"It could," Torin agreed. "Or it could tear us apart. Everyone will want a piece of it. The knowledge that a single weapon exists, capable of destroying the plastic, could drop us right back into war as both sides fight to control it."

"If the underground room is part of a complex, there could be multiple weapons."

"That could be worse."

"That's remarkably pessimistic of you, Gunny."

"Occupational hazard."

"Fuk me, that's deep!"

The team came to a halt behind Bertecnic's sudden exclamation.

Torin put on a burst of speed and stopped by another, lower wall. If the blocks of stone hadn't been heavy enough to fall through, the ground under them should be solid. Safe.

The hole beyond the wall was ragged, new, and Bertecnic stood at the edge of it. "It's solid," he said over his shoulder.

"Good." Torin pointed at the ground beside her. "Get back here."

"I don't . . ."

"Now."

His nostrils flared when he reached her, and he leaned in to inhale her scent.

Not the first time she'd been sniffed in the line of duty. When his hands reached for her waist, she snapped, "Enough."

He jerked back, smoothing the fur below his vest. "Sorry, Gunny."

She nodded, opened her mouth, and raced forward as a line

dropped into the hole from an overhanging branch. She snagged Ressk's pack as he dropped past, hauling him to the edge. "You were ordered to the plateau."

"I saw . . ." He gestured at the hole.

Torin released him, slowly, keeping her body between him and the dangling rope. "Ressk, he's not down there now."

"I know, I . . ." Ressk took a deep breath and let it out slowly, nostril ridges fluttering. "I know."

"We'll discuss this later." His shoulder muscles were rigid under her hand. "After we get Werst back. Depth?"

He sank to his haunches as though his strings had been cut, both hands gripping a fistful of dirt. "I pinged at ten point seven nine meters. Same artificial surface as in the other room. Excluding ceiling height, same dimensions. Less debris."

Less debris because the edges of this hole were holding firm. In spite of crushed foliage four or five meters out, there'd been no constant deterioration.

When she glanced back at Bertecnic, he shrugged. "Told you it was solid, Gunny."

"Do we check it?" Binti asked.

Dutavar sidestepped in the direction of the plateau. "The other one was empty."

"They had injured to remove from this one." Vertic glared Bertecnic still as he started back toward the hole. "I doubt they searched it as thoroughly."

"And the Artek felt technology start up right before Werst went through the ceiling. Mashona." Torin stepped out of the way as Binti came forward and dropped to one knee.

"They wouldn't give two shits about Werst," Ressk growled.

"One of their own went down," Binti reminded him. "And they think Werst's you, and they . . ."

"I know, Mashona! It doesn't help."

She wrapped her hand loosely around his ankle. "Yeah."

"Mashona, what do you see?" Torin didn't trust Binti's eyes more than the scanners. She trusted Binti's analysis of what she saw.

"Debris. Three empty sealant tubes. Blood. Scanner says Krai and Taykan. Nothing on the walls, but something *about* the walls. I can't tell from here if it's a color variation or shadow." She tapped the edge of her scanner. "Minor temperature variation, but—again—that could be the shadow."

"I could check." Ressk surged up onto his feet. "I know he's not down there, Gunny, but he was. We can't move on the anchor until after dark, and I need to do something."

Dark didn't happen for almost two hours. "Go on, then . . ."

He was on the line before she finished.

"Merinim, Freenim, with him. They climb as well as we do," she explained to Binti as the Druin shrugged out of their packs. "And they're a lot lighter. No reason for them to waste energy climbing up or the Polint to waste theirs hauling our asses out."

"You speaking of your own ass, Gunny?"

"I speak for every ass, Mashona."

Ressk had reached bottom before Freenim joined his bonded on the rope.

"You think we'll find a weapon?" Binti asked quietly, as Torin knelt by her side. "Down there?"

"I'm not ruling anything out. I think if we're going to get the rest of the hostages out of the anchor alive, I'll take all the help we can . . ."

Torin, EMP from the shuttle just took out the DLs on the edge of the plateau.

Is that why you had us keep watch, Gunny? Keeleeki'ka sounded more amazed than the situation called for.

"Basic precaution. No chatter on the implants."

Martin did what Torin would've. Just took him longer.

"Chattering, Warden Ryder."

Sharing vital information, Warden Kerr.

• ⸻◆⸻ •

Werst staggered into the wall at the top of the stairs, and leaned there, breathing heavily, nostril ridges half open. He'd have gotten more air had he opened them further, but he wasn't leaving his face that vulnerable.

"Oh, for fuksake." Malinowski slapped the wall above his head. "You've barely gone four meters."

"Go easy," Ganes called from inside the infirmary. "He lost a lot of blood."

"And unless he gets his ass downstairs, he's going to lose more. Sarge wants to see him now."

Werst caught a glimpse of her hand moving toward his shoulder out of the corner of his eye. When he bared his teeth, she snatched it back. Feigning dizziness that was uncomfortably close to the truth, he descended as slowly as possible, balancing on the thin branch between convincing the mercenaries he was too weak to worry about, and frustrating Malinowski to the point where she used a boot to speed him up.

The big common room was crowded. No surprise. Four days in Susumi had made it clear that three Polint could be a crowd on their own. Actually, where the fuk had Martin even found Polint? Last thing Justice needed was a Confederation/Primacy mercenary exchange program.

Not his problem. Not right now.

The overhead lights were bright, offering no advantage to Krai eyes meant for flickering shadow. Given the number of people, the room was surprisingly quiet.

The panels that sealed the six tall windows in the long outside wall while the anchor was in space had been replaced, but then, it was the first thing anyone with half a brain would do. Even the idiot gunrunners had managed it, although they hadn't done it well. Nor had they replaced the panels in any of the second-floor windows. Werst knew the infirmary window in this anchor was also unsealed, so it seemed Martin wasn't significantly smarter. To be fair, were Werst trapped in an anchor with a Strike Team approaching, he'd leave the panels off the second-floor windows, too. With the Strike Team's VTA also in play, the windows were a safer place to station shooters than the roof.

Plus, the second-floor panels were a pain in the ass to replace.

Sealed, the common room smelled like seven species had been locked in for longer than the mechanical air exchange could handle. Actually, fuk that. It smelled like Polint. Less pungent a smell than Dornagain at least. The black—Camaderiz—and the one who had the same reddish brown coloring as Bertecnic—Netro . . . Netrvoo . . . something. Fuk it. Netro. Netro and Camaderiz had folded their legs over by the long wall and were talking quietly. Tehaven, Dutavar's brindled brother, sat a little apart staring at his slate, lips moving. All three wore black vests and a variable number of knives. Three RKah leaned against the wall beside them.

Two thirds of the tables and benches had been pushed toward the long inside wall. Given the occupants of said benches were all Niln and Katrien—the first constantly tasting the air, the second visibly shedding—he'd found the hostages. No help there although no danger either. Big piece of the room he could ignore.

A table tucked in the outside corner held a Krai—Sareer—and two di'Taykan—deep blue would be Mirish, the pink Pyrus. Commander Yurrisk's crew. Sareer glanced over, her nostrils flared. If the commander's crew knew Strike Team Alpha the way Martin's seemed to, his cover was blown. Other species might have trouble telling Krai apart, but Krai sure as *chreen* didn't. She frowned and turned her attention back to her sah. Good.

Fuk, he could use a sah.

Commander Yurrisk, the Druin, and an elderly Niln stood by the large orange sheet of plastic-looking something. The commander's fingers and toes stretched and curled constantly. He swayed to the right, then to the right again. Anyone with eyes could see he was out on branches that couldn't hold his weight, one small step from a fall. The Druin beside him was an unanswered question. She didn't move like Merinim, so probably hadn't been a soldier, at least not recently. She didn't move like she was helpless either. She made him think of vines; flexible, mistakenly perceived as fragile. Vines could bring a tree down. The Niln was *Harveer* Arniz, soil scientist, one of the hostages. She appeared to be tired but unhurt, not at all cowed, and fascinated by whatever the *chreen* they were staring at.

Werst hadn't seen that much plastic in one place for years. Even inorganics, blameless as far as starting wars went, had been replaced. Could be academics didn't care. Could be they didn't know. He had the impression they lived isolated from reality, even more than most civilians. He didn't recognize the raised symbols on the sheet, but he'd gotten most of his education in the Corps, so subjects that didn't contribute to surviving the war tended to be glossed over—although instructors usually substituted the word *win* for *survive*.

They'd managed to hang the sheet from the ceiling about a meter out from the inside end wall. Most anchors Werst had seen used that wall as a pass through to the galley, but the panels remained up. One less exit to guard. One less exit he could use.

Martin stood back a meter, a meter and a half, from the

commander, Brenda Zhang beside him sucking on a pouch of water. He had a vague memory of Zhang declaring she had salvage rights to his weapons and that looked like his knife in her boot sheath. Good. When the fight started, if she tried to use it, the grip and balance would throw her off. Amateur. Martin was as big as he remembered. Maybe heavier, but that was a Human aging thing. He stood chest out, chin up, dominating the room with his height and weight. Still throwing it around, then. His face was red, the skin stretched tight, because they were near the equator and Robert Martin was too tough to need protection from UV. Idiot. How the fuk had he made sergeant? Although everyone but the hostages were armed, Martin was the only one with his hands on his weapon. Too many eyes on him for him to have his hands on his junk. Asshole.

Unlike Malinowski, both Martin and Zhang wore modified combats. On the one hand, Werst knew how to exploit the weaknesses of the uniform, but, on the other, civvies were easier to get teeth through.

Bad luck it'd been Trembley, the youngest and least trained of the Humans, who'd already been taken out of the fight.

The distance Martin kept from the commander, the Druin, and the Niln . . . who walked into a bar. He snickered.

"What's so funny?" Malinowski demanded.

"Your face."

"Fuk you."

The distance Martin kept looked . . . looked . . . Oh fuk, he'd lost the word.

The room tilted.

"Fall down," Malinowski muttered, "and I'll drag you over to him."

Toes flexing against the floor, Werst clenched his teeth against the return of the second protein shake he'd drunk before leaving the infirmary. Krai didn't vomit. Krai didn't waste food. His stupid body would not betray him, not now. A deep breath, cautiously drawn through nostril ridges mostly closed, and he straightened, took another step.

The distance Martin kept looked *supervisory*.

He was two meters out when Martin's gaze switched to him and his lip curled.

Werst maintained a steady *fuk you* expression.

"If you've finished your threat assessment, Warden Ressk, make yourself useful."

Hating the sound of his bonded's name in that *serley chrika*'s mouth, Werst showed teeth. "You might want to sound less *revenik* if you want my help." Fukhead.

"You might want to remember, I need your brain, not your fingers and toes."

"Yeah, because torture leads to clear thinking." Moron.

"Okay, I *don't* need . . ." Martin nodded at the huddle of scientists. ". . . all of them. So start protecting the innocent, Warden, or I'll thin the herd a little more."

Harveer Arniz hissed. Good for her. "What do you want me to do?"

Martin nodded toward the plastic. "What is it?"

"Orange. Looks like plastic."

"Listen tree fukker, I don't . . ."

"It's a data sheet. Or at least it looks like one." *Harveer* Arniz took Werst's arm and pulled him closer. "We have to start somewhere, so we're proceeding under the premise that it is what it looks like. I found it hanging in an underground room, and the decision was made to bring it with us . . ."

Werst thought he heard one of the hostages mutter, "Stupid decision."

". . . when we returned to the anchor after learning the Wardens had landed."

"This piece of ancient technology will lead us to the weapon," Commander Yurrisk announced.

"If we can learn to use it," the Druin said calmly. In Federate.

As Werst tensed, the commander visibly relaxed. "Which is why you're here, Warden Ressk. None of the scientists . . ."

"Hostages."

The commander tore his gaze away from the data sheet and turned to look at him with less crazy in his eyes than Werst had expected although the edges of his nostril ridges were quivering. "What?"

Reminding himself he was supposed to be Ressk, Werst kept both tone and expression respectful. Always a chance that if he reminded the commander of what he'd gotten himself into, he could reach the decorated Naval officer buried under the trauma. Failing that, respecting the enemy was often seen as weak. He could use that to his advantage. "As a Warden, I don't see them as scientists. I see them as hostages."

"You do?"

"Yes."

Commander Yurrisk sighed. "You can see them as a *vertak abquin* for all I care. They can't figure this thing out, but your reputation has convinced Sergeant Martin you can. He seems to have made a study of Strike Team Alpha. Can you?"

"Probably." Ressk's answer. Ressk's arrogance about this sort of thing. Ressk chubbing up in the presence of new tech. Although, odds were high that with his more unique education, Alamber would have better luck with this kind of thing.

"Get to it, then."

The data sheet was approximately three meters by two and a half. Werst had no idea why they assumed it was a data sheet. It was alien. It could be anything. Hell, it was plastic. It could be a fukking family reunion. They were called sentient, polynumerous molecular polyhydroxide alcoholydes on the official documents and there were lots of molecules on a sheet that size. Enough for speech if they decided they had something to say. Needed Gunny or Ryder in here to kick their shape-shifting plastic asses into a conversation, though.

The raised symbols were the same shade of orange as the background, visible only because they threw a thin, translucent shadow. As Werst watched, a symbol separated from a clump in the upper right corner and shifted down to the bottom left. Okay, sure, it looked a lot like a data sheet, but even he knew function didn't necessarily follow form. Weapons in the Methane Alliance looked like pudding.

"Well?" Commander Yurrisk demanded. "Well?"

Werst could feel the weight of the Druin's stare on the side of his face. Why did she speak Federate? He raised a hand in the universal symbol for *give me a fukking minute here* and walked around to examine the far side. Ressk wouldn't answer without all available information.

The back was a big orange sheet of plastic. Looked a bit like a tarp. None of the symbols showed through and he was all alone. He'd have made a run for it had there been anywhere to go, but there was SFA in the way of escape routes. He walked slowly, reminding himself of where the Polint's major blood vessels rose close to the surface, and emerged on the opposite side with every eye in the place locked on his position.

His lips curled. He intended to snarl. He wobbled instead and barely stopped himself from grabbing the plastic to keep himself upright.

Her gloved hand on the commander's arm, the Druin's inner eyelids flicked slowly across the solid black. If the tell held true across the species—or at least across his data set of three—she was considering new information. He was the only new information in the room.

Aiming for distraction, he turned and thrust his face toward the plastic, inhaling loudly. "Petroleum," he said, pulling back.

"Well done. That's what I told them." *Harveer* Arniz sounded snippy enough, he doubted they'd listened. "While your nose isn't exactly peer reviewed, it's good to have confirmation. No one else bothered to check," she added when he glanced her way.

The anticipation in the room had grown to an almost physical presence, waiting for his words of wisdom.

What would Ressk say? "I need to taste it."

"No." Commander Yurrisk left no room for argument.

Werst argued anyway. "Took a bite out of the plastic aliens on . . ." Fuk. Ressk hadn't been on Big Yellow. ". . . the prison planet. I know what they taste like. Fastest way to see if this is them. And we might get a reaction. Get enough of a reaction, and you can ask them where the weapon is."

"Too dangerous," Martin said before the commander could respond. "They're not going to want anyone to have the weapon. It won't be safe."

"We have to stay safe. The weapon will keep us flying." The commander brushed a hand over his scalp. "I need the weapon. We won't risk it."

So much for the possibility of a polynumerous, molecular, polyhydroxide alcoholyde distraction. "Fine. Then I need my slate."

"Your slate got smashed when you fell in the pit," Zhang called out. "Pretty sure it got busted up when it impacted with your ass."

That explained the pull/pain/pull of a Polint-sized bruise while going down the stairs.

•—♦—•

"Werst's slate . . ."

From the edge of the hole, Torin watched Ressk pick up the pieces of the slate and cup them in both hands. Nostrils flared, he brought his hands up to his face, and inhaled. The two Druin at the bottom of the pit with him examined the walls, giving him his space.

After a moment, Freenim shook his head. "I can't see the color discrepancy, Mashona."

"Half a meter to your right." Torin glanced over in time to see Binti cock her head and say, "No, wait, it's gone." Her head returned to the original angle. "And it's back. It's subtle."

"In your language that's another word for invisible, right?" Freenim slid sideways, eyes on the wall. "Here?"

"Another step. Yeah, there. You're at the longer edge. Parallel side's about half a meter over, stops about half a meter up and moves left toward the floor at close to a forty-five. Upper edge is about your shoulder height."

As Merinim moved to the parallel, Torin copied Binti's head motion. Still nothing. Not with eyes alone, not augmented by the scanner. "Any chance the angled edge could be shadow. Or lack of shadow?"

"Could," Binti allowed.

"Search for a connection between the underground rooms," Vertic said, all three Polint a safe distance away from the edge. "Stealth surveillance doesn't usually include moving from post to post in the open."

"Stealth surveillance?"

"Can you think of another reason a species with superior technology would hide under a city of primitives?"

Torin couldn't think of another reason why she'd do it, but unknown aliens were another matter. "There was a tunnel under the ruins by the VTA."

Just like that, then? You didn't think about mentioning it earlier?

"Thought it was a skinny cellar and I got distracted by getting the hell out of there."

We'll talk. I'm running subsurface scans around the VTA.

"The plastic can look like anything, Durlan." Torin heard Bertecnic pacing as he spoke, foliage crushed under foot. "Why hide? Those big beetles look pretty plastic anyway."

"We don't know these things were created by the plastic." A hand on Binti's shoulder and a nod directing her gaze back into the pit kept her from turning with Torin. "There's a lot of plastic in the universe. Most of it isn't sentient."

"The data sheet could be millennia old and yet it didn't shatter when rolled." Vertic's voice skirted *and I'm an officer* for the first time in days. "Precedent suggests that when a large and unexpected amount of plastic is found, the plastic is involved."

"Then why dig underground rooms?" Dutavar asked. "Why hide? The plastic could move about unseen."

Binti shifted, the motion enough to spill a fan of humus off the edge. "Why turn into a big yellow spaceship when their default color is gray?"

"To catch our attention." Torin stood and stepped carefully back from the edge to the line gouged in the dirt marking where the ground had scanned as solid. The line had been there when they'd arrived. "I can't believe I'm asking this: is anyone carrying plastic?"

"I am, Warden." When she turned, Dutavar held out a small open container. In it, a three-centimeter white cube. "It's organic," he said. "Pre-discovery of the plastic. Carried by all active military personnel."

Torin stared at it for a long moment. "Bait."

He nodded. "Bait."

Odds were the military bias against volunteering information had kept him from informing her of what he carried. Or, he could have thought it was none of her business. They had, until recently, been enemies, and had known each other less than a full tenday.

The piece was much smaller than the minimum density required for the plastic to achieve interactive sentience, but probably large enough to attract other pieces. The Primacy military was being proactive. Good for them. Or not. Torin honestly couldn't decide.

She could feel the anticipation as she reached for the cube. Even the ever-present background drone of insects seemed to have quieted.

Torin?

Craig could make her name sound like a hundred different questions. She appreciated the efficiency. "Nothing. No reaction." It felt a bit like foam. A bit like firm tofu.

"So you've come to believe these underground rooms were built by the plastic?" Vertic asked as Torin tucked the container into a pocket.

"I believe in not leaving large unanswered questions blocking my retreat. Particularly, if I'm about to be shot at. If the plastic reacts to me, we'll know."

"And if it doesn't?" Bertecnic asked.

Torin glanced over at him. "Hive mind, remember. And it's been in my head."

Binti grinned. "So you're saying that if it's smart enough to maintain an interstellar war for centuries, it's smart enough not to piss you off?"

"That's what precedent suggests."

Dutavar's vertical pupils were thin slits in gray eyes. "Warden, if there's a reaction, my government wants to question the plastic."

"It's a lot less satisfying than you'd think."

And I are having the right to the first interview.

"Only on our side of the line." She removed her pack and reached for the rope. "And if your government has come up with a way to restrain the plastic, I'd like to hear about it. They'll break into molecular components and can't be held." The assumption was that very high or very low temperatures would destroy them. The military was working on an electrostatic charge to hold them in place, but as none of the plastic they'd used in testing had risen up and complained, it all remained unsubstantiated theory. "Dutavar, secure and drop another line. If this goes wrong, we'll need to get everyone up at once."

Dutavar nodded. "On it, Warden."

Two meters from the bottom of the pit, she swung out far enough to miss the puddle of blood and dropped. The floor absorbed most of the impact. Seemed it wasn't only Krai bones that had kept Werst alive.

Freenim caught her eye, then nodded at his bonded. The Druin readied their weapons.

"Ressk?"

"I'm good, Gunny." He tucked the broken slate out of sight and stood. "Where do you want me?"

"By the lines. In case one of us has to get out fast."

Torin, for fuk's sake, be careful.

"Always."

I don't think that means what you think it means.

Shifting her weight onto the balls of her feet—her body refused to believe the plastic couldn't be fought—Torin held the white cube between the thumb and forefinger of her nondominant hand . . .

"You're lined up with the color change, Gunny." Binti's voice drifted down from above. "Dead ahead."

. . . and pressed it against the wall. White to gray.

Nothing happened.

Webbing creaked as one of the Druin shifted position.

Dirt pattered against the floor.

Binti called a sitrep back to the Polint.

Werst remained an injured captive.

By some freak chance a soil scientist might or might not have accidentally discovered a weapon to use against the plastic.

Still nothing.

"Gunny?"

"Wait." She pressed harder. Felt the cube give. No, not the cube . . .

"Good thing you're fast." Ressk caught her left hand and hit the tip of her finger with a shot of sealant before she could add more than a couple of drops to the puddle of blood on the floor.

Torin?

Ressk checked her thumb and released her. "Let it go, Ryder. It just clipped her. It's not bad enough for you to kiss better."

"Is that a door, Gunny?"

"That's a door, Mashona." Torin flexed her finger, the analgesic in the sealant already numbing the pain. "And we've got a passage heading toward the anchor."

• ——◆—— •

"Disable the communications capability on one of the confiscated slates," Commander Yurrisk interrupted Martin's refusal, his tone pure officer shutting down a lower rank. "He can use that."

Werst kept his lips over his teeth and hid his relief that the bullshit he'd been spouting had had the required result.

"He can wave his dick at it for all I care at this point," Martin muttered. "Malinowski, upstairs. Eyes on the plateau. Zhang, get Warden Ressk a slate."

"Which slate, Sarge?"

"Not Ganes'. Otherwise, I don't care."

The Druin in red, Werst still hadn't heard her name, handed him a pouch of water. "You need to replace fluids, Warden Ressk. You must be parched after that . . . speech."

She sounded amused.

That was . . . probably not good.

•—◆—•

"I've got a ping at 500 meters, Gunny."

"Not far enough."

"Could end at another door," Freenim pointed out. "Could end at a junction. There's only one way to know for certain."

Torin stared down the passageway, mentally flipping through options. If it brought them closer to the anchor unseen, the time spent on exploration would be worth it.

The size of the passageway was a problem. If it came down to life or death, she and Binti could squeeze through, but it was low enough and narrow enough movement wouldn't be comfortable, easy, or quick. Not a hope in hell of the Polint fitting, but if they were as fast as Vertic implied, they'd be better as a distraction out on the plateau.

Waste of resources stuffing a Krai *under* a jungle.

As far as exploration was concerned, that left the Druin.

"Would you be this hesitant about sending us after a potential weapon had we not so recently been enemies?" Freenim asked quietly beside her.

"*Been* enemies, not are enemies."

"You've put the years of war behind you so easily?"

Easily? No. However . . . "I allow you behind me with a loaded weapon. I'm more than capable of sending you down a passage created by the plastic a millennia ago."

He nodded, understanding her definition of trust. "But if we're the first to see what's down there, you're aware we may gain an advantage we could take back to our government."

Did he want the politics of her decision on record? Presit was always listening and when Presit was listening, Dalan was recording. "I'm *aware* we can't trust the plastic," she said. "The tunnel registers as inert and, in the end, that means absolutely nothing. It could get us closer to the anchor, but my experience says it will definitely fuk with us." She remembered the trip through Big Yellow, Crucible, the prison—the bodies she could lay directly on the plastic. In Big Bill's favor, that whole shitstorm had nothing to do with sentient, polynumerous molecular polyhydroxide alcoholydes. Except that the war the plastic had engineered had created Captain Cho and his crew. She thought about the war and about her dead. About the Primacy and their dead. "Worst case scenario, we can't break into the anchor, we come back here, and you go in. Best case, we let science deal with it after we're gone."

His inner eyelids crossed the glossy black in a slow blink. "And if the tunnel had been large enough for you?"

"Same answer."

"And if the weapon to destroy the plastic is down that tunnel?"

"We're not here to fight the plastic," Torin reminded him. "We're here to rescue the surviving hostages."

Freenim met her gaze. "The plastic was responsible for the deaths of millions on both sides of the war. If we have a chance to take their war to them . . ."

"The dead are dead." Torin fought to keep her hand from rising to her vest. "Here and now, I—and through me, you—am responsible for saving the lives of those hostages." Turning away from the tunnel, she raised her voice. "We're done, people. Let's move!"

Her implant indicated a private channel. **You're walking away?**

Craig. He sounded so neutral she wondered if he disagreed with her decision. Although, on a private channel, he had no reason for diplomacy. She watched Ressk reach the top of the pit. Beckoned Freenim and Merinim forward. Freenim hesitated, then nodded and obeyed. "You heard my reasons."

Still, big decision. Surprised you didn't run it by Vertic first.

"It's my decision, not hers." The two Druin went up attached to the same line, their weight incidental to the Polint pulling on the other end. Torin ran the other line through her belay loop, gave the signal, and walked the wall as she rose.

No shade on your decision, Torin, but there's an officer in the field.

"You've never cared about that."

You do.

"I did. Now, it's my field."

Later, when they'd wrapped up, they'd have to talk about the relief in his huff of air. **I love you.**

"Never doubted it."

Oh, come on, Boss! You expect me to clear two frequencies simultaneously?

Correction, semiprivate channel. "Yes, I do, because you're just that good."

Yeah, whatever. Don't do it again.

Vertic waited at the top of the pit. "So we're walking away from a potential weapon," she said as the ropes were rolled and stowed.

It was a statement, not a question. Torin settled her pack on her shoulders. "We're doing our job."

"If the plastic interferes . . ." When Torin turned to face her, Vertic shrugged. "We have clear evidence they were here. We have no clear evidence they all left."

"If the plastic interferes," Torin repeated, "we melt them into slag if we have to burn the jungle down to do it."

"Not so adverse to vengeance, then."

Torin concentrated on opening her fists, releasing the crushed webbing. "Not at all adverse to removing whatever keeps me from doing my job."

•—♦—•

"And again, nothing on this one I can use." Werst tossed the slate down with the other two. "I need programs that can run both a mathematical and a spectral analysis . . ." What the fuk, it sounded good. ". . . not games and music and letters home." He channeled Ressk's pissiest expression, the one he used when every new addition to R&D tried to treat him like a stupid grunt. "And I need a scan that analyzes tech signatures instead of extrapolating buildings from two rocks and a post hole." Credit where due, Nerpenialzic—whichever of the huddled Niln ancillaries he might be—kept clear directories.

"That's the third slate you've rejected," Martin growled.

"Yeah, I can count." Turning to face the big Human—having that much angry mass at his back made his fingers twitch toward an absent weapon—he saw movement from the corner of his eye, and paused. The top line of eight symbols on the data sheet—and while other patterns were random, that was definitely a line—changed, one after the other. A ninth symbol appeared at the end of the line, disappeared, and reappeared down in the lower right quadrant, symbols there arranging themselves around it, one disappearing to make room.

"What did it do just now?" Every word out of Martin's mouth sounded like a criticism. Asshole.

Werst showed teeth. "You were looking right at. Eyes not working?"

"If the next words out of your mouth aren't an answer, I'm killing a lizard."

And he would. He had. So far, the Katrien had been safe; Werst knew most Humans had a soft spot for small, fur-bearing mammals even when those small fur-bearing mammals were as irritating as fuk and reported certain statements totally out of context. Martin seemed true to species, but eventually he'd run out of Niln.

"It looks like it figured something out. The figuring." He pointed to the upper line. "The solution." He pointed to the new symbol. "I can't begin to understand what it means because I don't have my slate. Or a slate that's worth more than shits and giggles. You got that, right?"

The random Katrien who'd been assigned to record—older, female, high odds she was Dr. Tyven a Tur durGanthan, one of the geophysicists—jerked when she realized she'd been addressed. "I got that. Yes."

"Good. That's very good." He mimicked the tone Gunny used to reassure civilians caught up in unexpected violence. Not the tone she'd used on other Wardens and nothing like the tone she used on the Katrien she spoke to most frequently. "Don't worry about specifics. Keep the whole thing in focus, we can break it down later."

"I *know*."

No shit; it wasn't complicated. Unfortunately, he needed to kill more time and those useless instructions had taken care of . . . not a lot. Gunny couldn't move the team across the plateau until after dark and the days on Threxie were too *serley* long.

"You know what I need?" He kept himself from folding his arms as he met Martin's scowl. Folded arms would slow his reaction time in a fight. "I need Lieutenant Commander Ganes' slate. He's an engineer, so he might have something useful, and he's military, so he'll be using subroutines that actually make sense."

"You're not getting Ganes' slate."

"Why? If you disable the communications, it's the . . ."

"Sarge smashed it when Ganes tried to get a message out," Zhang said around a mouthful of half-chewed fruit. "Couple nights ago, he broke into where we had them locked up. Actually, had it in his hand when I caught him. Go me, right?"

"By all means, go you." Werst sighed. "I'm getting nowhere using the lame-ass abilities of civilian slates, so disable the communications on another military slate and let me work with an OS that makes sense."

Martin stayed quiet long enough Werst thought it might have worked, that he might have been able to add *military vs civilians* into his *us vs them* mindset. Which would put Werst firmly on the *us* side. Although . . . He glanced past Martin, past Zhang, to where Commander Yurrisk, his crew, and their random Druin sat together in the corner. Separate. Humans. Non-Humans. Ganes carefully isolated from his fellow scientists. If Martin hadn't added Polint mercenaries to his crew . . .

The data sheet shivered, a fine ripple rolling from top to bottom.

"Fuk me." Werst stepped back, trying to focus on the entire thing as all the symbols began to move. To regroup. It almost looked as if the movement of one symbol influenced the surrounding symbols, both in the place it left and the position it moved to. It reminded him of . . .

Of . . .

It wasn't the symbols, it was the movement . . .

"What?" Martin demanded.

And recognition slipped away.

Asshole.

Werst continued to stare at it for as long as he thought he could get away with, but the symbols had stopped moving, the pattern he'd almost recognized gone.

"Sarge asked you a question, tree fu . . ."

Yurrisk cleared his throat.

Zhang snapped her teeth closed on the next word.

Did she know Humans had picked up the teeth snap from the Krai? Werst doubted it. "I need a military slate to analyze that pattern."

"No."

"Why not?"

Martin scowled when Werst turned toward him. "Because I said so."

"If it helps us find the weapon . . ." Yurrisk began, rising to his feet.

"Warden Ressk can turn a military slate into an incendiary device,"

Martin said, cutting the commander off. "We can't risk him destroying the data sheet. It isn't safe."

Werst gave himself a mental kick. Marines were taught to how to overload the power source of their slates—although they were also taught it was a trick best kept in reserve until losing fingers seemed like a fair trade. He'd never known anyone who'd done it, but if it also worked with civilian slates . . .

The three rejected slates had been tossed onto a stool too far away for him to snatch one up without being seen. He needed Martin to keep his attention on the commander for one more . . .

Nope.

"Fine." With any luck, Martin hadn't seen him move. He waved his arms around anyway. "Why should you trust me? Even though I'm standing here because I'm the only one dirtside with a hope of getting anything like information out of this. Have you considered your second-best hope is Lieutenant Commander Ganes? He should have a look."

"No."

"He's wasted in the infirmary."

Martin snorted. "He's a doctor."

"He's not that kind of doctor."

"You're alive because of Ganes."

"Proves my point. He can think outside the box. This . . ." Werst cocked a thumb at the plastic. "This was never in a box."

Martin's brows drew in. "What are you talking about?"

"Two brains are better than one and all that. Bring the commander . . ."

"I said, no."

"Because you're an idiot!" Although he'd deliberately provoked Martin's reaction, Werst barely resisted the urge to intercept the blow, climb Martin's arm, and tear his throat out. As the big Human's knuckles made contact with his cheek, he threw himself back onto the rejected slates, and dropped flailing in the midst of the mess, sending pieces of stool and slates in all directions. Using his body to block Martin's view, he slid one of the slates into the deep pocket on the right leg of his overalls and rolled under the lower edge of the plastic, seemingly propelled by the toe of Martin's boot that had barely touched his ribs.

"Are you trying to destroy the data sheet?" he rasped as he stood. It took a lot longer than it usually did. Gravity was being a *serley* fuk. Once on his feet, toes spread against the movement of the planet, he clamped a hand over the healed patch on his throat. Still holding. Good. "Do you know," he panted, "what would happen if you slammed me into it?"

"No," Martin snarled. "Do you?"

"No, and that's the point." The flare of pain from the bruise on his left shoulder blade made his arm feel weak. His fingers spasmed, so he curled them into a fist. "We could lose everything on it." Still clutching his throat with his right hand, he stumbled around the edge of the plastic farthest from the debris. "Am I bleeding?"

"Nah." Zhang leaned in. "You're good."

Harveer Arniz had already begun to clean up the mess, tossing slates and stool pieces into one jumbled pile. Martin hadn't noticed the missing slate in his enjoyment of the elderly Niln crawling around the floor at his feet. Asshole.

"I can't do this . . ." Breathe. ". . . without the proper equipment." Releasing his throat, Werst waved his right arm, drawing Martin's attention. The flush rising on Martin's face suggested he didn't know how to react to violence ignored. Even money said his usual response was more violence. Unfortunately for his stunted interactive abilities, he needed to keep "Ressk" not only alive, but functional. Sucked to be him. If he could keep Martin off balance until the Strike Team arrived, it would make everyone's job easier. "Failing that, I need to taste it." It hadn't tasted like anything the last time. "At least we'd know if this was them, *the plastic,* not some random debris left behind by other aliens entirely."

Martin frowned. "What other aliens?"

"Exactly."

One of the older male Niln jumped to his feet. Tail straight out behind him, he snarled, "I have had enough! You . . ." A slender gray finger stabbed at Werst. ". . . are no more likely to understand this artifact than a *firnacin* would."

"Yeah?" Bracing himself, Werst spread both hands in the classic *bring it* position. A fight would waste some time for sure and, the way he felt, an elderly professor might stand a chance. "I've had a lot more contact with the plastic than you have, buddy."

A number of vowels got lost in the hissing, but the Niln—*Harveer* Salitwisi now Werst had gotten a better look at him—managed an impressive volume and a string of complicated insults Werst barely understood. He yelled back every derogatory thing he'd ever heard Ressk say about the nonmathematical sciences; not loudly, his throat wouldn't allow it, but with as much enthusiasm as he could muster. Three of the ancillaries were trying to get Salitwisi back into his seat, but he kept struggling free and finding more things to complain about. By the time Werst risked losing his voice completely, he'd begun tearing into the entire situation, hands and tail waving counterpoint.

Harveer Arniz waved a piece of stool and yelled, "You tell him, Salit!"

Zhang was laughing.

The red on Martin's cheeks had darkened to purple.

He pulled his KC around into position and stepped forward.

The three Polint stepped forward with him.

Zhang stopped laughing.

Fuk. That was a little more off balance than Werst had intended. He shoved the Katrien—still recording—toward her people. *Harveer* Arniz caught her at the last moment before impact, but the Niln's attention remained on the far corner . . .

. . . where the Druin had her fingers spread over Commander Yurrisk's chest and her mouth to his ear.

The commander shoved her away and jumped to his feet. "Sergeant Martin!" His voice filleted the layers of sound. His nostril ridges were closed, his teeth exposed, and he was terrifying in the way of those beyond consequences. His eyes were dead. His teeth were . . . cracked. With his lips back that far, Werst could see the shadow where a triangular piece had broken off. Krai teeth didn't break. Werst could see the blood and hear the screaming. What the fuk had happened on the *Paylent*?

The noise fell off as Martin turned. Werst suspected *Harveer* Arniz had slapped her hand over her colleague's mouth, but he didn't dare turn to check. Didn't dare take his eyes off the commander. Had to be ready for anything because anything could happen. The commander had gone to the place in his head where the war continued. Where everything was a dark and terrible absolute. Sliding, sliding, sliding . . .

Werst had seen a Marine break during a battle. Took twenty-seven rounds to stop his charge and no one looked close enough to be sure they'd all come from the enemy.

"You will not discharge your weapon inside, Sergeant. Not among my people. It isn't safe." The dead gaze flicked to the hostages. "You will sit and be quiet and stay safe." One more flick to Werst. "Do you think this is funny?"

"No, sir."

"Find what that thing says about my weapon. Now."

The threat, the blood and screaming, the flaying, the devouring—it was all in the delivery.

"Yes, sir."

He frowned at Werst's response. "I don't . . . I don't know you! How did you get in here?" He dropped his head between his shoulders, took a step forward, and stopped, a red-gloved hand on his arm.

"He's the Warden who fell in the hole. He was injured with Gayun." The Druin stood at the commander's side, her voice calm, matter of fact. Where had she come from? Werst hadn't seen her move. "He's Krai."

"I can see that!" More a protest than a confirmation.

"There are no Krai among the enemy."

"Of course . . . no Krai . . ."

"He's here to solve the data sheet."

"When we sell the weapon, we can keep flying."

"Yes."

He met Werst's gaze. The *Paylent* had returned to haunting the back of the commander's eyes. "Find my weapon, Warden."

"Yes, sir."

The commander straightened. Tugged at the hem of his tunic. Swayed right, then right again. Swallowed. His nostril ridges slowly opened. "Sergeant, stand down. Remember, you won't be paid until my weapon is found."

Zhang had put her back against the wall. A muscle jumped in Martin's jaw, but he moved his hands away from his weapon, spreading his arms with exaggerated emphasis.

Commander Yurrisk was like keeping a bomb around. Most of the time, it was a metal cylinder taking up space, no big deal. Every now and then, it began to count down and then all anyone could do in its immediate area was hope it didn't explode.

Werst had to admit that in Martin's place, on the planet with the weapon, with the ship in orbit, and Zhang available to fly her, he'd have shot the commander just to keep from waking up and finding himself ankle-deep in blood.

Why hadn't Martin?

•—◆—•

Torin could see the jungle thinning up ahead, actual open space between the trees, the perpetual dusk under the heavy parts of the canopy broken by broad bands of sunlight. "That's far enough." She pitched her voice to Dutavar at the front of the line. "If they can see us, they can shoot us. We'll base here. Ressk, you and Bertecnic set out perimeter pins. High and low. They've Krai on their side; we can't assume the canopy is safe."

Bertecnic circled back around, but Dutavar stayed where he was, arms folded, facing the plateau. When Freenim moved toward him, Torin shook her head. Freenim met her gaze for a moment, then tipped his helmet back and rubbed at the red dimple in his forehead. "We've got time for food before we lose the light."

Torin nodded in turn. "I could eat."

As she joined Dutavar, she heard Freenim giving Merinim and Mashona their orders. He asked for Vertic's assistance. Good chance he'd be having a talk with his bonded later about the difference.

Dutavar stared out at the anchor.

"If it's possible," she said, "your brother will be your responsibility. If it isn't . . ." And this was the corollary she hadn't mentioned before. ". . . I need to know you'll follow orders."

He cocked his head and turned to meet her gaze, vertical pupils narrowed, gray eyes speculative. "You've left the question of my loyalty a little late, haven't you, Warden?"

"I've left you enough time to make up your mind. You didn't have enough information back on the *Promise*; if I'd asked, I wouldn't have believed your answer." She hadn't had enough information either, and he had no reason to believe *her*.

"And if I told you my duty to my family comes first?"

"I'd remind you our best chance of success comes as a team. And then, if I had to, I'd take you down."

"Incentive to lie."

"You think I wouldn't know?"

He tried to look away. Frowned. And finally said, echoing Torin's early words, "If it's possible, my brother will be my responsibility. If it's not possible . . . try not to kill him. He won't surrender. Honor demands he fulfill the terms of his contract and if he has to be killed to prevent that, our mother will not be pleased."

"I'll do my best."

"Then I will as well." She nodded toward the anchor. At this distance it looked small and fragile. A tiny rectangle of safety on an alien planet. "Do you see anything?"

"No. Scans say the plateau is empty."

Torin's scanner showed a flock of ground-feeding birds, four tiny mammals, seven lizards, and the ubiquitous unnumbered insects. But she allowed his point. Not far from the edge of the jungle, she could see a hole and a pile of dirt. Beyond that, on the forty-five, a fabric shelter holding a few pieces of equipment. Not weapons. Her scanner's database held no matches. The old Navy VTA was a half kilometer from the anchor which was, in turn, significantly closer to the edge of the cliff than to the jungle. Martin had locked in the lower shields, activating the defensive grid that kept her scanner from reading anything more than a spaceworthy rectangular cube. There was a faint heat signature in one of the upper windows. Probably the closest shooter Martin had to a Mashona—without Mashona's eyesight or her skill.

Of course, she now had access to the enemy who'd once cracked anchors open. "Your opinion, Santav Dutavar."

"I don't understand why he's keeping the Polint inside."

"They're safe inside."

"They'd be safe outside."

Not necessarily. *They* had Mashona. "He could be using them to intimidate the hostages, or to intimidate Werst . . ." She wasn't sure Werst could be intimidated. ". . . or he doesn't trust them out of his sight."

"When we're bought, we stay bought. It's a matter of honor."

"Good to know. He could be using them to intimidate and control Commander Yurrisk and his crew."

"Martin is working for the commander."

"Commander Yurrisk can't afford to keep his ship flying, so he can't afford mercenaries. Martin's working for the buyer of that weapon."

"Why do they need Yurrisk, then?"

"He has a ship. It puts another layer of separation between the buyer and the seller. And it muddies the water when Justice gets involved."

"My people could also be here to make the water dirty."

Close enough. "Vertic thinks they're here to take down Strike Team Alpha."

Dutavar huffed out a thoughtful, "Good strategy."

"Not a guarantee."

"As you say, Warden."

She gripped his arm. "Go get some food. It's going to be a long night."

"Are you . . . ?"

"In a minute. If I stare long enough, I'll drag Martin outside by force of personality."

"That's . . ." His mane lifted. "You're making a joke."

Torin raised a brow, then shifted just enough to refocus on the anchor. Dutavar snorted, then started back toward the jungle.

"That was impressive, Gunny." Firiv'vrak's approach had been silent, but the odds were high only one thing on 33X73 smelled like cherry candy. "Would you have carried out your threat to take him down?"

Glancing down at the Artek, Torin smiled. "I don't make threats."

TEN

"**WE ASSUMED A SMALL** exploratory expedition, wiped out by a native weapon. Perhaps even accidentally. We were wrong."

Werst, sitting cross-legged in front of the plastic sheet, weight carefully off the bruises on his left side, froze as he heard the slap of Commander Yurrisk's feet against the floor, felt the air currents shift as the commander's arm waved over his head. As it had become clear that "Ressk" had no answers, the commander had grown agitated, muttering and pacing, Qurn's low voice a constant background hum.

"This data sheet changes everything. The plastic were here to observe the native population. They were discovered and destroyed, but not before compiling a warning that was never retrieved. The natives would have given the weapon that saved them a place of honor. In a temple or a shrine. Protected from the elements."

"It wasn't put in stasis," Werst muttered, rolling his eyes.

"I'm not crazy." The commander's breath ghosted against Werst's ear as he leaned in close. "I know the weapon won't work, not after all this time. That doesn't matter. The buyer will take it as is. Pay me for it. Keep us flying. Save my ship. Save my crew."

Werst suspected the odds were higher that the plastic, when discovered, had wiped out the native population, destroyed the weapon, and left a scorecard behind. Fixated on keeping his ship and crew . . . Werst huffed out a breath. He might as well use what was clearly the relevant word. Fixated on keeping his ship and crew *safe*, the commander had done minimal research on the planet as a whole. He didn't know all the planetary populations had disappeared around the same time. Werst could see how genocide wouldn't have occurred to the scientists, but it sure as shit should have occurred to Robert Martin.

He closed his teeth on a grunt of pain as Commander Yurrisk squeezed the bruise on his shoulder. "If the plastic continues to hold their secrets safe, we need to find the temple by other means."

"It's a big city," Qurn reminded him gently, tugging his hand free.

"Pop quiz: How can we locate specific buildings in under the trees?" *Harveer* Arniz crossed to Werst with a bowl of food. From the Katrien stores, given the smell. "We can't," she continued as his stomach growled. "How many times do we have to go over that?"

Werst heard boots approaching and when Martin's foot appeared, arcing through his peripheral vision, he gripped *Harveer* Arniz's tail with his foot, and dragged her clear.

The food went flying, at least half of it onto Werst's lap. *Harveer* Arniz clutched a handful of his overalls and pressed unhurt against his side.

Martin seemed satisfied with the mess. "You don't eat until you get me some answers."

"Your finger looks like a sausage," Werst said, lips off his teeth. "And you wouldn't like me when I'm hungry."

"I don't like you now."

"Out of his way, Sergeant. Your assistance is not required." Commander Yurrisk's voice was ice and iron. "Send the *harveer* back to her people." The iron had gone, but the ice remained.

"She should clean up the spilled food first." Qurn seemed permanently set to calm and supportive. Or to subtle manipulation. Werst hadn't yet determined which. He'd bet the Primacy had different laws about AI, though. She'd aced that creepy, serene robot thing popular last season on the vids. "When the spilled food has been removed, Warden Ressk can return to work."

"Doing what?" Zhang asked. When everyone turned to stare, she shrugged. "Just asking. What's he good for? If he can't do shit without a slate, and you won't give him one he can use, it'd make more sense to shoot him, right? Take an enemy out before the fight starts and all that."

She wasn't wrong. And she should keep her mouth shut.

"As long as we hold Ressk, we have leverage for negotiations with the Wardens."

Everyone, including Zhang, turned to stare at Martin.

"What?" He spread his hands. "We're going to want to leave the anchor eventually."

Commander Yurrisk nodded. "We can trade him for the weapon."

Which sounded reasonable except Gunny didn't have the weapon, wouldn't go looking for the weapon, and had been instructed not to negotiate with the hostage takers. Martin had to be aware of the first two points and suspect the third. What was he playing at? Seemed he wanted "Ressk" alive for more than his tech abilities. Why?

Too many unanswered questions. Time to double up.

"Commander Yurrisk, sir." Werst stood, only mildly exaggerating the amount of pain he was in, and stared at a point over the commander's left shoulder. "When the symbols shifted, I thought I saw a pattern I recognized."

"You knew the symbols?"

"No, sir, but the *pattern* looked almost familiar. Like leaves against the sky." The commander was Krai, he'd understand that. Werst saw a fine tremble run through his body and remembered too late that Commander Yurrisk's injury denied him the trees.

The commander swayed right, then right again. Swallowed. Shook his head. Turned and vomited into the metal bowl Pyrus held ready. Spat. Straightened. Sareer handed him a pouch of water as Pyrus took the bowl away as quickly and unobtrusively as he'd brought it.

Werst had also forgotten one of the most common side effects of vertigo. Krai didn't vomit. Krai didn't waste food. He felt his face heat, embarrassed for Commander Yurrisk, wanting to shield him from the non-Krai in the room.

"Unless there's something on the wall I should know about, look me in the face, Marine."

To Werst's surprise, the commander's eyes were clear and focused. As though physically hitting bottom had reset his mind. "Sir."

"Will you know the pattern if you see it again?"

"Yes, sir."

"Then back to it, and keep me informed."

"Yes, sir." He turned back to the plastic, aware of Qurn studying him. Seemed like she had a thing for Krai. He could understand that.

As he sat, carefully avoiding the last of the spilled food, Commander Yurrisk informed Martin that he wanted Malinowski to report any movement on the plateau so that they could begin the negotiations.

"She was sent up there to shoot . . ."

"Her orders have changed. There are six Wardens on a Strike Team. We have one. Leaving five to face us. If their pilot stayed with the shuttle, four. As long as we remain in the anchor, we're safe. They have to negotiate."

No, they didn't. Werst let the conversation fade to background as he eased himself back down, crossed his legs, and slid the slate out of his pocket.

Harveer Arniz moved closer, tossing a wet cloth into the empty bowl. "Those are my overalls, Warden, and you've made a mess of them."

"Hey, you made the . . ." Her expression wouldn't have been out of place on any Corps DI. "Sorry." He picked a soft square of noodle off his thigh and ate it, sliding the slate into the matching pocket on her lower leg, mouthing *Ganes* before saying, "At least they're waterproof."

Inner lids flickered across her eyes. "Not my point."

Before she turned to go, she touched his cheek with her tongue.

• —♦— •

"We're hard to see at the best of times if we're not specifically being targeted. In the dark, with no one aware we're here, I could stroll to the VTA."

"We."

Firiv'vrak's antennae flattened. "You're not coming with me."

"I am." Keeleeki'ka's translation sounded smug, and she smelled of acetone. "I'm learning your story now."

"No . . ."

Torin cut her off. "Unfortunately, the reasons for not leaving Keeleeki'ka on our VTA still stand."

Even the clack of Firiv'vrak's mandibles sounded peeved. "We have to humor her because of her political position."

"Close enough."

"I'm standing right here," Keeleeki'ka muttered. But the smell of acetone began to fade.

Torin shifted in order to meet as many of Keeleeki'ka's eyes as possible. "You will obey Firiv'vrak in the field. Agreed?"

"Agreed."

"Firiv'vrak, you will treat Keeleeki'ka as a comrade in arms. Agreed?"

The biscuit Firiv'vrak held cracked. Neither Artek had eaten much. They ate their own dead, so Torin assumed they'd been living off the land. "Agreed. Although I'm not a story."

The smell of acetone grew stronger. "Everyone's a story."

I are agreeing and as I are recording you eating and talking and not accomplishing anything, we are going to consider this the extended personal interest segment.

"It can't all be thrilling runs through the jungle, Presit."

Trust me, Gunnery Sergeant Kerr, except for when you are having been attacked, the running are not being especially thrilling. Dalan are having to edit extensively. I are thinking, montage.

They'd settled to eat on the cleared road where they had full three-sixty visuals, however limited in distance some angles might be. Overhead, the break in the canopy allowed them to see the sky. Torin checked her cuff. Still sixty-one minutes to sunset, twenty-three minutes after that to full dark. Unless this turned into a siege, one way or another they'd be done before the moons rose. She turned back to the Artek, who smelled of heated milk and were ignoring each other. "You're sure you can get into the shuttle if Beyver's locked the door behind him?"

Firiv'vrak waved her arms. "Not my first boarding party, Gunny. That's a CFN223 VTA. I'm sure I can get into it."

"Good." For values of the word relevant to present need. The past could stay in the past—and considering the weight she continued to carry, Torin was aware of the hypocrisy of that thought. "Once in, secure Beyvek before he gives the alert. Then shut down the implant search pattern so that Alamber can return to cracking the mercenaries' slates."

"Do we even need to disrupt their communications?" Binti gestured with a protein stick. "They're all inside. They can shout."

Once I get into a slate that contains the anchor codes, I can control the anchor's defenses.

"He can open the door," Bertecnic snickered.

If the door prevents us from accessing the interior, it's a defense.

"Yeah, but I thought that was Werst's job."

I can let him open the door. Make him feel useful. But once I'm into the emergency evacuation protocols, I can pop the window shields.

Vertic leaned in, although Alamber was eight kilometers away in the other direction. "What good will that do?"

Time it right, and it's mercenary flat packs.

You are not being permitted to be killing indiscrimin . . .

"He knows, Presit. We can use it as a distraction. Firiv'vrak, when you've taken the shuttle, open a private channel to Craig."

Fair go that, between us, we can get anything in the air.

"We don't need it in the air." Every pilot Torin had ever met assumed that if it flew, they could fly it. Given what she'd seen both Craig and Firiv'vrak do, in this case, they might be right. "We need to know if there's anything on board we can use against the anchor and you need to hold it in case of an attempted escape—where hold it means lock the hatch. Craig can talk you through how to keep it secure in case of an attempted breach."

"I have successfully breached the CFN223 . . ."

"So you said . . ." Past in the past. ". . . but this time you're on the other side. Now, about the anchor. The upper windows are uncovered, but if we approach across the plateau, they'll pick us off. Cover enough for an Artek is not cover enough for the rest of us."

Just one shooter so far, Gunny. Ressk waited at the edge of the plateau, keeping watch. Closer to Werst.

"Glad to hear it." She caught the pouch of coffee Binti threw her. "Precedent says mercenaries don't spend their money on helmets."

"Can't use a helmet to kill someone," Freenim noted.

Torin caught herself waiting for Werst to disagree. She doubted anyone else had noticed the pause. "Probably the reason. This lot seems typical." There'd been no visible helmets in the hours of images the DLs had acquired. "We'll assume there's still scanners in the VTA and they haven't been sold for fuel or food or air, but the Artek throw a minimal heat signature." It had given them an advantage infiltrating Confederation positions during the war. Sh'quo Company hadn't run into a lot of them, but Gamma Company in Sector Nine had specific, Artek-generated profanity. "Once the Artek are on board, they won't be a problem for the rest of us. The anchor has external security cameras. Once we're in range, we're committed, so they have to be taken out."

"That's me," Binti acknowledged.

"But they know we're coming." Merinim scooped a brown

gelatinous glob out of a pouch and licked it off her fingers. "They'll have prepared."

"They don't know we have Primacy assistance."

"So they think there's five of you out here?" Dutavar shook his head. "No offense intended, Warden, but they have a secured location. What do they think a single Strike Team can do?"

"So far everything we've been sent to do." Binti spread her arms. "We're just that good." She shrugged as Torin cocked a brow in her direction. "Well, we are."

Slate balanced on her thigh, Torin pulled up the map of the plateau and surrounding area. Adjusting the angle, she zoomed in on the cliff. "Durlan, could the Polint move at speed along this ridge . . ." She traced it with a finger. ". . . from the edge of the jungle as far as this diagonal crack?"

Vertic stretched her upper body out toward Torin's leg and cocked her head, vertical pupils open wide. It wasn't a large image. "I wouldn't risk the section three meters lower or anything closer to the waterfall, but that upper ridge, that's easy enough."

"Can it be done while carrying Ressk?"

She took another look at the ridge. "Given his flexibility, yes."

The infirmary window?

"The infirmary window," Torin agreed. "It's on the far side of the anchor, and they know we're coming in from the jungle."

Freenim nodded. "They don't think we can reach it without being seen."

"That, and smarter people than Martin have forgotten to include the infirmary in their battle plan, fixated on how they'll need it later. Ressk will slip into the second floor and deal with the shooters unless Mashona's already neutralized them. Once they're out of the picture, Ressk, get to Werst. Dutavar, Bertecnic, you'll be making the run."

"No," Vertic protested, rearing back. "I won't be left behind. *And* I've already carried Krai, just like Bertecnic. All three of us go, or Dutavar stays."

Torin met her gaze. "Dutavar has military kit. That means military strapping and that's safer for Ressk. He's the best shot of the three of you and the security cameras on the rear of the building need to be taken out. His coloring also provides the best camouflage, although Bertecnic is dark enough that he's unlikely to be seen. You're bright."

"Bright?"

"Gorgeous," Binti told her. "But visible. Any small amount of light will just bounce off all that gold."

"It's possible you were never chosen for a night infiltration during the war," Freenim said quietly. "You'd have been sprayed dark. My unit durlan, who had similar coloring, hated it, and opted out when she could."

"Dutavar flickers." Firiv'vrak rose up so her eyes were even with Vertic's. "Although as our vision combines multiple images, that could be us. In the dark, Bertecnic blends. You glow."

Vertic sighed, settling her bulk back down in the cradle of her legs. "But I'm not staying behind because of my coloring, am I?"

"No," Torin told her, the memory of Vertic's rank, the habits of a lifetime keeping the cutting edge off her voice. "You're staying behind because I say so. There's three young male Polint in the anchor. What happens if you call to them?"

"Call? Unless they're standing in the upper windows, we'll have to amplify my voice to be heard inside, so . . . nothing. Even if Alamber sends it through their slates, nothing. But," she continued before Torin could speak, "face-to-face is different. In the presence of a female who hasn't gathered, at best, biology will negate the contracts, they'll switch sides and turn on my enemies to impress me. It's more likely, though, that they'll be confused and, therefore, easier to take down." Vertic shrugged and unwrapped another food pack. "As I said . . ." She nodded at Dutavar, who tossed his head, his mane up. ". . . not many of our males make Santav Teffer."

"Then you need to be out front where they can see you. You two . . ." Torin pointed at Dutavar and Bertecnic. ". . . will take them from behind."

I always miss the good stuff, Alamber muttered.

Vertic ignored him. "How will they see me if they're inside?"

"If you're right," Torin told her, "and they're here to take me down, I'll present myself for the taking to draw them out."

"Present yourself?" Freenim asked in the dry, matter-of-fact tone common to senior NCOs addressing officers they felt were about to commit stupidity.

Torin didn't appreciate it being used on her. "I'll offer to negotiate for the hostages."

"Your Justice Department has ordered you not to negotiate."

"The mercenaries don't know that. They're trapped. It's logical we'd offer them a way out."

"And then you'll betray them?"

"Yes." She raised a brow when Freenim laughed.

"You didn't pause before you answered," he explained.

"Makes more sense they'd shoot you, though." Binti shrugged as all attention turned to her. "Well, it does."

"Not a problem." Attention turned back to Torin. "Ressk will be in the anchor by then and will have taken out their shooters. And if he hasn't . . ."

So little faith, Gunny.

"If he hasn't . . ." Torin repeated, then fell silent as another piece fell into place. She was good, she knew that, but there were Marines who were better. Martin could have stacked the deck with numbers alone and had an easier time of it than going to the trouble of finding and hiring the Polint. "If I'm killed by members of the Primacy, that could crack the peace."

Vertic ran a claw between two flagstones. "Think highly of yourself, Gunny?"

She are not needing to—although I are also believing she does. Gunnery Sergeant Torin Kerr are being the face of the peace for much of the Confederation. Her death may not be starting the shooting again, but it are going to be affecting the current talks, there are not being a question of that.

And who wants the war to continue, Craig muttered. **Those who manufacture the weapons. Who manufactures the weapons in this sector . . .**

"Anthony Justin Marteau." The military industrial complex was huge, but MI made weapons and weapons couldn't be repurposed.

I know you liked him.

"I've liked assholes before."

He laughed. **True enough.**

"And it's all circumstantial so far."

No smoke without fire.

"That's not . . ."

Torin, let me have my moment.

"The moment is yours, Warden Ryder." Torin turned her attention back to Vertic. "Once the Polint emerge to go after me, you advance and scramble their perceptions while Dutavar and Bertecnic come around the building and join the fight. Freenim, Merinim, and I will slip inside before the door closes again." She turned further until she faced Dutavar and Bertecnic. "How fast can the two of you cover the distance to the cliff if you're out in the edge growth where it's open enough to run?"

Dutavar rubbed at one of the larger patches of orange fur on his hip. "It's about five kilometers . . . eight minutes."

"Give or take," Bertecnic agreed.

Both Artek made a speculative sound, and Keeleeki'ka clicked, "Fast."

Freenim shook his head. "We're lighter. Closer to the weight of the Ner. It should be one of us instead of Ressk."

"Can you get up that diagonal crack, balance on Dutavar's shoulders, then jump for a second-story window?" He frowned and Torin added, "I don't ask rhetorical questions, Durlave Kan."

"We can." He blinked. Exchanged a speaking look with Merinim. "But not easily." The corners of his mouth twitched up. "Point taken. Ressk's the better choice."

"Dutavar hasn't carried Krai before." Bertecnic emptied a handful of nuts into his mouth, chewed, and swallowed. "We can switch harnesses. I can do it."

"You're too big for my harness."

You're not wrong.

"Alamber." Torin pulled the heat tab on a second coffee pouch. "Thank you for offering, Bertecnic, but Dutavar carries Ressk."

Bertecnic shrugged. "Happy to make the run without being weighed down by Ressk's fat ass, Gunny."

Ressk's muttered announcement that he was going to tie Bertecnic's tail in knots was indistinct enough, Torin ignored it.

"Mashona finds a perch . . ." Binti flicked a sloppy salute in Torin's direction. ". . . and the rest of us make for the tent."

"Tent's lousy cover, Gunny."

"If they don't know we're there, they won't shoot at it. Incentive to get to it quickly. Also, as you say, a tent's lousy cover, they won't expect us to use it—we're known to be smarter than that. It'll put us closer to the anchor, coming in at an angle they won't expect."

"If everything goes in our favor." When Torin glanced over at Freenim, he laughed again. "Never mind. Plans. Combat. Improvise. It's a theme."

"Once Ressk's inside, his implant will be blocked. How do we coordinate aggression?" Vertic shifted her foreleg to the left and began digging on the other side of the flagstone she'd uncovered. "What if the slates are blocked as well?"

"Then we trust Ressk to do his job . . ."

I take out the shooters, I free Werst. Werst goes Marine Corps on their asses.

". . . and we act accordingly."

Torin glanced at her cuff. "Sunset in seventeen." She swept her gaze around the team. "Pack up. Firiv'vrak?"

"Heading out, Gunny." She folded in her arms and rose. "I'll send Ressk back and wait on your word to go."

"*We* will send Ressk back," Keeleeki'ka corrected, scrambling into place by Firiv'vrak's side. "And your story will go on."

"Joy," Firiv'vrak muttered as they disappeared into the underbrush, leaving a lingering odor of wet dog and cinnamon.

•—◆—•

Martin paced across the common room adjusting his path to intercept any movement, enjoying the scramble out of his way. His reputation, like that of most bullies, had been built on air; if not constantly reinforced, it disappeared. He was, however, staying away from the hostages, and showing more control than Werst had expected him to, so . . . points for being a mature asshole.

"Sergeant!"

When Martin stopped in front of Commander Yurrisk, Werst shifted to keep both of them in his peripheral vision.

"I want Lieutenant Beyvek back in the anchor where it's safe."

"I told him to return after blocking their implant frequency. Seems he hasn't managed that yet." As the commander opened his mouth, Martin cut him off. "Disrupting the Wardens' communications will keep us *all* safe."

"You don't think the Wardens can work around such a minor disruption? These Wardens?"

Credit where due, the commander's dismissive tone made Martin's sound petulant.

"I think we need to cut them off from their VTA so that they can't put it in play at the last minute. And, yeah," Martin cut off the commander's response, "they've got slates, but the time it takes to activate a com unit is time we can use."

"That's not a good enough reason to risk a life, Sergeant."

"How is he at risk?"

"He's not here. With me." Yurrisk unclipped his slate. "I'm bringing him back."

"At this time, it would be more dangerous for him to cross to the anchor." Qurn gripped Commander Yurrisk's arm and spoke loudly enough to be overheard. Werst had no doubt the volume increase was deliberate. She wanted him to hear her. "If the sniper you told me of is in position, he'll never make it. Beyvek is in a VTA behind a secured air lock."

"I should have kept him here." The commander's lips were off his teeth when he turned his attention back to Martin. "You said he'd be back in minutes!"

"He should've been, but we grabbed the wrong Warden. That di'Taykan of theirs is clearly better with tech than the . . ."

Werst heard *tree fukker* in the pause. Wondered what would happen if Commander Yurrisk heard it, too.

". . . one we have."

The commander's eyes narrowed, and he stared at Martin for a long moment. "If anything happens to Lieutenant Beyvek, you will answer for it, Sergeant Martin."

"As you say." Martin pivoted on a heel and stomped back toward the plastic. "And what do *you* think about grabbing the di'Taykan?" He slammed his knee against Werst's shoulder.

Werst rocked sideways, sucking air through his teeth at the sharp flash of pain. He'd fought through pain before. A twist and a snap and he could hamstring the asshole. Couldn't bite through the combats covering his legs, but he could crush and tear and not have to fill his mouth with Martin's blood. Tempting.

"Evidence suggests he's not as useless as you are."

Did Martin expect Ressk to straighten and declare, *"I'll show you who's useless,"* then solve the mystery to prove Martin wrong?

"So who fuks the di'Taykan? Everyone or just Kerr? Got to be everyone, right? He's di'Taykan. Of course . . ." Martin slammed

Werst's shoulder again. Werst ground his teeth and thought about yanking tendons off bone. ". . . Humans are amazing in the sack, so maybe Kerr's enough for him. She fuk you, too?"

"No time," Werst snapped, nostril ridges flaring and closing as he breathed through the line of fire that spread out from the continuing impact of Martin's knee with his shoulder. "She's too busy cleaning up Human stupidity."

"What?"

"But she appreciates you removing the apostrophe," Werst sneered. "Misplaced apostrophes really piss her off."

"Yeah, well, fuk her. Yeah, fuk Gunnery Sergeant Torin Kerr and her whole shit-doesn't-stink life." His finger stopped just short of Werst's chest and was withdrawn significantly faster as Werst curled his lips back off his teeth. "She needs to pace herself. We've changed more than the apostrophe."

Bam. They hadn't suspected Humans First because of the species purity party line. Looks like they were wrong. "So what's with the Polint?"

"You ever see them fight?"

"I have."

"That's what's with the Polint."

The scientists had no fight in them so it seemed Martin had planned ahead in case a Strike Team showed up. The Polint were weapons. Insurance. That made Martin smarter than Werst had thought.

•—◆—•

"What do you think you're doing, lizard?"

I'm helping bring you to justice for the deaths of Dzar and Magyr, you ki seewin. Arniz held out the covered dish. "I'm taking your young Trembley a bowl of soup. His body requires nourishment to heal."

"Why should you care? You're not Human." Martin stomped across to her, the metal in his boot heels ringing against the floor. Was the noise supposed to frighten her? Please, she'd heard forty-seven firsters leave a lecture hall on the last day of classes. "Lift the lid off the bowl, lizard."

She sighed and lifted.

"Taste it."

"It's Human food."

"And too good for you. Taste it."

"Are you assuming I've poisoned it?"

"Taste it," he repeated, smiling.

It seemed Sergeant Martin wasn't aware the Niln could consume seven substances fatal to Humans. When Dr. Ganes joined the expedition, they'd all been required to learn what those seven substances were. To think she'd once been appalled at the concept of accidental death.

Dzar had been conscientious about memorizing the list and ensuring none of the seven were included in their stores. Shortsighted of her, as it happened.

Holding the bowl against her chest, Arniz tucked the cover under her elbow, and pulled a spoon from the narrow breast pocket on her overalls. The soup package had been labeled chicken noodle although the makers of the soup had been civilized enough to use a nonflesh-based chicken substitute.

"Well?"

She forced herself to swallow the spoonful of broth. "It's disgusting."

"You're disgusting," he said conversationally. "Sterilize the spoon before you give it to him. What do you want?"

The di'Taykan with the pale pink hair—Pyrus—had risen. "I'll take the food up. I want to see Gayun. He shouldn't be alone."

"He's in stasis."

"Sergeant . . ."

"No."

"Sergeant." Yurrisk's repetition of Martin's rank didn't sound like the same word. It sounded like *know your place* and *don't make me come over there.* "You can't open the pod, Pyrus." His voice gentled. "Gayun won't know you're there. Let *Harveer* Arniz take the food up. You two . . ." He waved Mirish closer to Pyrus. ". . . stay together. Watch each other's backs."

Arniz could hear grinding from Martin's mouth, and his jaw made small movements back and forth, but he unclamped it to say, "Zhang."

"Sarge."

"Accompany the lizard upstairs. Move Trembley into the infirmary, then spend . . ." He checked his cuff. ". . . a moment or two with Ganes. He hasn't had enough *Human* contact."

"He's in medical. I hate medical, Sarge."

"I don't care. Move Trembley, see Ganes, then take up position in the window of the room you moved Trembley from. We've got plenty of ammo; they have to worry about precision, you don't."

"Sucks to be them." Zhang headed for the stairs. "Come on, lizard."

"Why do you hate medical?" Arniz asked as they began to climb. Knowledge was power. She'd never realized that so viscerally before.

"It smells funny."

The height of the risers was a compromise between the many species who might use an anchor. Slightly high for Arniz, slightly low for Zhang. They climbed three steps side by side.

"You smell funny, too," Zhang added on the fourth.

At the top of the stairs, the mercenary shoved her toward the infirmary. "In you get," she ordered as Arniz stumbled and nearly dropped the soup. "Stay out of the way. Malinowski! Help me move Trembley."

"The bunk's on rollers," Malinowski's voice came from the room the Niln had used for their night nest. Arniz's tail lashed. She'd probably touched personal belongings. It wasn't enough the Warden wore her overalls? "Move him yourself."

Zhang rolled her eyes until whites showed all the way around. It was a fascinating feature of Human eyes. "Lazy cow!"

"Fuk you, Zhang. There's a Strike Team out there I need to shoot."

"Like you could hit them without using a ship." She turned to Arniz. "Navy gunner. A terrible shot if the target isn't half a kilometer long. What are you still doing out here? Go."

Arniz ducked a second shove and stepped into the infirmary. Dr. Ganes moved back to give her room. He'd been standing just inside the door, watching. Maybe listening to voices from downstairs. How well did Humans hear? She held up the covered bowl. "I've brought food for Trembley." Which was when she realized she had no food for Dr. Ganes. Who was a colleague.

He must have read the realization off her face because he smiled. "It's all right, *Harveer*. There's protein shakes up here if I get hungry. I'm fine."

Across the hall, Zhang cursed at Trembley's bed. Arniz moved closer to Ganes and slid the slate out of her lowest pocket. "The young Warden sent this."

"It's Dr. Lows'." The Katrien slate looked tiny in his hand. "Did he mention what he wants me to do with it?"

"We didn't have time for an extended conversation. I assume he thought you'd know."

Ganes opened his mouth, closed it again, and finally muttered, "He must need help with the block. It's the only tech we discussed. Still, if he had to send me a slate, I wish he'd been able to send mine."

"*Se tenis tin arramani ki haliven.*" Arniz spread her hands. "If wishes were . . ." And frowned. "Never mind. It loses a little in the translation."

•—◆—•

Martin was now the only Human in the common room. He paced—no, he prowled. Every movement said *look at me, I'm in charge.* Or maybe *notice me,* Werst allowed. There was a lot of psychology tossed around during Warden training; why people did what they did. Werst didn't care about why, only what, but some of it had stuck. Bullies often felt they could only be seen by being shitheads. Understanding didn't make them any less a shithead, though.

As he prowled, Martin watched Commander Yurrisk and his crew as much as he watched Werst or the hostages. His hands were in constant movement over his . . .

Not his. Werst knew the gouge on the barrel. The discoloration on the butt. He half rose, then settled again. There was nothing he could do about Martin carrying his weapon. Did he think Martin would apologize and return it? At least Werst would know where it was when the fighting started.

Martin was waiting for something.

For dark, Werst assumed. For the attack.

"Be ready," he told the Polint. The Polint looked unimpressed, but all three stood, stretched out the stiffness in their muscles, and began checking weapons. As well as the heavy machetes, they'd all strapped on multiple knives. Netrovooens had a leg sheath—redundant bordering on ridiculous given it hung eight centimeters above claws Werst had seen used to disembowel a Marine.

Be ready for what? Martin could hold off the Strike Team indefinitely from inside the anchor. And as long as he had the hostages, he had the upper hand in any negotiation.

Too bad people safe in government offices had decided they didn't negotiate with hostage takers.

They weren't here. Fukkers.

"You, Ressk, I should've let you bleed out. You, Tehaven, lock him in the storage room behind the kitchen." He smiled down at Werst, showing teeth, but spoke to Tehaven. "If he decides to be a hero, rip him apart."

Werst held both hands up where Martin could see them and flipped him off.

Martin had been Corps. He knew the Krai gesture. "Fuk you, too. You and you . . ." Martin pointed at Pyrus and Mirish. "Get the data sheet down. We don't want to chance it being destroyed in the fighting," he added as the commander rose to his feet.

"There will be no fighting, Sergeant. The anchor is safe. We're safe in the anchor."

Commander Yurrisk had believed he'd gotten his people to safety on the *Paylent*, only to find he hadn't. The word *safe* haunted the commander. Werst suspected he could put most of the important parts back into the redacted report.

"Suppose the hostages panic? There's your map to the weapon, gone."

"I need that weapon." He swayed to the right. And again.

"I know."

Martin's reply sounded more like *no shit* to Werst.

Commander Yurrisk's bristles made a harsh *shunk shunk* against his palm as he rubbed his head with both hands. "Pyrus, Mirish, take the sheet down. Carefully. Let it roll."

After four days in Susumi, Werst recognized a Druin frown when he saw one. Seemed Qurn didn't believe the sheet to be in any danger. Could be she wanted to keep studying it herself. Could be taking it down had been Martin's idea and she trusted him as far as she could spit a *vertak*. Could be she'd attacked an anchor during the war and knew how unlikely anyone would get in without heavy artillery.

Well, anyone but Gunnery Sergeant Tor . . .

Fuk. Martin expected Gunny to breach the anchor. That's why the Polint were prepping. In close quarter fighting, even these kids would be deadly. Mashona would lay down a covering fire, driving Zhang and Malinowski back. Gunny, the Druin, and Ressk would come in, raised to the second floor by the Polint—who'd be left outside. Even if they could get to the windows, their big asses couldn't get through it.

Gunny, the Druin, and Ressk—they'd restrain Zhang and Malinowski. They couldn't free Ganes. They'd slip down the stairs, expect surprise to be on their side, and be met by the Polint. They'd be unable to fire weapons because of the hostages. They'd . . .

"Get up." Tehaven's hand engulfed his shoulder, thumb pressing into the edge of the largest bruise. "I don't want to miss the fight."

Would Gunny realize it was a trap? Better question, would that stop her?

•—◆—•

"Coming through!"

Trembley was pale, teeth clenched as Zhang shoved his bed into the infirmary. One end bounced off a machine of some kind. Arniz couldn't identify it, she was too busy scrambling out of the way.

"Be careful," Ganes snapped, swinging the autodoc clear at the last moment. "You may need this equipment later."

"You're not the boss of me," Zhang told him cheerfully, parking Trembley's bed by the window. "Give the kid his soup, lizard. Then get back downstairs with the rest."

"Not a *kid*. And you're a crap driver," Trembley muttered.

"And you're a crap patient."

He rolled his eyes and she slapped his leg. "Ow!"

"Yeah, yeah, you'll live. So, Ganes . . ." Giving Trembley's largest toe a final squeeze, she crossed to Ganes and leaned up against him. They were close to the same height, both significantly smaller than either Martin or Trembley, Ganes' deep brown skin contrasting beautifully with Zhang's lighter tones. Arniz hadn't previously been aware of the color variety Humans exhibited. The differences among her own species, and there were many, were significantly more subtle. Although, subtle was not a word anyone would apply to Zhang, she realized, as the young female continued to breach Ganes' personal space. "Martin said you needed human contact. Got time for a quickie? Sex is easier than conversation, am I right?"

He removed her hand from his groin. "You hit me in the face with your KC."

"You were trying to get to the communications equipment. Opposite sides, no hard feelings."

"And yet I'm not interested." He sidestepped, cleared the piece of equipment Zhang had him backed against, and moved away.

Why didn't Ganes try to overpower her? But then, he couldn't leave the infirmary, could he? Overpowering her would be the end result, and he'd be punished for it. Observation suggested Trembley no longer blindly followed Martin's philosophy, but Arniz doubted he was ready to take up arms against the sergeant. She, herself, had no idea of how to work a gun. Wasn't sure if she could fire at another sentient being even if Ganes taught her how. She thought of Dzar falling. Of Mygar falling. Perhaps she could fire at Martin.

Zhang didn't seem bothered by the rejection. She grinned and spread her arms. "Your loss. Oh, hey, drugs . . ." When the cabinet door refused to yield to either her fist or her weapon, she shrugged and headed for the infirmary exit. "You . . ." A finger with a blackened nail pointed at Arniz. ". . . soup and get out. Come on, lizard, a little hustle. Me and Malinowski, we've got Wardens to shoot."

Arniz crossed to Trembley's bed. She'd never really internalized the size of the infirmary, the multiple shades of gray made it difficult to get a handle on the dimensions. The size and the contents of infirmaries were legislated lest people not take their health and safety seriously, but—if she'd thought of it at all—she'd have assumed a Human-sized bed would have filled any free space. It didn't. Room remained for both people and equipment to maneuver. If her estimate of the dimensions was correct, there was room for another Human-sized bed if needed. In a world less restricted by the government, they could have had another lab.

"You look thoughtful." Trembley sounded rough, but he tried to smile. Or perhaps it was a grimace.

"You look horrible." She cocked her head. "Although not as horrible as you did, so that's some progress at least. I brought you food." The broth was urine yellow, the mushy noodles almost white, and the chunks of beige were allegedly protein. It had tasted mostly of grease. Trembley's midsection made noises under the bandaging. "Can you sit up?"

"I think, I . . ." He sucked air through his teeth when he moved, his expression as much embarrassment as pain.

Then Ganes was there, one large hand between Trembley's shoulders, supporting him until he could move a triangular form into place. "Don't undo my work."

"Yes, sir. I mean, no sir."

Arniz gratefully passed Trembley the soup. The smell had begun to curl her tongue. She rubbed the spoon against her sleeve before she handed it over.

"Move it, lizard! Wardens to shoot!" Zhang called from the door.

"That's Ganes' bed you're in," Arniz said, patting Trembley's bare arm with her fingertips. The skin was warm and damp. "Now you have something in common to talk about. Remember, though . . ." She held his gaze. ". . . he's Navy, so try to get along."

"Well, I don't know . . ." This curve of his mouth was definitely a smile. ". . . if he's *Navy*."

She touched his cheek with her tongue. He tasted of pain. The younger races were tougher than they looked if they made these kinds of life and death decisions all the time.

"Good one, lizard." Zhang struck her shoulder as they left the infirmary. "Gotta say I like you reminding Trembley that Ganes was Navy."

She hadn't.

She'd reminded Trembley he'd been a Marine.

•—◆—•

"You don't look much like your brother up close," Werst said cheerfully, scanning the corridor for a way out. The walls were too smooth to climb and too far apart for him to work his way along the ceiling. Given the way he felt, he was just as happy not to have to make the attempt. "I mean the variegated fur, yeah that's him, but the shape of your face, I don't see it."

"Shut up."

"Your chin's pointier."

"Shut up."

"Your mane's darker. And shaking us doesn't do shit," he added, hanging off Tehaven's hand as the Polint flung him around. He'd have bruises on his right shoulder later to match the left, but it was worth it for the look on Tehaven's face. It also let him use his feet to check if the door they were passing was locked. It was. "Look, kid, not my species, but doing the dirty work for a *serley chrika* like Martin isn't going to score you any points with your females."

"Shut up!"

Werst let his knees take up the shock as his feet slapped back down onto the floor. He kept his left heel raised, but it still hurt like fuk. "Dutavar explained the whole 'trying to prove your worth now there isn't a war on,' but, honestly, I wasn't listening."

"He thinks his way is the only way!" At the end of the corridor, Tehaven jerked him into the kitchen, empty of everything but the smell of food. "I can fight. I don't need a uniform. I don't need his *control.*"

"You don't need the crap this is going to land you in either, kid." This close, Werst could see the bright green PCU in Tehaven's ear. Confederation built—the Confederation had its colon in knots when it came to privacy laws and hidden tech, thus the size and the color. It explained how Martin made himself understood.

Tehaven reached out his other hand and yanked open the heavy door that kept the storeroom secure during the anchor's drop to dirt. "Stop calling me . . . *Dupoht!*" He froze. "How do you know my brother?"

"Dutavar? We served together."

"That's not possible . . ."

As confusion loosened his grip, Werst twisted free, dove between the Polint's front legs, and used the webbing under his stomach to propel himself out between his back legs. He snapped his teeth as he went by. Polint balls didn't exactly hang free, but they were obvious enough.

Tehaven leapt forward.

Werst slammed the door behind him and locked it. Then he sagged against the polished metal to catch his breath, fingertips against the scar on his throat. Still seemed to be holding. The door shivered under his back as Tehaven threw himself at it, but the storage room was too small for him to get a good run. The door had been designed to survive drop impact, and Werst only needed a few minutes.

There were vents in the kitchen ceiling.

• —◆— •

Torin lost sight of Dutavar and Bertecnic a lot sooner than she'd expected. She knew the success of any part of a plan didn't guarantee success of the rest, but it never hurt to start a fight on a positive note.

"I wonder if we'd be harder to see at dusk," Firiv'vrak said thoughtfully by Torin's knee. "As the light changes, as mammalian vision adapts, we'd be another shadow against the . . ."

Dirt sprayed up no more than two meters out from the point where the ancient road left the jungle.

"They're shooting at shadows," Torin noted.

Firiv'vrak's antennae flattened. "So they are."

"Take them when you can, Mashona. Keep them alive if possible."

Roger, Gunny. No clear shot yet. Shooter's being careful.

Another shot. Another spray of dirt.

Correction, shooters. There's a second one two windows to the right of the first.

The last of the light faded, and the sound of the insects intensified.

"Firiv'vrak. Keeleeki'ka. Go."

They dipped on and off Torin's scanner, proving the ground wasn't as flat as it appeared.

At the cliff, Gunny. Heading down.

"Roger, that. Ressk?"

Not so much climbing as . . . Garn CHREEN!

"It's okay," Torin said as Vertic's ears flattened. "He's enjoying himself. And he should do it silently," she added as Ressk whooped. "I'd rather they didn't hear you coming."

Sorry, Gunny. Teeth together.

"At least he's not dwelling on the state of his bonded."

"There's that. All right, Durlan, wait here until the Polint emerge. You can cover the distance fast enough there's no reason to expose yourself before it's necessary. You're a big, bright target."

"So I've heard," she huffed. "And if the young males don't emerge?"

"New plan."

Vertic's tail flicked. "Can you share this plan?"

"Not yet." Torin grinned. "It depends what emerges in . . ."

The plateau lit up like midday under a white dwarf, each blade of grass standing out in sharp relief. Torin couldn't see the Artek, but they were out there. Exposed.

"Mashona!"

On it, Gunny. One. **I haven't got anything . . .** Two. **. . . big enough to take it with a single shot.** Three. Four. **But if I keep hitting the same spot.** Five. Six. Seven. Eight . . .

And it was dark again.

Rounds from the anchor hit up in the trees.

Torin blinked and tried to recover her night sight. "Remind me to make yet another requisition for impact boomers."

With pleasure. I'm moving left. Two, three trees over.

"You think they can hit you?"

At this distance with what they're firing? The way they're firing? Not a chance. But if I get a better angle, I can hit them.

"Gunny!"

Torin followed the line of Vertic's arm. One of the shooters had begun tearing up the dirt about two hundred meters from the shuttle.

• —◆— •

"What the fuk are you doing, Malinowski?" Martin threw his slate down on the desk in the anchor's communications room.

"I saw a bug!"

"Yeah, well, this armpit of a planet is all about bugs." The security feed from the plateau side of the anchor filled the desk's surface with blurry shadow, a pixilated mass rising up into the air on the other side of the darker mass of the shuttle. "What idiot installed such a worthless piece of shit?"

Peering through a small air vent over Martin's head, Werst silently echoed the question. The NOD had a crap image intensifier, low luminous sensitivity, and enough visual noise only experience let him identify the fuzzy mass as flying dirt.

"Not a bug, *Sarge, a . . . FUK! I'm hit!"*

Werst's lips curled back off his teeth. One for Mashona.

"How bad?"

"Through the fleshy . . ." She sucked a breath through her teeth, loud enough the slate picked it up. "*. . . fleshy part of my arm. The new one. Damn, that's going to void the warranty."*

"Seal it. Mashona's better than we thought. I'm sending the di'Taykan up. They get there, you and Zhang come back down."

"You said the Wardens wouldn't shoot to kill. Looks like the sniper'll put a hole in whatever she can hit."

"Good thing we've got non-Humans to waste, then."

The vent was smaller than the palm of Werst's hand. He had Martin alone, and couldn't get to him. Couldn't squeeze enough of himself through to do any damage. That *pistol* they'd taken off the gunrunners would have come in handy in such close quarters. Two shots at the base of his head—bam, bam, severed spine. As he followed Martin out of the room, slithering through the rigid tubing, thankful the bruising was on his back not his front, Werst was starting to see the pistol's attraction.

At the edge of the common room, the tubing narrowed as it ran vertically up toward the second floor and broadened as it dipped and ran along the inside wall just above floor level. About five centimeters past the junction, a bundle of cables came through the lower curve.

He couldn't go up. His skull wouldn't fit, let alone his shoulders. He'd have to follow the cables.

So the vents on the roof were out. Fine. Cables running in such a convenient location meant there had to be an access hatch somewhere close. Or it was an idiotic design. He wasn't ruling that out.

"Pryus! Mirnish!" As Werst passed a small oval vent, Martin barked out the di'Taykan's names as though he had the right to command.

"Mirish!"

He ignored the correction. "Get upstairs. Send my people down."

"Sergeant . . ." The commander sounded unimpressed.

"Malinowski's been hit. Zhang helping her. We need shooters in those windows to keep everyone safe."

Not hard to see how Martin, the waste of oxygen, had been manipulating the commander all along.

"Go, then. But Mirish, Pyrus, fire defensively only. Our position will be stronger during negotiations if no one dies."

Two of the hostages were already dead.

Werst stretched out his arms, grabbed two handfuls of cable, and dragged himself through the barely adequate space, the slick fabric of the Niln overalls all that made movement possible.

He ignored the bruising. Ignored the pain.

Imagined passing both on to Robert Martin.

•—♦—•

Beyvek has been neutralized. Am in process of shutting down search program.

He didn't fight back, Keeleeki'ka added. **He had a weapon and time to fire, but used neither. I am securing him. There is urine.**

"Artek can be startling if you've never seen one before," Freenim said quietly.

"And a hell of a lot more startling if you have," Torin noted. "Lieutenant Beyvek was Navy. Mashona?"

I'm set.

"Freenim, Merinim, let's go."

Although 33X73 was a MidSector planet, its rotation had their particular piece of it pointing away from the core. The stars were scattered enough their light was neither help nor hindrance. Torin locked her scanner on the tent's position. Three meters and one thirty-five degrees off her zero, Merinim followed. Freenim had their six.

Search program requires command codes. Keeleeki'ka, ask the prisoner . . .

Beyvek is nonresponsive. How does he breathe with his nostrils closed so tightly?

I've got this. Craig sounded amused. *_Don't mean to skite, but I've rebuilt more control panels than the yard at Ventris. One on one, Firiv._*

Torin ran at full speed, crouched low, KC in her left hand, pack humped high on her back, her silhouette as non-Human as she could make it. The tent, reflective in sunlight, absorbed what little light there was, and she was almost on it when her scanner flared.

The ground to the right dropped six meters.

As the soft dirt crumbled and her right foot went out from under her, she threw herself to the left, landed on her knee, pivoted, and subvocalized, "Pit at my ninety!"

Then she threw herself forward, grabbed Merinim's wrist and yanked her hard enough that her next two strides were on air. Impact knocked Torin's breath out, but she got her arms around the Druin and rolled them away from the edge.

"Seriously?" Merinim sighed into Torin's chin. "Another one?"

"This one reads as dirt all the way down. I assume the archaeologists are responsible."

"That makes all the difference." She rolled off Torin and up into a crouch.

"Tent," Torin said, eyes on the anchor. "Now." Merinim slid in under the fabric. As Freenim raced up, she indicated he should follow. Then she gave them a ten count. Had Craig dropped out of sight like that, she'd appreciate a moment.

"Where the fuk is Tehaven!"

Werst dragged himself a few painful centimeters farther along the conduit, his breathing fast and shallow.

"Might be eating."

One of the Polint. Werst couldn't tell which.

"Eating? Who said it was snack time? You, Camaderiz, go haul his hairy ass out of the kitchen!"

It wasn't so much a tent as Torin defined it as a fabric shelter over an assortment of equipment—some partially disassembled. In the center of one of the two longer tables, a violet light was flashing. Torin recognized two screens and what might be a microscope. Nothing else. The tables could be folded, the stools had broad feet so they wouldn't sink into the dirt, and none of it would be any help cracking the anchor. The actual entrance to the shelter was ninety degrees off the anchor—also no help—so the three of them dropped and peered out under the side opposite from where they'd entered.

And we've dummied it out. Wholesale destruction saves the day. Scan is off. You're on, Alamber.

Already locked and blocked our implant frequency . . .

"He's fast," Freenim murmured.

Not if you need me to go slow, Alamber purred. **Well, hello there. Thanks, Keeleeki'ka.**

"For?" If Keeleeki'ka was improvising, Torin wanted to know about it.

She just sent me the codes to Beyvek's slate. From Beyvek, I can get into every slate he's had contact with. Relevant to this mission, that's his crew and Martin. Once I'm into Martin, I'm into the rest of the mercenaries. This won't take long now, Boss.

The Artek hadn't mentioned computer skills. "Keeleeki'ka?"

I carry the story of Inwetermin who said to the government, that information is ours and if you keep it from us, I will take it.

Uh, Gunny, should we be worried that Inwetermin's familiar with Confederation codes?

"Not our job, Mashona." Torin's scanner marked two of the upper windows empty of glass, but the angle was too acute for her to pick up a heat signature. "Not our job."

• —◆— •

"You useless, four-legged idiot!"

"It's not my fault!" Tehaven's voice, halfway between a growl and a whine, cut clearly through the wall. "The Warden said he knew my brother!"

"How?"

"He said they served together, but that's complete shit."

"He said? He answered you?"

"Yeah. Then he went after my balls and locked me in the storeroom."

"I don't care if he tried to remove your balls with his teeth!" Martin wasn't so much shouting as screaming. He was pissed. Werst snickered as he dragged himself forward another half a meter, Martin's reaction easing his pain. "Your PCU has a translation program!"

"So?"

"So that's how you understood him. How did he understand you, huh? How? I knew that tree fukker was hiding something."

Delusional dimwit. Heartbeat pulsing in his throat, Werst sucked air through his teeth as his fingertips touched lapped metal.

"Malinowski! Zhang! Get to the roof. He's Krai. If he's in the vents, he'll have gone high!"

And he would have, had the vertical not been as big around as his dick.

•—◆—•

I'm in, Boss. A few minutes more and I'll have control.

"Can you patch me through? Slate to slate?"

How much of you?

"Voice only."

No problem. Audio takes almost no space.

Torin crawled forward on her elbows until her hips were clear of the tent, then drew her legs in under her.

"You're certain they won't shoot?" Freenim murmured behind her.

"I'm certain they're lousy shots."

"One out by animal attack, one into the pit with Werst, one hit by Mashona, one down in the shuttle—four casualties, eleven mercenaries remaining. They can't all be lousy shots."

Technically, five mercenaries and the crew of the *DeCaal*. "They can't all fit in those two windows either."

Gunny, we're in back of the anchor. Dutavar has used the Artek weapon to take out the camera.

"The camera? Singular?" The Artek weapon fired silently, but she'd expected half a dozen cameras.

Scientists seem to trust walls. Infirmary window is unshielded, lights are on.

"Proceed with caution."

She grinned as Ressk grunted, **Where's the fun in that.**

Okay, Boss, I'll have control of the anchor in no more than five, but you're in now. You talk, they'll hear you.

• ——◆—— •

"Hey, Dr. Ganes? Doc? There's something crawling on the window."

"Insect, attracted to the light." Ganes crossed the infirmary to stand by Trembley's bed. The young man had been plucking at the covers since *Harveer* Arniz had left—the outward sign of an inward turmoil. Ganes had left him to stew, but was willing to interact should he have something to say. Young and stupid was, after all, a correctable condition. "See the way the wings pick up the light?"

"It's . . . huge."

"Wingspan of approximately eleven centimeters. The mandibles are prominent, but we're still not sure if it's a predator or . . ."

A long-fingered hand wrapped around the central body and, with a panicked flutter of wings, the insect disappeared.

"Not good!" Trembley groped for a weapon that wasn't there.

The circular end of a narrow rod clicked against the window. Clicked again. Faster until the sound became a continuous soft burr. Fractures ran across the theoretically unbreakable glass and a moment later the sheet collapsed into pieces, each no more than five millimeters square, small enough it sounded like rain when it hit the floor.

"Well, that worked. Points to R&D." The Krai climbing into the infirmary flashed teeth. "Commander Ganes?"

"Dr. Ganes."

He paused, bare feet three centimeters above the floor. "Didn't think that through . . ."

Ganes pulled the extra blanket off the end of the bed—his extra blanket, his bed—and used it to sweep the glass aside.

"Thank you, Commander." His finger slid over the trigger guard as he shifted his grip on his weapon. "If it isn't Private Emile Trembley."

"He's no threat. He's been injured." Ganes leaned out the window and saw two large shapes heading for the corner of the anchor. He didn't think they were Dornagain, they looked a little more like H'san. "Warden Ressk?"

"How do you know . . . Werst!" Nostril ridges shut, he ran for the stasis pods. "Where is he?"

"Your bonded is downstairs. Weak from blood loss, but walking and talking. I repaired the hole in his throat and unless he's done something extraordinarily stupid, it should have held."

Warden Ressk blinked. "Your bedside manner sucks, Commander. Doctor."

"I've been told."

"Attention in the anchor. This is Strike Team Lead Warden Kerr. Surrender the hostages immediately."

"What's she doing?" Trembley demanded.

"Negotiating."

"That sounded more like an ultimatum."

"Yeah, that's what I said. Stay in here." Without waiting for a response, Ressk slipped out of the infirmary and across the hall to the Niln sleeping chamber.

"Who's that?" Ganes heard Malinowski snarl over the sound of two sets of boots descending the stairs from the roof.

• ——◆—— •

"Go!" Martin yelled.

Arniz watched the Polint race for the air lock. Watched Qurn struggle to keep Yurrisk by her side. His nostril ridges were closed, his teeth exposed, and unless the Druin was a lot stronger than she looked, she wouldn't manage it for much longer.

"Once the hostages have been surrendered, unharmed, we will discuss the multiple infractions committed against the laws of the Confederation."

Like the murder of innocents. Like the murder of Dzar. Like the murder of Magyr.

• ——◆—— •

Torin raced for the air lock door as Vertic ran across the plateau roaring a challenge. One of the Polint running toward her stumbled. Dutavar and Bertecnic charged out from behind the anchor.

Everyone stay clear of the windows!

The shields fell first, then the glass, leaving six large openings behind.

Torin changed course.

• ——◆—— •

There were more than three Polint outside the anchor. Had Martin been hiding more Polint in the shuttle?

"New plan!" Turning from the windows, eyes narrowed, Martin lifted his weapon and pulled the trigger.

Arniz began moving before the screaming started.

• ◆ •

"Too many!" Fingers white on the edge of the mattress, Trembley swung his legs out of bed. "The Warden's outnumbered!"

Ganes watched Malinowski charge into the Niln sleeping chamber after Warden Ressk, Zhang at her heels. If she'd remembered she was holding a KC in one hand, the Warden would already be dead. As it was, a Krai in close combat against two Humans and a di'Taykan wouldn't stand a chance.

Trembley staggered across the infirmary in underwear and bandages.

"Where the hell are you going?" Ganes demanded, grabbing his arm.

"To help." Trembley pulled free.

"Who?"

"The Warden!"

Ganes had no choice but to trust him; he couldn't follow him out the door.

• ◆ •

Werst forced the access hatch open to see *Harveer* Arniz hanging on Martin's arm, wrapped around arm and KC both, preventing him from shooting again. Her tail flailed at his groin. She hadn't made contact yet, but it was only a matter of time.

Some of the hostages were screaming, some weeping. Most were bleeding.

Head, arm, and one shoulder were out of the hatch. His implant was still fukked even though he could see the gleam of Vertic's fur outside the anchor.

Martin stopped trying to shake Arniz off and reached around to his back with his free hand. "This wasn't meant for you," he snarled. His hand dwarfed whatever he'd pulled from his waistband. Werst didn't recognize what he held until he fired four rounds into Arniz's body. Then he fought harder to get free as Martin flung her across the room, arms, legs, and tail flopping.

• ◆ •

Mirish moved quietly down the hall to the door of the Niln's nest room, any noise her boots may have made drowned out by sounds of the fight. She raised her weapon, took a moment to aim, and fired.

Watching from just inside the infirmary, Ganes had nothing to throw, no way to take her down before she fired again. By the time he found something, the Warden and Trembley would both be dead.

He was screaming when he slammed into her, his hand half a meter behind him on the floor.

• —◆— •

Werst had nearly worked himself free when Commander Yurrisk threw himself toward one of the open windows, nostril ridges closed, teeth bared. One step, two. He seemed to hang for a moment, then he dropped. And twitched.

Martin reached down, grabbed an arm, and flung the commander up over one shoulder. He fired the KC hanging from his other shoulder one-handed, then ran for the window.

Qurn sprinted across the room, scooped up the rolled data sheet, and followed.

Teeth clenched, Werst was halfway across the room, sure he could catch her given the weight of the plastic, when his implant pinged.

Out on the plateau, he saw Tehaven rear. Saw Qurn maneuvering her awkward burden through the fight. Druin were stronger than he'd thought—stronger than Freenim and Merinim had let on, and he'd have words to say about that later—but their spines were much like Human spines. He could break Qurn's neck if he could get close enough.

He could . . .

He ran for *Harveer* Arniz instead, tonguing his implant as he dropped to his knees beside her. "Gunny, Martin is heading for the VTA. He has Commander Yurrisk."

• —◆— •

Torin changed course again, tense muscles loosening at the sound of Werst's voice. "Has Commander Yurrisk?"

The Commander's unconscious. We need Ryder in the air ten minutes ago. Martin fired on the hostages before he ran.

On my way!

"How many down?"

Most of them.

The shuttle's medical facilities, as good as they were, wouldn't be enough.

Torin tongued her implant. "CC 882Alpha Override."

She passed an Artek fighting a Krai. Had to be Sareer, the others were accounted for.

Strike Team Lead Warden Kerr, contact by implant is against regulations. That code should not be in your possession.

Alamber had thought differently. Torin had agreed with him. "Get down here, now." The red Polint, Netro-whatever, was on the ground with Vertic's forefoot on his throat. "We have multiple civilian casualties."

Regulations keep this one in orbit until . . .

"Now. Or their deaths are on your head."

You are not able to accuse this one of . . .

A deeper, familiar voice broke in and Torin remembered a Dornagain rising up out of a well, an enemy in each hand. Finds Truth Through Inquiry, like others of her species, knew where the lines were drawn. **Detaching in ten, Warden Kerr. We're on the way.**

Martin was almost to the VTA.

Torin lengthened her stride.

"Gunny!" Firiv'vrak came up beside her, one arm stretched out, flexible digits holding an oval shape the size of Torin's palm. "Boarding pass. You didn't see it and I never gave it to you."

Torin snatched it out of the air as Firiv'vrak put on speed, her wedge-shaped body aimed at the running Druin carrying the plastic roll.

• —◆— •

Werst had his palm pressed over a sucking chest wound when Merinim dropped to her knees by his side, sealant in hand.

"I've got this," she said. "Ressk is upstairs!"

• —◆— •

Torin saw the Druin brace herself and swing the roll as Firiv'vrak caught up. The blow flipped the Artek over onto her back. A second blow slammed her into Torin, cutting her feet out from under her. Head tucked in, Torin landed on her shoulder, rolled, snapped her helmet off leaving the strap tangled in Firiv'vrak's legs, and got back onto her feet. Her uniform stiffened around her knee.

Martin, Commander Yurrisk over his shoulder, had reached the VTA's ramp. The Druin wasn't far behind.

She could hit them, hit both of them. Full auto, she couldn't miss. But she wasn't Binti Mashona. She couldn't guarantee she wouldn't kill them.

So tempting.

She hit the override on her cuff, turning the support for her twisted knee off, gritted her teeth, and ran.

• ◆ •

Dr. Ganes and Mirish fought silently, rolling about on the hall floor. Ganes seemed to have the upper hand so Werst leaped over them, slamming Mirish's head into the floor with a foot as he passed.

He landed on Malinowski's shoulders, wrapped his feet around her neck, drove his elbows into her temples, and jumped clear as she began to fall. Ressk had Zhang half buried under bedding, but she continued to fight.

The di'Taykan, Pyrus, had curled up in a corner, head down on his raised knees. He was mumbling in Taykan, the same words over and over, and Werst suspected the fight had sent him back to the *Paylent*. His weapon was nowhere in sight. Not good, although it would have to do for now.

Zhang caught Ressk under the ribs with a boot heel. He grunted and fell back.

Werst took his place and grabbed her ankle. Yanking her in close, he blocked a blow to the side of his head, and, as she flailed, choked her out.

Ressk sat up, nostril ridges slowly opening. "I had that."

"You're bleeding." His right sleeve dripped onto a pillow.

"Not mine. It's . . ." His eyes widened and he scrambled up onto his feet and off the nest.

Werst followed to find him on his knees beside Trembley, two fingers pressed into the young Human's throat. "Ressk . . ."

"It was a single shot!"

"The back of his head's gone."

"Fuk."

• ◆ •

Too late for the ramp, Torin jumped for the shuttle door, feet splayed awkwardly, boots magged. She slapped the thing Firiv'vrak had given her against the control panel, hundreds of tiny, flexible filaments slipping between the cracks.

Chasing Martin was low priority—way below rescue the hostages, avoid an incident that could cause another war, and arrest as many of the mercenaries as possible. But C&C was on its way down. If any Confederation/Primacy interaction on 33X73 had caused a renewal of war, that renewal had to have been planned in advance, and if she was with Martin, that would prioritize pursuit.

Although she didn't want Martin as much as she wanted Martin's employer. Someone had paid to send a crew of mercenaries after the ancient H'san weapons, playing on Major Sujuno's desperation. Someone had paid mercenaries to accompany the *DeCaal* to retrieve another ancient weapon, playing on Commander Yurrisk's desperation.

The correlations could be coincidence.

Could be the same someone.

Someone she needed to stop before more of the broken were further damaged, before more innocents died. Before the war started again.

Gunnery Sergeant Kerr! Vertic hadn't forgotten how to command. **Let it go! That's an order!*

Torin had spent most of her adult life following orders. Good orders. Bad orders. She'd taken comfort in knowing she didn't have to be responsible for the larger picture, that she could deal with the details that allowed her to complete the mission and bring her people home alive. A comfort she sometimes missed.

Gunnery Sergeant Kerr!

"Warden Kerr," she replied, as the lights turned green and she forced the hatch open far enough for her to slide into the air lock. "And I'm doing my job."

ELEVEN

TORIN FOUGHT TO KEEP the outer hatch of the air lock open, but without something strong enough to stand up to the force of its closing, she didn't stand a chance. She could remember arguments among Marine air support—the dangers of a door that couldn't be operated manually versus the dangers of a door that could. In the end, instinct had won as much as reason. A closed door was safer.

She stepped back as the clamps engaged and the light turned red again, turned, and felt the deck begin to vibrate under her boots.

Martin had started the engine.

Swearing, she dove for the inner hatch. Locked.

She'd left Firiv'vrak's boarding pass attached outside.

The sudden surge up into the air almost dropped her to her knees. Hand slapped flat against the bulkhead, she managed to stay standing long enough that sitting became her choice. Knees up, boots magged to the deck, the pressure of her back against the bulkhead would keep her secure.

They'd cracked her jaw when she made Gunny and the implant upgrade that had made it possible for her to order C&C to the ground should have made it possible for her to contact her team even inside an ex-Navy air lock, on her way off planet. Should have. Didn't.

The block had been lifted off Werst's implant when the Druin-in-red had carried the plastic data sheet out of the anchor. Carried it all the way to the shuttle.

"Fukking plastic." She let her head fall back and bounce once. Once more for the amount of trouble the plastic had been responsible for—where trouble meant not only centuries of carnage, but the discussion she'd be having later with Craig. He'd understand the situation, he understood the job; the lack of communication, not so much.

The inside of the air lock wore familiar patterns of wear—scuffs and dents in the matte-gray metal made by boots and equipment and people in too much of a hurry to be as careful as they should. The air smelled like the air on the plateau with undernotes of sweat and grime and age. Part of her found it reassuring, familiar.

The greater part of her wanted to write up the maintenance crew.

Unless the shuttle was in worse repair than it appeared or Martin was an idiot—both possible—he had to know she was there. On the one hand, he couldn't space her. Safety protocols would keep the outer hatch closed so long as sensors read life signs.

On the other hand, when the inner hatch opened, he'd be ready for her.

On yet another hand, at least he hadn't gotten away.

• ——◆—— •

"What happened to your voice?"

Werst touched the scar on his throat. "Got put back together by an engineer. Everything works. You okay?"

"I'm good. He . . ." Ressk nodded down at Trembley. ". . . got that one . . ." He jerked a thumb at Malinowski. ". . . off me. He didn't have a hope in hell of taking her out, but he gave me enough time to get a couple of solid hits in on the other one."

There was a bruise rising on Ressk's cheek. Werst sketched the edges with his thumb. "*Chirtric dirin avirrk* to take on all three."

"Fuk you." But he was grinning, so Werst counted it a win. "It was just Pyrus at first, the other two came out of nowhere. If Trembley hadn't . . ."

"Yeah."

"He was injured."

"Yeah." Werst held out a hand for half of Ressk's zip-ties and knelt to secure Malinowski. "Deal with Zhang."

Ressk shoved her over onto her back. "What do we do with Pyrus?"

The words the di'Taykan continued to repeat sounded like denial.

"Hang on." When Werst got to the window, the fighting was over. "Mashona! Where's Gunny?" He couldn't raise her implant.

"On the shuttle. She went after Martin."

Of course, she did, Craig muttered as he set their VTA down on the plateau.

Her helmet are showing only dirt.

Of course, it is.

"Fine. Plan B. How sane is Sareer?" Held by two Artek, she wouldn't be happy. Werst didn't care about her mood.

"How are we to determine sanity without a basis of comparison?"

He thought that was Keeleeki'ka. At this distance, in the artificial light spilling from the anchor, it was hard to tell. "Mashona!"

"Fine. Whatever. It's not like I don't have things to do." Binti jogged over, bent down, and jumped back as Sareer snapped at her. "She's fine. Why?"

"Pyrus is back on the *Paylent*."

Sareer's gaze snapped up to meet Werst's. Turned out she had an extensive profane vocabulary. In multiple languages.

"True what they say about sailors," Ressk muttered.

Binti grinned although Werst wasn't sure if it was at Ressk's comment or the profanity. Now the implants were working again, it could've been either. "I'll bring her up."

"What do we do with Trembley?"

Werst grabbed a piece of fabric out of the nest and dropped it over the corpse. "Gunny's got body bags. When she gets back, we'll treat him like a Marine."

Both Commander Ganes and Mirish were unconscious, sprawled across the doorway. They rolled the commander carefully onto his back and, while Ressk secured Mirish, Werst retrieved the commander's severed hand.

"Snack time?" Binti asked coming up the stairs behind Sareer.

"He's Navy." The wrist had been cauterized. Good. Well, good unless all the heat had cooked the interior. It didn't smell cooked. Werst ran for the infirmary, tossed the hand into the empty stasis chamber and hit start. The things were supposed to be idiot proof.

Back out in the hall, Binti and Sareer stood waiting for Ressk to get out of their way. He kept spraying sealant on Ganes' stump—although it had to have cauterized, too—and refused to move.

"It wasn't supposed to go this far," Sareer whispered as Werst crossed to stand behind his bonded.

He closed his hand on Ressk's shoulder, thumb stroking the skin of his bonded's throat. "It never is," he said.

•—◆—•

Arniz blinked awake, not entirely certain of where she was. Her unintentional naps weren't usually painful. Though, to be fair, the pain was distant. Muted.

Inner lids slipped across dark eyes on the pale face above her. The clothes were black, not red. Another Druin. A different Druin. "Who are you?" Her voice sounded strange, slurred. Had she been eating *rizkins*? She didn't think so. Her mouth tasted of copper, not mint.

The Druin cupped her cheek, gently moving her head. "I suppose that right now, I'm a Warden."

"I see." She didn't. This Druin wasn't speaking Federate. The voice Arniz understood came from the Druin's slate. Also, her tail hurt. A lot. She shifted and hissed.

"It may be broken. I'm sorry, I don't know anything about tails."

"Why would you?" Frowning hurt, too. "Did we crash?"

"No." The Druin blinked again. "You were shot."

And it all came rushing back. "Martin."

"Yes."

"That *ki seewin*!" Arniz assumed the noise the Druin made was a chuckle. "One of your people was with Yurrisk."

"I know." The Druin's hands were bloody. The air tasted of death.

Arniz tried to reach out, but couldn't move her arm. "Why?"

"Why was one of my people with Commander Yurrisk? We don't . . ."

At first Arniz thought the roar was in her head. When the Druin prevented her from thrashing, she realized it was a shuttle landing very close to the anchor. It took her a moment to recognize the higher pitched foreground noise as Salitwisi yelling about the planet being a Class 2 Designate.

She sighed. And had trouble breathing in again.

". . . Niln with internal dam . . ."

"More than . . ."

". . . many dead?"

". . . two stasis pods upstairs." Arniz knew that voice. It was Ressk. The Warden. "Get Arniz into one. Get Lows into the other. The dead can move on."

"In the sun," she whispered. The Druin leaned in. "Her name is Dzar. Put her in the sun."

"We can do that. We're going to immobilize and move you upstairs now."

The spray tasted of jasmine. A Human-sourced plant originally, it had become a noxious weed in northern parts of the Niln homeworld. Too many of them loved the scent to be thorough about eradica . . .

"Because we have a landing pad that was specially designed for landing on. I don't care if you're Wardens. I wouldn't care if you were the original inhabitants come back from an extended vacation! The shuttle goes on the pad!" Salitwisi sounded one extended vowel away from hysteria. Arniz wasn't entirely unsympathetic.

"I miss being unconscious," she sighed. The stretcher shifted. She didn't remember being put onto a stretcher. "Can you put me out again?" she asked the Druin as Salitwisi declared he and only he would keep his hand on his ancillary's wound.

"Maybe when we get you to the infirmary."

On the other side of the room, Salitwisi turned every sibilant into a hiss. "Because Hyrinzatil is my ancillary, that's why, you uneducated brute!"

Arniz winced and hoped whoever Salitwisi'd insulted had taken it personally and not as a reference to the Younger Races as a whole. An extended lecture on social prejudices would be all they needed right now. "Can you put *him* out?"

"Not my call."

"Pity."

"The stump's cauterized and I'm pumped full of pain killers," Lieutenant Commander Ganes slurred. "I'm fine."

"Yes, sir. This is your bunk." Not like Trembley was going to need it again. "Lie down on it." Ganes was an adult and an officer and if he said he was fine, he was talking out of his ass. Werst had no intention of arguing with him. He backed him toward the bed, let gravity put him down, then grabbed both feet and swung them up before Ganes could protest.

The moment he was horizontal, Ganes blinked twice, tried for a third, and failed to get his eyes back open.

"Yeah, you're fine." Werst lifted the stump up onto the commander's stomach—the sealant covered the burn, if not the lingering scent of cooked meat—checked pulse and respiration, hoisted a bag filled with medical supplies over the less bruised side of his back, and headed for the stairs.

He'd taken the commander's hand from the pod and tossed it into a stasis pouch . . .

"Looks like a lunch bag."

"I find one tooth mark on my hand and I kick your ass."

"You're welcome to try, sir."

. . . then he and Ressk had helped move the two dead—Trembley and a young Niln named Dzar—into the small room next to the nest. They'd left the prisoners where they lay. They didn't deserve any better.

Mashona and Ressk had escorted Pyrus and Sareer downstairs. Ressk had gone reluctantly, nostril ridges flared to draw in as much of Werst's scent as he could, but he'd gone. Werst may have done some scenting as well. Led away with his wrists secured, Pyrus wept silently, but he knew where he was. Might have been why he wept.

Werst had stayed behind to take care of Ganes.

As he reached the top of the stairs, Malinowski, still in the nest room, ran out of Federate profanity and switched to what Werst assumed was a Human language.

"We have another three pods incoming with C&C." At the bottom of the stairs, Ryder reached out and stopped a stretcher from ascending. The Katrien's head rolled limply left. "Dog's bollocks to hump them up and then down again, even with the AG."

"Seems like a design flaw to have the infirmary on the second floor," Freenim agreed, backing up to give the stretcher room to turn.

"And they're all flawed the same way," Werst told him, halfway down.

Freenim blinked. "Why?"

"We blame the H'san."

"That's what we do now," Ryder added, speaking quickly, shifting his weight from foot to foot. He pointed toward the closest common room wall. "Line the three who need stasis there—shorter run out to the C&C shuttle."

Guiding the stretcher back into the common room, Freenim shook his head. "There's going to be more than three."

"Internal injuries get podded. External get sealed. The *Warner-Lalonde*'s heading in, full Med-op by tomorrow. We . . ."

Everything with any potential to be medically useful has been unloaded. Alamber broke in. He was on the group channel, but it was clear to anyone with half a brain he was talking to Ryder. **Go.**

Ryder caught Werst's gaze. "I . . ."

Werst snapped his teeth. "Go!"

They'd needed all capable hands on injuries while medical supplies were being unloaded. Field dressings weren't a problem, not for anyone who'd served in either Confederation or Primacy military, but these were Niln and Katrien, and Ryder was the only one who'd begun the civilian EMT course. Seemed salvage operators came in a multitude of species. Who knew? Gunny'd be all over the rest of them finishing it now.

He clearly wanted to be at the shuttle controls, ready to fly as soon as possible, but he'd done his job.

"Ryder finally going after Gunny?" Ressk asked as Werst spun around. "Your situational awareness sucks by the way."

"I fell in a pit."

"My point."

Werst moved close enough for touch. "How'd he get the Navy to move so quickly?"

"Sicked Presit on them."

"Ouch."

Presit was nowhere in sight, but Dalan moved among the injured, camera up, red light announcing his mobile invasion of privacy. He skirted Salitwisi, still applying pressure to a stomach wound on a younger Niln even though death had stopped the bleeding.

Strike Team Alpha's job had been to free the hostages and three of them were dead. Well, five, counting the two Martin had taken out before they'd hit dirt. It could have been a lot worse. Not that the gunny was going to accept that.

Of course, she was still on the job.

"Why aren't you with Ryder? There's four of them," Ressk continued when Werst turned. "Beyvek's in the shuttle, and Martin will have cut him free. Ryder's still shit at close combat and I'm not saying Gunny won't have taken all four out, but a little competent backup won't hurt once he's got a grapple on them."

Ryder didn't pull a trigger. That was the deal. He didn't carry the dead.

"Can you fight?" Ressk demanded suddenly. "*Chreen!* Your injuries . . ."

"I can fight." He could do what he had to. He'd have time on the shuttle to rest.

"C&C will be here before you hit vacuum." Ressk pressed his forehead against Werst's, then pulled away. "Go."

So he went. The air in his lungs, air Ressk had breathed out.

The Polint had taken care of their own injured. Only fair, Werst figured; they'd caused most of the injuries.

Vertic had two of the mercenaries at her feet, the black and the red, Camaderiz and Netr-something. Dutavar knelt by his brother. Tehaven was on his side, breathing heavily.

"How many times do I have to say it! Your way isn't the only way!" Tehaven snarled as Werst ran by.

Dutavar grabbed his brother's wrist. "This is the wrong way."

"You're the wrong way!"

"Bertecnic!" Vertic's sudden bellow made Werst's blood throb against the scar. Ressk was right, his situational awareness was shit. He stumbled, caught himself at the last minute, and tried to move a little faster. His stamina was shit, too. And Ryder wouldn't wait because Ryder didn't know he was coming. Might not wait even if he knew. What the fuk did he think he was doing, charging off like he was the only one on the team who could throw a punch . . .

"Need a lift?"

This time when he stumbled, a big hand hauled him up into the air. He twisted around it, grabbed a footful of Bertecnic's vest, dropped onto the Polint's back, and held on with all four extremities. Distance that would have taken him another five, maybe ten minutes to cover, disappeared under Bertecnic's undulating stride. Wasn't Werst's first ride on a Polint. He didn't enjoy it any more than he had the first time.

"Durlan says to bring Gunnery Sergeant Kerr back."

"Damn right."

"She says the chain of command is a twisted mess without her."

Durlan Vertic currently commanded five fully grown male Polint. If she decided to take over, she couldn't be stopped. Twisted mess seemed accurate to Werst.

The ramp had started to withdraw when they reached the shuttle, so Werst jumped clear before Bertecnic had fully stopped. He slid into the air lock as the outer hatch closed and into the cabin as the inner swung shut. Then immediately into a seat as Ryder engaged the engines. Once they were high enough for the dampeners to kick in, he unbuckled and moved up to the copilot's chair. Only to find it already occupied.

"Are you having invited him?" Presit demanded, punching Ryder in the arm.

"He invited himself."

Ryder had the shuttle on the kind of angle that meant Werst had to hold on to the chair back to keep his balance. "But you invited *her*?"

"She also invited herself. I didn't want to take the time to toss her out."

"I are going where the story are being and Gunnery Sergeant Kerr are providing a career's worth of story." Under the self-serving justification, Werst heard concern. Which didn't negate the self-serving justification as there was SFA the Katrien could do once they caught up. "Dalan are fine recording on the ground," she continued, waving a dismissive hand, enameled nails gleaming in the light. "I are knowing better than to be trying to get anything but grief out of a trauma situation. It are all in the editing."

"It?"

"Yes, it." Her tone suggested that if he didn't know what *it* referred to, she wasn't lowering herself to explain.

"You might want to park it," Ryder growled, eyes locked on the board. "There's a storm in the upper atmosphere. I'm going straight through and it's going to get . . ."

The shuttle moaned, twisted, and slid about three meters left.

". . . bumpy."

Torin braced herself as the shuttle dropped into place on the *DeCaal*, deck shuddering as the clamps engaged. It'd been neither the worst ride she'd ever taken, nor the most dangerous. That honor went to the trip from Big Yellow to the *Berganitan* in an HE suit, strapped into Craig's salvage pen, blood running down her arm to fill her glove. In comparison, this ride was a welcome breather after a long day. "First-class ticket," she muttered, standing and stretching as the engine powered down. She checked her KC. Turned to face the inner door. If Martin planned on leaving the VTA, he'd have to go through her.

"*Warden Kerr.*"

"Robert Martin."

"You know who I am." He sounded smug.

"I know enough. Throw down your weapons and open the door. I'm willing to accept your surrender."

Several seconds of silence followed.

"Because the great Gunnery Sergeant Kerr only has to ask," Martin sneered.

"Warden Kerr." She smiled. "And I didn't ask."

The silence extended.

"You're trapped in the air lock," he said at last.

"You're trapped in the VTA," Torin replied. "It's a matter of perspective." If Craig grabbed air the moment the medical supplies had been unloaded, he wasn't far behind them. She had to keep Martin distracted until the grapples were deployed.

"What do you think is going to happen when the inner hatch opens? You can't kill all four of us before one of us kills you."

Four? Martin, the Druin, Commander Yurrisk, and . . . Beyvek.

"Except . . ." He was laughing now. Laughter was good. Laughter took up time. He could chortle evilly for the next hour as far as Torin was concerned. *". . . you're not allowed to kill us, are you?"*

Only technically true, and Justice had worked hard to fill the Strike Teams with those who'd do what Justice wanted. Torin had been responsible for enough death—she touched her vest—that she'd prefer not to add to the total. "I can do anything I want to you, as long as I complete the paperwork. If I happen to only kill you, I can cope."

"You'll die right after I do."

"Perhaps. You'll be dead. You'll never know." Stalemate. The bully Werst had known wouldn't trade his life for hers. "And I hate doing the paperwork; dying would let me avoid it."

"You'd . . ."

"We haven't time for this," another voice snapped. Not Krai, had to be the . . .

The light in the air lock went off. The clamps holding the inner hatch shut disengaged.

Not smart. When the door opened, those inside the VTA would be silhouetted in the light. Torin hadn't wanted to do it this way, but needs must. She took two long steps to stand tight against the bulkhead. A white line opened along one edge of the hatch, painting the floor just past the toes of her boots.

She slipped her finger through the trigger guard.

The lights came back on.

Temporarily blinded, she felt a sharp pain in her cheek and sudden cold spread out across her face. She squeezed the trigger as her knees buckled. Heard shouting, hit the floor.

Oh, yeah, she knew this feeling.

•—♦—•

"There are being a saying among my people that a straight line are not always being the shortest distance between two points. You are understanding what I'm saying?"

"It would've been faster to go around the storm." Ryder's teeth were clenched so tightly Werst was impressed he managed to get the words out.

Presit gave a satisfied chirp. "That are what I'm saying."

"Werst?"

He threw the diagnostic up into the air. "Another half an hour to match orbit and you'll need to dead-eye the hookup when we reach the *Promise*. Sensor array's completely out."

"Please," Presit sniffed. "Warden Kerr are being fine. Warden Kerr are always being fine. And I are having noticed she are having turned her helmet scanner off."

"Gunny lost her helmet dirtside," Werst told her.

Presit sniffed again. "Typical. She are never thinking of my visuals."

•—♦—•

Torin hadn't entirely lost consciousness. She felt herself lifted onto a stretcher. Saw Martin carry Commander Yurrisk out of the air lock slung over a shoulder, one hand dangling level with Martin's ass. Saw a whole lot of orange plastic go by and felt the hair lift on the back of her neck. The sight of so much plastic in one place made her hands twitch to hold the alleged weapon that had brought Martin to 33X73.

Her head wobbled on her neck as the stretcher rocked, the generator creating a weaker field along the left. Felt the contact the stretcher made with the side of the hatch. Felt it tip. Felt the surge of adrenaline at the prospect of falling, helpless.

A flash of red, barely seen over the curve of her cheeks, lifted Torin's foot back onto the padding as the stretcher leveled out at the last moment. She could hear the generator whine, long past its time for a maintenance overhaul. It became easier to keep her eyes open. She still couldn't move.

Up above, the interior of the ship showed the same signs of wear and grime as the air lock and loops of multicolored wire hung between the bulkheads. It resembled the interior of Salvage Station 24 and Torin considered the comparison to be a compliment—the salvage operators had children to keep safe. It was obvious Commander Yurrisk had done all he could to keep the *DeCaal* flying.

The control room looked to be in better shape. Not good shape, but better. Teeth gritted, she rolled her head to the left and saw four seats original to an Aggression class ship: OIC, helm, weapons, and communications. All four looked to be fifty percent duct tape. The weapons console and what looked like half of communications had been removed.

Qurn bent to lay the roll of plastic on the floor, realized there was no room, and propped it in a corner. "Fine, they wanted her dead," she said, carrying on a conversation Torin had clearly missed the beginning of. "But you can't tell me that your leaders wouldn't prefer Gunnery Sergeant Kerr alive and converted to your way of thinking."

"You think she'd *convert*?"

"Why wouldn't she? Or don't you actually believe your way is the right way for your species?"

"I don't have to keep *you* alive," Martin snarled and Torin realized two things.

One, Martin was Humans First. No question.

Two, for the first time since the team had left the station, she could hear the silence between words. The Druin didn't need a translator. She spoke Federate.

Three things.

She could move.

She hit the deck on one knee, rolled under the stretcher, shoved it at the approaching Druin, and felt the unmistakable pressure of a KC muzzle against her spine. When she stood, the pressure moved from the base of her neck to the top of her ass, but it didn't let up. Odds were high a round would go through her uniform at this range. If she caught a break and it didn't, the impact could still shatter vertebrae.

"You said she'd be out for hours, Qurn."

"If she were Druin, she would be. As she isn't . . ." Qurn spread her hands. "The reaction is variable between species. Variable within species as well."

"Don't care. You." Martin pointed at Qurn. "Cover her. You." When the finger came her way, Torin raised a brow. "Get out a zip-tie. Beyvek, secure her."

"The zip-ties are in my pack," Torin said in her *dear lord, you're an idiot, but it's in my best interests not to actually say that* voice. "My pack's dirtside."

He searched her face, checking, she assumed, for the lie. She'd been a senior NCO in a company that luck of the draw had dropped into more hard combat than any other two; Robert Martin wasn't nearly skilled enough to find anything but what she wanted him to find.

"Tie her with this." Qurn detached a decorative red braided cord about a meter long from her sleeve.

"Yeah," Martin scoffed, "as if that'll . . ."

"It'll hold her." Qurn's voice, on the other hand, said *you can trust me. Our interests align, and I would never lie to you.* It was remarkably effective and Torin made a mental note to ask Freenim about subharmonics.

Qurn tossed the cord to Lieutenant Beyvek, who caught it one-handed—the pressure of his weapon not letting up. "Hands behind your back."

In his place, she'd have folded her forearms across the small of her back and secured them wrists to elbows. The last couple of years had taught her that was the only way to ensure certain people remained secured. Without her experience, Beyvek crossed her wrists and wrapped the cord in a diagonal pattern. Unfortunately, the Krai still built with rope and the more traditional had kept the old skills of net making alive. The binding felt like Beyvek had come from a traditional family. She could flex her fingers, there was no chance of her circulation being cut off, but regaining her freedom wouldn't be easy.

When Beyvek slipped out from behind her, he gave her a hard shove, pushing her back into the remaining rear corner. Storing her out of the way much as Qurn had done with the plastic. She watched the Druin slide a metal rectangle into a fold of her robes, noted that

Martin had dropped Commander Yurrisk into the OIC's chair, swiveled it around to face the board, and settled himself at the helm. Then she took a moment to examine Beyvek.

He had a rising bump over one eye, the bruising feathering back into his mottling, numerous small cuts on both hands, and a split nostril ridge—souvenirs of his fight with the Artek. He also wore no discernible expression . . . emotions either shut down or crippled. Given that he'd followed Commander Yurrisk from the *Paylent* and the Artek did most of the Primacy's boarding, Torin bet on the latter.

"In case you missed it, Lieutenant . . ." She kept all censure from her voice. Pure senior NCO to junior officer—respectful of what they could be, there to support what they were. "Martin here is a member of Humans First."

His gaze tracked up to her face. "So?"

"You're not Human, sir. Martin considers you a lesser species, not worth his time."

"He saved the commander."

Ah. "How is Commander Yurrisk, sir?"

Beyvek's nostril ridges shut, the split seeping blood. Emotions surged free—anger, terror, guilt, pain—and were quickly shut down again. "An enemy came through the air lock. How do you think he is?"

"I don't know, sir." Although she could make a good guess. "The report from the *Paylent* was almost entirely redacted."

"Twenty-five of us fought our way free and the commander stayed on our six all the way to the engine room. Did you know there's an air lock between the rest of the ship and the engines in case of a blow-back?" His tone remained conversational. "Double hatches. But they got through. Three of us died immediately. Seven of us were injured. They kept coming. There were pieces. In pieces. People in pieces. The commander . . ." Anger. Terror. Guilt. Pain. Gone again. Torin could see his throat work as he swallowed. "Use your brain, Gunnery Sergeant. I know thinking's against Marine SOP, but try. The commander kept us alive. Now, I'm keeping him safe."

"And the others from your crew?"

"They'll catch up."

"Did Martin tell you that?" She couldn't work even a millimeter of give into the cord. "You can't trust Martin."

"Sergeant Martin saved the commander. Brought him to the VTA."

"Commander Yurrisk needs more help than Martin can give him." Torin met Beyvek's gaze. "I can get him that help. I can get all of you help."

No one, not the most underexposed of the Elder Races, would consider Beyvek's flash of teeth a smile. "Help? We got *help*."

"Yeah, some military therapists can be next to useless," she agreed. "But none of us . . ." Emphasis on us. ". . . are military anymore."

He stared at her for a long moment as his nostril ridges slowly opened. "Once you're in," he said quietly, "you never get to leave."

Her arm twitched against the hold of the cord and she squared her shoulders against the weight she still carried as he turned his back on her. He wasn't wrong.

A rustle of fabric flipped her gaze over to Qurn in time to see her peel off a glove and touch the plastic with long, pale fingers. The plastic remained inert. She turned toward Torin's scrutiny—Torin was ninety-nine percent certain she'd bit back the sudden intake of breath punched out of her lungs when flesh came in contact with the plastic—and said, "Your teammate, Ressk . . ."

She rolled the name out so dismissively, Torin would've bet high that she hadn't bought Werst's impersonation of his bonded.

". . . believed he saw a pattern in it."

"Spill your guts, why don't you," Martin grunted, attention on the board. The *DeCaal* hadn't been retrofitted with a hard light display, keeping Torin from seeing what he was doing.

"It makes no difference." Qurn drew the glove back on.

Torin raised a brow. "You're going to let him give an artifact with so much potential to a third-rate terrorist organization with delusions of grandeur?"

Qurn looked up at her, her minimal features making her face almost as expressionless as Beyvek's. "I'm not letting him do anything."

"Damned right." Martin leaned back, grabbed the commander's right wrist, and slapped his palm down on the screen.

Around the time of the *DeCaal*'s commission, electronic field readers had been installed on ships below a certain size in an attempt to prevent the Primacy from using captured vessels against Confederation forces. One member on each ship's crew had been designated as the key to unlock the ignition sequence. They had to be alive and reasonably healthy in order to match the same electronic field recorded originally. The field couldn't be faked or duplicated. Unfortunately, battle killed indiscriminately and when the seventh ship lost its key in the middle of a deployment, Parliament declared the attempt too flawed to continue paying for.

The Corps had given the Navy a lot of shit about it—they'd refused to allow Parliament to code their weapons the same way—but, to be fair, everyone Torin knew in the Navy had called it an asinine idea.

Either the *DeCaal* had already been in the scrap yard when the feature was recalled or Commander Yurrisk had reinstalled it. Fifty/fifty chance.

"Martin didn't save Commander Yurrisk," Torin said. "He needs him to unlock the ignition sequence. He's using the commander to escape."

"He got him off planet." Beyvek shrugged. He was better at it than a lot of Krai. "That's good enough for me."

She needed more time. Craig had to be close.

Torin leaned back against the bulkheads, the angle of the corner giving her bound hands enough room to work, the pressure keeping her shoulders still and attention off the attempt. Still no give. The bulkheads also kept her upright when the occasional wave of dizziness swept over her as her system worked at flushing the last of the Druin tranquilizer. "Since we're all going with you, care to share a destination?"

Martin snorted. "We're going to a future without the Elder Races fukking us over."

"In what way?" Both Martin and Qurn turned to stare at her. Beyvek didn't bother. "I'll make it easier," Torin said in her most patronizing tone. "We're no longer at war. How are they fukking us over now?"

Her most patronizing tone had been designed to get a response. Martin surged up out of his chair, remembering at the last instant he needed to keep the commander's palm against the screen. "We died in their war!"

"Yes, we did." She'd grant him the *we*, he'd worn the uniform.

"And what did we get out of it?"

"An extended life expectancy, a presence on multiple planets, and membership in a civilization worth preserving."

Qurn's mouth twitched.

"I don't do rhetoric," Torin added. "Destination?"

Martin returned his attention to the board and the bars of light flashing across it. "When we're moving, I'm gagging you."

"You don't care he's withholding information?" she asked Qurn.

"Not really."

"Lieutenant Beyvek . . ."

"He saved the commander."

"Yes, sir, he did," Torin agreed. "Now, where's he taking him?"

Beyvek glanced back at Torin, raised his weapon, then turned to face the back of Martin's head. "Where *are* you taking him?"

"To safety."

"You're not a Susumi engineer," Torin pointed out, "This ship's canned equations are Navy. The Wardens can track them." Ressk and Alamber could track them. The official tracking program was still being discussed in a Parliamentary committee. "If Martin does the math, we'll all become the kind of statistics that discourage amateurs from jumping. We all die. The commander dies."

Beyvek's shoulders stiffened, and his finger slid through the trigger guard. Naval officers understood the dangers of decompression better than most, but weighted against his instability . . . Torin bet that if any area of the ship still had self-patching up and running, it would be the control room and, besides, that close even the Navy could hit a soft target.

Her foot in his back would fling him forward, taking him out while keeping his aim more or less on Martin.

But then Qurn was there at Beyvek's side while Torin's eyeballs twitched and she fought to keep her knees from buckling as her brain insisted the room had begun to spin.

"You need Sergeant Martin to help keep Yurrisk safe." Soft, convincing, Qurn leaned in, her hand over his, and lifted his finger back to a resting position, pressing—no, stroking—it flat. "Remember how Martin carried our commander to the shuttle? He won't endanger him now. Will you, Sergeant Martin?"

"We're going back to our entry point and through, a simple reverse equation," Martin growled. "My people will meet us on the other side. They'll have original equations Justice can't track."

"And Sergeant Martin will take Warden Kerr with him, leaving the three of us on the *DeCaal,* so we can gather up the rest of the crew. Isn't that right, Sergeant."

"Yeah, that's right."

"Safe," Qurn said close enough to Beyvek's ear, it twitched. When he nodded and the tension left his arms and shoulders, she murmured, "Well done."

Torin wouldn't have believed Martin had he said the H'san liked cheese, and she sure as hell didn't believe he had any intention of giving up the *DeCaal*. Humans First had lost their fleet, they needed another. Hell, she was pretty sure Qurn didn't believe him either. Beyvek did. Or he believed Qurn, if not necessarily Martin.

"Keep watch." Qurn squeezed Beyvek's shoulder and crossed to communications, pulling off her gloves as she dropped into the seat.

Martin's fingers staggered across the board where Craig's danced, but Torin knew she was nearly out of time. She could feel the engine vibrations through the bulkheads and deck. Craig had to be close.

If puking would slow things down, she'd happily spew the contents of her stomach over the deck.

"I'm reading a second ship." Qurn had removed both gloves to work the station.

"Who asked you?"

"A *second* ship," Qurn repeated.

"It's C&C. Full of Dornagain." Martin threw up both hands, then slapped them hurriedly back down as port engines fired and the ship lurched. "They land when the danger's over and apply protocol. They're harmless. Toothless. Useless. Besides, Wardens respect personal property." Torin could hear the sneer in his voice. Mashona, back on 33X73 could probably hear the sneer in his voice. "Human Wardens are dogs licking the feet of the Justice Department. Dogs with no bite. You're not allowed to blow up civilian ships, are you, Warden Kerr?"

Torin remembered the *Heart of Stone* and, as the *DeCaal* left orbit, muttered, "Not anymore."

Puking did not slow things down.

• ——◆—— •

"Dumb-ass rules," Werst muttered, ignoring the ladder from the shuttle lock and dropping straight down to the deck. At this point, pain was relative. Ryder's longer legs had already taken him to the first hatch. "If we'd blown their ship up when we got here, they'd be shit out of luck right now."

To Werst's surprise, Presit slid down the ladder's outside supports like a vacuum jockey on alert, hitting the deck seconds behind him. "And you are being a good enough shot to be sure you are not blowing up the Susumi engines and irradiating the entire system?"

"It's not shitting through the eye of a needle, there's three square meters to avoid." To his further surprise, she kept up to him on the flat. "Even you could hit it safely."

"Then Justice ships are having to be armed, and where are the line being drawn between the Wardens and the military?"

"You read the minutes of the Parliamentary committee." Werst had endured a compulsory assembly where Commander Ng had shared their conclusions. If Ressk hadn't had an illegal copy of the new *Band of Jernine* on his slate, he wouldn't have survived.

Silvered fur rose and fell and disappeared into the air filters as Presit shrugged. "Of course, I are having read it. *The Strike Team's primary responsibility remains the control and capture of mercenary groups during violently illegal activities. As the budget contains no provision for boarding parties, we find it to be preferable that the occasional mercenary vessel escape rather than the Wardens' ships be armed creating a third fighting force.*" Quote over, she snorted. "I are wondering what they are thinking you are doing if they are thinking you are not fighting?"

"Politicians think?" Three strides ahead entering the control room, Werst dropped into the copilot's seat as Ryder started the engines. "And Martin's not getting away."

"The *DeCaal* are an Agressive class ship."

"Yeah?"

He heard her sigh from the seat behind him. "*Promise*, while being a fine ship, are not being fast enough to catch it."

Ryder drew both hands, fingers spread, across the board, and *Promise* flung herself out of orbit. "We've a fair go," he growled.

• ——◆—— •

"The Wardens are in pursuit."

"C&C," Martin began.

Qurn cut him off. "Not C&C. The *Promise* has left orbit, has corrected for rotation, and is following, all engines on full."

"Too bad you never got around to replacing the rear guns." He slapped the commander's dangling foot. Yurrisk showed no sign of regaining consciousness. Qurn seemed unconcerned, but as that could mean she didn't care as much as it could mean he was in no danger from the drug, Torin wasn't reassured. Pushing the commander's foot out of his way, Martin swiveled the chair around to face Torin in her corner. "Looks like your team doesn't want to lose you, Warden Kerr."

"Yeah, well . . ." Torin shrugged. ". . . I owe Warden Ryder money."

"What happens if we space you?" He sounded as though he honestly wanted to know.

"I die. They keep coming. You go to rehabilitation with extra bruises." She swallowed. Carefully. The vertigo and accompanying nausea hadn't completely faded. With the *DeCaal* down to bare bones, she'd thought it strange the puking protocols were still in place in the control room. It hadn't been necessary to untie her so she could clean up the vomit. Although, given Commander Yurrisk's vertigo and the *DeCaal's* starboard wobble, maybe it wasn't all that strange.

When Martin returned his attention to the board, he snickered. "We've doubled our lead. That cobbled-together bucket can't catch us."

• —◆— •

"They're fast." Werst checked the *DeCaal's* lead—again—and found it had more than doubled.

"They are still not responding when I are hailing them. That are being very shortsighted. If they are not responding, they are not knowing I are offering them an exclusive venue to be telling their story."

"What do they care?" he asked. "They'll still be going to rehab."

Ryder's thumb drummed against his thigh. "If they make it to rehab."

Presit sniffed. "Rehab are being weightless against the exposure I are offering them."

As she began listing the ways Martin and his people would benefit from that exposure, Ryder threw up a new screen of equations and separated out the bottom line. "They'll be one point nine seven seven million kilometers away by the time they reach the jump site."

That wasn't good, Werst acknowledged with a grunt, but it wasn't the end of the line either. "So we follow the jump. Not the first time you've had to do that."

"Or we get ahead of them."

Werst's nostril ridges closed. "You're not."

"Only choice."

"I are not understanding," Presit pointed out petulantly, having finally realized no one was listening to her.

"Micro jump," Werst told her, strapping in as Ryder began working the equations.

"Oh."

He unstrapped and squirmed around. "Oh? That's it? That's your response?"

"Of course not. I are going to be needing your data to be proving to my network the jump are having occurred."

"You're weirdly calm. This has never been done before! No one knows how time will pass in Susumi on a jump that short."

She combed her claws through her whiskers; first the left, then the right. "I are having ridden the Susumi wave of a Primacy battleship with Craig Ryder at the helm of the *Promise*." Then, in case Ryder thought that might be a compliment, she added, "Riding the first micro jump are being an excellent story to be adding to my portfolio. There are very likely more awards in my future." Without being told, she pulled the crash harness over her shoulders and buckled in. "I are ready. Proceed."

"Holy fukking shit!" Martin reared back, the chair under him shrieking a protest at the abuse.

Susumi alerts, proximity alerts, and radiation alerts screamed out warnings as the *Promise* appeared suddenly, emerging from Susumi space only a few thousand kilometers away, riding the exit wave ninety degrees across the *DeCaal*'s trajectory. Caught in the lateral disbursement of Susumi energies, the *DeCaal* rocked hard to port.

Torin rolled with the motion, threaded her legs through the loop of her arms, and, on her way to her feet, drove her head into Beyvek's nostril ridges. He gasped, snorted blood, and dropped. She saw Martin rise, swing his weapon around toward her, then all she saw for a moment was a flurry of red fabric. As it cleared, Martin flinched, slapped his cheek, and collapsed.

In the sudden silence, Qurn bent over Commander Yurrisk, one bare hand wrapped around his throat.

"Back away." Torin's hands remained tied, but they were in front of her now, holding the comforting bulk of Beyvek's KC and at this distance she couldn't miss.

"I'm checking his vitals!" Qurn snapped, sounding rattled for the first time. "I've never used the drug on a Krai."

"Have you used it on a Human?"

"You're fine."

Relatively speaking, she was. Torin slung the KC's strap over her head, pulled a zip-tie from her side seam, and dealt with Beyvek before pulling a second tie and moving to Martin. She secured Martin's wrists and ankles with the ease of frequent practice—although she'd never practiced bound—while the bulk of her attention remained on Qurn. Torin neither liked nor trusted unanswered questions, and this was a chance to resolve at least one.

When Qurn suddenly stepped back into the minimal space between the seats just as Torin leaned in to check the board, her robes tangled with the weapon swinging forward off Torin's shoulder. It might have been an accident, but Torin didn't think so. A power play had been inevitable. They danced for a moment, Torin keeping Qurn's hands busy in case she intended to try for another shot of the tranquilizer. The instant they gained enough floor space, she grabbed a double handful of fabric, lifted, and put Qurn down a half meter away, releasing her and regaining the KC in one extended move. "You're not a member of the Confederation, I'm not certain what your rights are, but cooperating is in your best interest right now. Who the hell are you?"

"I'm an agent of the Primacy government." Qurn's hand began to slide beneath her robes. Torin cleared her throat and she froze. "I have identification."

"And I have a need to see both your hands."

"Of course. That's not a problem." She held them out in the universal gesture of *I'm humoring you because you have a gun.* Her colorless white skin had picked up pink tints from the red of her clothing and Torin wondered if Qurn's fashion statement was not only a way to keep from leaving DNA evidence but also used to direct attention away from the minimal facial features of the Druin. In a less overt outfit, no one would recognize her. "My government," she continued calmly, "sent me into the Confederation to determine if the peace is legitimate. We know so little about you. I observe. I interact. I've been with Yurrisk and his crew for the last four months."

"You've got a good handle on the language."

"I'm very good at my job."

If she saw it as a job, then she wasn't a fanatic. Point in her favor. "And you went along with Martin because . . . ?"

"Martin. Yes." Qurn shook her head. "Humans First is a potential fissure in the peace process. If Martin took me to their leader, even as his prisoner, I'd have gained important information."

"Which you'd have shared with the Confederation."

She blinked. "Probably not."

"Odds are high the leader would have had you killed."

"I can be quite persuasive."

"Wouldn't matter. You're not Human."

"We have speciesists within the Primacy." She blinked again, dismissively, looking for a moment very much like Freenim. "I know how to work within their parameters."

"And what were Commander Yurrisk's parameters?"

There was honest emotion on her face when she glanced over at the unconscious body. "In the beginning, he was a way to an end. Broken, desperate, and easy to manipulate. As time passed, I grew fond of him. I found I could keep him stable. I don't often see the results of what I do, and it was good to be useful." Qurn lifted her minimal chin, daring Torin to comment—although Torin doubted she realized she'd done it. No matter. No one who'd served with the di'Taykan worried much about interspecies relationships. "He needs help," Qurn added when it became clear Torin wasn't going to judge.

"I know." Torin nodded toward Beyvek. "They all do. They'll get it."

"How can you promise that?"

"How can I not?" Torin shifted the muzzle of the KC to follow Qurn's step sideways.

Qurn stilled. "I don't blame you if can't give me your trust."

"Good." Torin let the KC hang, then let the metal rectangle she'd lifted from Qurn's robes slip out of its nest in the knotted cord and into her hand. Aimed it and pressed the slightly raised stud on one narrow end hoping she'd found the trigger.

Qurn flinched, a hand rising to her cheek, eyes opened impossibly far.

"Trust isn't given," Torin said as the Druin collapsed. "It's earned."

On the screen, the *Promise* had begun to come around. Her implant still out, Torin gave the com station a quick once over. Half a dozen blinking lights suggested Craig or whoever he'd brought with him, were trying to get through. Torin assumed she'd be able to figure out the communication controls in time, but as long as the *DeCaal* continued to race toward a preset jump point, she had a more important problem.

Leaning over Martin's body, Torin studied the board. She'd been infantry, boots on the ground, and the only ship she'd ever flown was the *Promise*. The *Promise*'s controls were unique. Back when she and Craig had been trying to make a go of it as CSOs, Craig had insisted she learn to fly for safety's sake if no other reason. Those lessons were very little help when facing a traditional Navy configuration.

Two things worked in Torin's favor. First, Martin's piloting skills were no better than hers and the navigation program had been doing most of the work. Second, vacuum was unforgiving, so the most important functions were designed to be obvious—in case of situations similar to the one she found herself in.

Vaguely similar, given the Druin, the big roll of orange plastic, and the red cord still binding her wrists.

A slide to the left shut off the stern engines.

A tap on the bar to the right fired the bow engines—the computer controlling the duration required to counteract their forward momentum.

As they slowed to a stop, Torin slid the power bar to the smallest non-negative integer.

The vibration under her boots faded.

The lights went out.

The air stilled.

"Shit."

•—◆—•

"*. . . Kerr . . . communi . . . not . . .*"

"That's a slate signal. Werst . . ." Hands in constant movement over the board and through the hard light projections, Ryder flicked a screen to the right. ". . . match it."

Werst, who'd been told in no uncertain terms to keep two readouts out of the red while they dropped the energy from the micro jump, leaned back and reached out with a foot.

". . . just in . . . try . . ."

All three of them flinched at the sudden burst of static and Werst slammed his nostril ridges closed as a new cloud of silver-tipped fur wafted past his face. Turned out Presit hadn't been as calm about the micro jump as she'd wanted him to believe. Surprise. He scowled at the frequency fluctuations. Should've sent Ressk up with Ryder. Or Alamber. He wasn't tech . . . "There!"

"Promise, *this is Warden Kerr. Are you receiving?"*

The tension went out of Ryder's shoulders. If Torin was on coms, she was in control of the *DeCaal*. "Receiving. Torin, are you all right?"

"I'm fine. Martin and Lieutenant Beyvek have been restrained. Commander Yurrisk and Qurn, the Druin, are unconscious. My implant isn't working, and I've powered down all systems along with the engines."

"On purpose?"

"Yes, I shut off life support on purpose."

"Hey, your warrior ways are strange to me."

"Ass."

"It are adorable how you two are flirting even though we are still likely to be dying!"

"You brought Presit?"

"Not so much brought."

"Likely to die?"

"The theory about a micro jump accumulating more energy than a longer jump turned out to be accurate. We're having a little trouble getting free."

"A little trouble?"

"I'm on it."

"And you're the best."

"Why was I the only one concerned about something that had never been done before because even Susumi engineers considered it to be too dangerous?" Werst grumbled as he finally got the two readouts to lock in the green.

"Because Craig Ryder are the only thing Gunnery Sergeant Kerr . . ."

"Warden Kerr."

Presit huffed out an audible breath before continuing. ". . . Warden Kerr and I are agreeing on."

"Likely to die," Torin repeated.

"That are fact, not opinion."

The alarm was not unexpected.

"Nothing to worry about," Ryder muttered as the *Promise* tilted sideways. "Just the Polint's quarters breaking off."

"Just?" Werst matched Ryder's tone.

"Everything's sealed, that's what the internal locks are for. We'll pick it up later."

"If we are surviving . . ."

"You're surviving."

"That are not being . . ."

"You're surviving."

The alarm shut off. Werst's ears continued to ring in the sudden quiet.

The ancient pilot's chair crackled as Ryder leaned back and rotated his wrists. "And we're out."

Presit snorted. "You are so wanting to be saying 'I told you so,' aren't you, Gunnery Sergeant?"

"Yes, I am."

Presit's laughter sounded close to hysteria, but Werst wasn't sure his sounded much better. He felt like he'd been running on adrenaline for a tenday. Or two.

"We're bonzer now, Torin. Out of the energy stream, heading your way." Craig flipped a timer up into the air. "How long before you're showing a noticeable drop in temperature?"

"Slate says three hours and forty-seven minutes."

"You may get a bit chilled, but it'll be easier to grapple and link air locks than talk you through bringing the *DeCaal* back up."

"The board's locked to Commander Yurrisk."

That explained why Martin had grabbed him. Loser.

"Like I said, easier."

"The plastic's probably blocking your implant, Gunny. Mine came back when Qurn humped it out of the anchor, yours dropped out when the air lock closed."

"Range?"

"Variable."

"That are being not only a lot of plastic, but the first technological artifact of the plastic we are having found. Do not be touching it . . ."

Werst saw a small, shiny black finger poke Ryder in the arm. Presit was up and moving around.

". . . until both Dalan and I are being present. This are needing to be recorded on more than a helmet camera or a slate."

"Yeah. No danger of that. Listen, Craig," Torin continued before Presit could respond, *"let Justice know Humans First will be waiting at the other end of the DeCaal's jump coordinates."*

The corner of Ryder's mouth closest to Werst twitched up. "I'd be stoked, but you've bunged the power and we can't actually pull the coordinates."

"Yet."

"What about the *Warner-Lalonde*?" Werst asked. "They're in-system."

"No. If the Navy goes, there'll be shooting and a debris field."

"Justice it is, then." The *Promise* tipped to port. Ryder waved Presit's questions silent and to Werst's surprise, she obeyed. "Besides, the *Warner-Lalonde*'s deck-deep in our medical emergency. When they've stowed the injured, you want to match up so they can transport the plastic and the prisoners?"

Werst glanced over at Ryder as the silence lengthened.

"Torin?"

•—◆—•

Torin turned the helm's chair and stared at Martin, a zip-tie holding him to the remains of the weapons station. He was conscious and struggling, muttering of revolution and revenge.

"I don't think that's a good idea," she said at last. "The captain of the *Warner-Lalonde* is Human."

•—◆—•

Two high-level Justice bureaucrats and a Parliamentary undersecretary waited at the intake desk when Strike Team Alpha and their prisoners arrived at Berbar Station, the *DeCaal* towed in by the *Promise*. C&C had remained on 33X73 waiting for the arrival of a larger team and dealing with the hostages who'd refused to leave.

"Dead and injured from a university expedition and ruins that may or may not have been left by the plastic all piled up together on a Class 2 Designate." Analyzes Minutiae to Discover Truth had sighed deeply. "I don't even want to contemplate how many Ministries we'll be dealing with before this ends."

Not Torin's problem.

The approaching Niln was. "Strike Team Commander Kerr, you are to release the Druin known as Qurn into our custody." They handed her their slate.

Torin read the documentation while Wardens escorted Commander Yurrisk and his surviving crew into one room for processing, Martin and his mercenaries into another—although Camaderiz and Netroovens, the two minimally injured Polint, could only be detached from Vertic after she swore she wouldn't leave the station without them. The documentation held very little of substance beyond proof that Parliament could move quickly when it wanted to and that while they had no evidence Qurn wasn't her actual name, they weren't taking her word for it either. Good thing; her words had been limited. Although they'd tried, singly and collectively, neither Freenim nor Merinim could convince her to expand on the information she'd given Torin on the *DeCaal*.

She was an agent of the Primacy government.

She didn't seem surprised by her welcoming committee.

"We also require the weapons you removed from her."

Or by that.

Torin smiled. "You'll have to sign for them."

"Of course."

"Alamber."

He stepped forward and handed the locked case to the Niln's companion. "You know, it'd be great if we could . . ."

"No."

"Really? Come on, now, that's not . . ."

"No."

Torin caught Alamber's eye, and he let it go. It would have looked strange if he hadn't protested at all, but he'd put minimal effort in. During the four days they were in Susumi, Alamber had analyzed everything in the case, and Torin would be willing to bet he and Ressk were halfway to at least one prototype. She'd carefully avoided finding out for certain.

"Perhaps we'll meet again, Warden Kerr."

She offered her own neutral *perhaps* as the-Druin-known-as-Qurn was escorted out. At heart, she'd always be a soldier and soldiers distrusted spies no matter their allegiance. Turning away from the hatch, she saw Commander Yurrisk staring through the glass of the intake room, eyes locked on the hatch where Qurn had disappeared. If, as she'd said, Qurn had grown fond of the commander, he, in turn, had grown dependent on her. Their interactions on the way back to the station had been recorded, access available only to their court-appointed therapists. The commander's crew had been considered a threat to themselves and had been interacted with accordingly. Torin could see more therapy training in the Strike Teams' future. Martin's group, on the other hand, had been eyes on the whole trip. Brenda Zhang had sulked. Jana Malinowski had seethed. Robert Martin had declared he would never betray the cause.

"We're done dying for the Elder Races, and if we're for the garbage now they no longer need us, we have to fight back."

Right off the pamphlet. He wasn't entirely wrong.

"Can you sit on your ass and do nothing while our dead are disrespected?"

Even given the number of cylinders she carried, the living concerned Torin more. Werst had carried Trembley out.

Torin never met Private Emile Trembley, but she'd known him. She'd known a hundred like him. She hadn't wanted Werst or any of her people to carry that weight, but she wouldn't take away his choice.

"Warden?" Approves of Redemption, one of the intake officers, handed her a slate. Torin checked that the details had been entered properly—mostly for the young Dornagain's sake—and signed off on the prisoner transfer. "Robert Martin wanted me to tell you that if you are not for us, you're against us."

Fair enough. Although being against murdering, delusional, self-absorbed fukwads was a low bar to clear.

"When I say us . . ." Approves' ears flattened slightly. ". . . I am, of course, repeating the pronoun as he used it himself."

"I got that. Thanks."

Craig leaned into the conversation and smiled. "You can tell Robert Martin that he can . . ."

"Warden." Commander Ng stood at the hatch. "Debriefing in conference three. Now."

Presit and Dalan were already set up in the conference room. Presit waved, and Torin knew that behind the mirrored lenses, her eyes were gleaming. She'd spent most of the return trip interviewing the prisoners and had been first off at the station, dragging Dalan behind her and announcing she had people of her own to inform of her return.

"Just be ignoring me," she said as they shuffled chairs away from the table to give the Polint space.

Werst muttered, "I wish."

Commander Ng activated the table. He couldn't have made it clearer that the Justice Department intended to keep their own records. "The Primacy wanted their people to return back across the line before debriefing, citing undue influence," he began, sinking into his seat. "Then both sides wanted you debriefed separately, Confederation and Primacy. That's not going to happen. Team debriefing, then individual as required. Standard operating procedure." Exhaustion lurked behind Commander Ng's eyes, and Torin wondered how long and how hard he'd personally had to fight. "We're streaming this to a Susumi satellite for full transparency and have already sent your after actions through. Begin when you arrived at 33X73. Limit your adjectives."

Even without adjectives, enough time passed that Alamber had sagged against Binti's shoulder by the time Commander Ng began filling in the details external to the team's report.

"Robert Martin approached Commander Yurrisk on Corlavan Station and told him his employer wanted a rumored weapon retrieved. Once the commander had the weapon in hand, Martin's employer would pay more than enough to refit his ship."

"And keep his people safe," Werst grunted. "I guarantee those exact words were used."

"Martin and his people would come along to take care of any necessary intimidation. Martin has admitted to Presit a Tur durValintrisy . . ."

"I are a celebrity. People talk to celebrities." Presit combed her whiskers. "I are also very good at my job."

"Martin admitted to Presit a Tur durValintrisy . . ." Commander Ng tried again. ". . . that his orders as a member of Humans First were to secure the weapon, then have the Polint kill the hostages."

"He paid for their services," Vertic said as attention shifted to her. Bertecnic fidgeted on her left, Dutavar—who'd rejoined them when the *Warner-Lalonde* transferred his brother to medical on Berbar—stood motionless on her right. "They'd have honored the contract."

"A culturally difficult position," Ng replied, saying, as far as Torin could tell, nothing at all. "We can assume Humans First intended the eventual discovery of the bodies to destabilize the peace."

"I thought they were tired of Humans dying for the Elder Races?" Binti muttered, shifting Alamber into a more comfortable position.

"War would restore the status quo. Make them feel powerful again."

The room smelled suddenly of chili. "Humans," Firiv'vrak declared, "are weird."

The commander spread his hands. "Not arguing."

Craig leaned his chair back—theoretically impossible given the design, but he always found a way—and folded his arms. "A return to war would require an increase in weapon production. And we can all dux out who that benefits."

It certainly gave Marteau motive, Torin acknowledge silently. "Fair point," she said aloud. Craig's brows rose. "You were right."

He grinned. "Damned straight."

"And we should dig deeper."

Ng cleared his throat.

"*We* the Justice Department should dig deeper. Not *we* personally," Torin clarified.

"Glad to hear it."

"Martin had a pistol." Hands flat on the table, Ressk leaned in. "We've taken down only one other person with a pistol, and he also had stolen guns."

"And what dung beetle makes guns," Craig muttered.

"Still very circumstantial, but . . ." Ng raised a hand, before Craig could protest. ". . . it's adding up. I'll put in the request."

"Request?" Freenim steepled his fingers. "I understood you to be in command of the Strike Teams."

"Justice needs to consider if there's evidence enough to warrant an invasion of privacy," Torin explained.

"We can't kick the door down and force a confession," Craig added when Freenim blinked.

Vertic nodded. "Because the door is steel and most likely trapped."

"That's . . ." It had been a long day. ". . . close enough."

Alamber made a very Polint-sounding hum, clutched the edge of the table, and pulled himself upright. "What are the odds Humans First intercepted a message sent from an archaeological site to a university by accident?"

"Higher odds that they're searching for specific terms," Ng allowed. "Why?"

His hair flicked slowly back and forth. "That means a meat network. It's harder to hack people."

"You think a network of people searching for information relative to Humans First could keep the search a secret?" Torin asked him.

"No one has more than one piece, Boss. They don't know about each other. They don't know enough of the secret to tell."

The front legs of Craig's chair hit the deck with a crack. Half the people at the table flinched. "Shite. Sorry. It's just, paying for that kind of a meat network . . ."

"Stop saying that," Werst grumbled. "I'm starving."

". . . would take a lot of money."

"You've made your point, Warden." Ng was too self-possessed to glance at Presit, but the implication was there in the angle of his head. The last thing they needed was Presit digging into Marteau Industries before the Wardens could. "Martin also admitted that if the Wardens found out about the situation on 33X73, the presence of members of the Primacy there with him would ensure Strike Team Alpha would be sent to intercede. This is why Martin chose the Polint. They had the best chance of defeating you. It didn't occur to them that you'd have Polint of your own."

Nothing they hadn't worked out themselves. Torin caught Vertic's expression and bit back a smile.

"It should have occurred to them," Freenim said, throwing up both hands. "You don't bring . . ." The translator paused for the first time in days. ". . . jelly to a knife fight."

Jelly?

"They had a lot on their minds," Torin told him while the commander coughed. Craig was the only Confederation member trying not to laugh, but then his poker face *was* that good.

Werst snorted. "Humans First aren't very smart even without the apostrophe."

"They're narrow-minded," Torin corrected. "And bigoted. And . . ."

"Assholes," Werst declared.

"Granted. But that doesn't mean they're stupid. If we keep thinking of them as stupid, we're going to have trouble."

"More trouble." Freenim met Torin's gaze.

She nodded. "More trouble."

"According to Martin, Humans First believed Commander Yurrisk would understand the Primacy killing Strike Team Alpha where he might have objected to Martin's Humans doing it." Ng tapped Martin's file closed as though he were tired of looking at it. "Leftover emotional connections from the war."

"Would Commander Yurrisk understand killing the scientists?" Vertic asked.

"Unless the commander is willing to speculate, I assume we'll never know."

Werst drummed his fingers against the table until Ressk closed a hand around his wrist. The scientists had been shot with Werst's weapon.

"Why did they need the *DeCaal*?" Binti wondered. "Humans First has ships. I mean, not a lot, not anymore, but why involve someone from outside the club?"

"We assume they were trying to keep their involvement less . . . traceable." The file Ng threw up into the air listed times and ship registrations. "Jump buoys keep records."

"Jump buoys." Craig leaned back again, his opinion of those who feared Susumi space clear. Torin resisted the urge to keep tipping his chair until he landed on his ass, but only because she'd need him unbruised if they ever finished debriefing.

"What *about* Commander Yurrisk?" Ressk asked.

Ng squared his shoulders. "Rehabilitation will give the commander and his people the help they need."

"He has a lot of people," Torin pointed out.

"Four. Gayun didn't survive his injuries."

Torin raised a brow.

Everyone at the table looked at the commander. At Torin. And back at the commander again.

"Justice is not concerned . . ." He closed his eyes as though suddenly aware of what he'd said. "I'll request we look into it."

"That's all I'm asking."

"I doubt that, Warden. I very much doubt that."

* * *

"Alamber . . ."

He stepped further into the bedroom, turning from a silhouette in the open hatch to a pale, dark-eyed, emotionally compromised teammate. "You were captured."

"Technically . . ."

Craig cut her off with a kiss on the back of her neck and slid toward the far edge of the bed. "He has a point."

Torin sighed and moved with him, beckoning to the di'Taykan. "Get in."

Nine hours later, after too little sleep, Torin found herself back in DA8 with her entire team—minus Dutavar who'd already boarded with his brother. Grouped together at the edge of the arm, they watched a Ciptran disappear into the Primacy's lock followed, two at a time, by eight Dornagain.

"That's just mean," she muttered as the pairs of Dornagain slowly undulated forward, highlights gleaming in rippling fur, heads ducked low so as not scrape the upper deck.

She felt Craig shrug, his shoulder bumping against hers. "It's an official delegation."

"Not very balanced."

"They're balancing against the entire Primacy, not merely a single team. Pile enough Dornagain on one end of a lever and you can move a galaxy."

"That's . . ." She bumped back into him. ". . . pretty accurate actually."

Both Ciptran and Dornagain were too large to intimidate and almost impossible to enrage.

"I'm surprised they're not sending your military species." Vertic had moved almost silently to stand beside her. "I'm surprised they're not sending you."

"I don't think Parliament trusts Humans right now," Torin told her.

Craig snickered. There was no humor in the sound. "*You* don't trust Humans right now."

"My people will be pleased about the Ciptran." The scent of cherry candy wafted up from knee level.

Keeleeki'ka snapped her mandibles together. "Our people."

The cherry scent grew stronger. "Whatever."

As the outer hatch closed behind the last pair of Dornagain, Representative Haminem took three steps away from the air lock and huffed out an impatient lungful of air. "Well, come on, you lot. We're already behind schedule."

"Better get used to that, mate."

He blinked at Craig. "I'm sure we'll be able to accommodate their speed. We manage quite nicely with the Lindur."

"Lindur?"

"Trust me . . ." Firiv'vrak rose up until her head was even with Torin's waist. ". . . you don't want to know. They squish."

"Most people squish compared to the Artek." Torin held out her hand, and Firiv'vrak stroked both antennae tips across her palm. Proof of trust. The antennae were fragile, and Firiv'vrak's outer mandibles could take Torin's hand off at the wrist.

Keeleeki'ka had the beginning of a new design carved around the edges of the triangular break in her carapace. "Presit won't allow me to hold your story, but I would if I could."

"Thank you."

Firiv'vrak reached back and tried to push Keeleeki'ka over. "It's not that great an honor, trust me. She's like nits under your belly plates."

Until Firiv'vrak's story was complete, Keeleeki'ka's people, whom the government wanted to appease, would protect her. Torin didn't trust governments either, so she found that reassuring.

Vertic stepped forward as Haminem shooed the two Artek into the air lock.

"The *politician* . . ."

"Ouch. Medic," Werst muttered.

". . . has assured me Camaderiz and Netroovoens will be released to the Polint in a very short time." Her ears pricked forward. "In turn, I assure you they'll be punished for what they had a part in. I have offered to hold them in my honor."

"All of them?"

Vertical pupils dilated slightly. "Perhaps not Dutavar as he plans to continue in the military. Tehaven, however, wants to be as far from his mother as the law allows."

"Still, four . . ." Binti nodded like she knew what she was talking about. ". . . decent beginning of a family."

Metal protested as Vertic drew her left front claws along the deck. "If I get them all back."

Torin touched her tongue to her implant. "Alamber."

A hatch opened down the corridor and Alamber emerged, his hair waving languidly back and forth, one hand dragging Bertecnic out behind him, the other adjusting his masker. "Just saying good-bye, Boss."

On his way down the corridor, Bertecnic lurched into the bulkheads. Twice.

Vertic sighed and gave him a shove toward the ship. "We may not see each other again, Gunnery Sergeant Kerr. Stay safe."

"I'll do my best, Durlan Vertic. You, too."

Merinim stood with Craig as Torin and Freenim stepped slightly to one side. "Any idea of what's going to happen when you get back?" Torin asked quietly. If their government had been willing they not come back at all . . .

"Arguments. Extended discussion." Freenim blinked. "Yelling. The usual."

"If there's trouble . . ." He had her codes.

"Don't be a stranger." She had his.

If two senior NCOs, retired, couldn't hold things together, it didn't deserve to be held.

Twenty-seven hours later, Presit fell into step beside her as she crossed the common. Torin shortened her stride.

"Qurn are having disappeared into the halls of Justice. Her existence are being dealt with government to government."

"Her existence?"

Presit waved a dismissive hand, nails glittering gold under the artificial light. Torin wondered where she'd gotten a manicure on Berbar. "Apparently, her existence are not a sure thing. I are not finding anyone leaking, but I are not saying I are not having noticed ripples in the smooth flow of information."

It took Torin a moment to parse that. "You'll keep investigating."

"I are not investigating." Presit wrinkled her muzzle showing small, white teeth. "I are, as Keeleeki'ka would be saying, looking for a story."

"The plastic is gone," Torin said, two steps later. "Our R&D wanted a look at it and didn't get one before the Ministry of Alien

Interference grabbed it. Better the military's all over them than us." The news had been filled with high-level military staff officers demanding they be given access to 33X73 and just as many scientists demanding they be kept away. Torin had already received two communications from General Morris and passed them both on to Commander Ng unopened. "I was a little surprised they didn't want either Craig or me to touch it."

"Maybe they are wanting to translate the data first, before it are becoming a being capable of further communication."

The emphasis on *communication* made it sound like *lying*. "Maybe. Werst told them the pattern he thought he recognized looked like infantry maneuvers. They weren't impressed."

"Are you wanting to impress them? Is that why you are wanting to be touching it?"

"No. That's not why."

"Maybe they are not wanting to risk that much plastic coming to life. It could be becoming a *very large being* capable of communication."

"Maybe they had political reasons I'm not aware of."

Presit snorted. "Maybe water are wet."

• ◆ •

Arniz watched the insects swarming over Dzar's body, wings flashing in the sun as they fed. She'd be sunwarmed bone in no time. Finally at rest. Hyrinzital lay on a bier next to her, but Katrien burial rites had called Lows and Mygar home.

She brushed her fingers lightly over the force field the Wardens had raised to keep the animals away, tasting ozone on the air. The largest Dornagain, Analyzes Minutiae to Discover Truth, had physically moved between members of the Ministry of Culture and the Ministry for the Preservation of Pre-Confederation Civilizations and had talked them around to allowing the platforms—her argument slow and ponderous enough that both ministries had surrendered more than agreed.

There was another anchor on its way and more scientists, funding easy to get when a potential plastic civilization had been uncovered. No one spoke of the potential weapon. She doubted anyone had forgotten it. The military had become strident about gaining permission to land. There was rumor of a H'san joining the new team.

Potential plastic civilization sounded ridiculous.

Harveer Salitwisi and Dr. Ganes were gone as well. One too broken to continue, one promising he'd be back. How long did it take to regrow a hand? Arniz had no idea. She vowed to be kinder to him and wondered if she'd remember.

She was staying until the end.

The first plastic had been found in a latrine and the Ministry of Alien Interference didn't know the difference between a feces and a fragipan.

•—◆—•

He allowed the Wardens access to his system without argument. With grace, even. They had a warrant, but he had nothing to worry about; he knew the limits of the law and what they could find.

The di'Taykan with the pale blue hair and the dark lines painted around his eyes was a bit of a surprise. He knew they'd allowed the Younger Races to try their luck beyond the harnessed violence of the Strike Teams. There was one in each of the C&Cs, but he hadn't been informed of any in data forensics.

"It's all yours, Warden di'Crikeys." The Rakva stepped away from the desk, bright yellow crest displayed.

Alamber di'Crikeys. In person, in the flesh, he didn't look much like the publicity pictures.

His presence explained why the Rakva Warden had been pleased to surrender the console.

He had nothing to fear from the Justice Department. They had no idea of where to begin. Of what to look for. Warden di'Crikeys, on the other hand, had learned his skills from criminals and this was, quite possibly, the only legitimate job he was qualified to hold.

He watched the di'Taykan settle into the very expensive chair and sigh happily as it formed around him. "Do you need me to stay here while you work?"

Warden di'Crikeys grinned at him, eyes lightening. "Planning to distract me? I'm good either way."

If only Wardens Kerr and Ryder had been able to see where their loyalty should lie.

So unfortunate Strike Team Alpha hadn't been destroyed on 33X73.

He should have ignored Gunnery Sergeant Torin Kerr, ignored the stories that built her up into a worthwhile foe, and concentrated on removing Alamber diCrikeys.

He stopped by his collection room on his way out, slid the ancient piece of H'san ceramic into a padded case, and tucked the small, pink plastic horse into his pocket.

Warden di'Crikeys wouldn't find it all—he'd personally coded the plans for the pistol and destroyed the printer after the run—but there'd be enough. A connection. A payment. An assumption. Not all admissible in court, and his position would offer a certain amount of protection, but, in the end they'd force him from the shadows.

The hell they would.

No one forced Anthony Justin Marteau to do anything. He'd leave the shadows by his own choice.

"I thought working with the Primacy team, with Vertic, would remind you of everything you missed about the Corps." Craig pulled his shirt off and tossed it toward the chute in the corner. Every pilot on the station wanted to buy him a drink, but he been pacing himself and was still remarkably sober. "I don't want you to go back."

"I'm not going anywhere."

"Not without me."

"Not without you."

"Aces. I've been thinking about the ruins on Threxie. Do you think we'll ever find out what happened to the natives?" He turned to see Torin sitting on the edge of the bed, frowning. "What?" When she didn't respond, he prodded her with a bare foot. "Torin?"

"Remember the animals in the ruins?"

"Not likely to forget, am I."

"When I was fighting, before you tranquilized it, I got a good look at its front paws."

"When it was trying to claw your face off?" He sat down beside her.

She leaned into his warmth. "The toes were long, and there was an extra digit on each foot."

"A dew claw. Sure. Lots of animals have them. You grew up in the country, you should know that."

"No, this was a visible thumb." Holding her left hand up at eye level, she moved her thumb back and forth. "It was fused to the rest of the paw. Obvious. Useless."

"Okay." He waited. He was good at that.

"The plastic was in our heads," Torin continued at last, still staring at her thumb. "Who's to say it didn't get into their heads . . ."

"The animals' heads?"

"The animals' heads," she repeated. "It got in and changed them at the cellular level, changed them until they weren't a threat. Bred them down. One generation, tool users. The next, animals."

"Because they'd discovered a weapon?"

"Because they'd discovered a weapon."

He leaned back on his elbows and stared at the line of her back for a moment. Or two. "You have a suspicious mind, luv."

She took a deep breath and let it out slowly. "Yeah."

APPENDIX

THE CHARACTERS OF *A PEACE DIVIDED*

Wardens

Strike Team Commander Lanh Ng	Human

Strike Team Alpha

Torin Kerr	Human
Craig Ryder	Human
Binti Mashona	Human
Werst	Krai
Ressk	Krai
Alamber di'Crikeys	di'Taykan

Strike Team Ch'ore (partial list)

Ranjit Kaur (Cap)	Human
Sirin di'Hajak	di'Taykan

Strike Team T'Jaam (partial list)

Doug Collins	Human
Gamar di'Tagawa	di'Taykan

Strike Team Beta (partial list)

Porrtir	Krai (pilot)

Dornagain

Analyzes Minutiae to Discover Truth
Finds Truth Through Inquiry
Vimtan Is In the Details
Many Pieces Make a Whole
Approves of Redemption

Niln

Vesernitic
Nubaneras

Support

Dr. Deyell, R&D	Rakva
Dr. Erica Allan, therapist	Human
Dr. Verrir, therapist	Krai
Paul Musselman, bar owner	Human
Sarrk, political aide	Krai
Feerar, Chief Ag. Scientist	Krai

* * *

Primacy

Vertic (gold)	Polint
Bertecnic (chestnut)	Polint
Santav Teffer Dutavar (variegated)	Polint (santav teffer = master corporal)
Firiv'vrak	Artek
Keeleeki'ka	Artek
Durlave Kan Freenim	Druin (durlave kan = staff sergeant)
Merinim	Druin

* * *

Civilians

Anthony Justin Marteau, CEO Marteau Industries	Human
Orina Yukari, Marteau's aide	Human
Representative Haminem	Druin
Presit a Tur durValintrisy	Katrien
Dalan a Tar canSalvais	Katrien

* * *

Named Scientists/Hostages

Harveer Arniz, soil scientist	Niln (harveer = doctor {PhD})
Dzar, soil scientist ancillary	Niln
Dr. Harris Ganes, engineer	Human
Harveer Salitwisi, archaeologist	Niln
Hyrinzatil, archaeologist ancillary	Niln
Harveer Tilzonicazic (Tilzon), xeno-botonist	Niln
Dr. Tyven a Tur durGanthan, geophysicist	Katrien
Magyr, geophysicist ancillary	Katrien
Dr. Lows a Tar canHythin, geophysicist	Katrien
Nerpenialzic (Nerpen), geophysicist ancillary	Niln

* * *

Mercenaries

Sergeant Robert Martin	Human
Emile Trembley	Human
Brenda Zhang	Human
Jana Malinowski	Human
Camaderiz (black)	Polint
Tehaven (variegated)	Polint
Netrovooens (bay)	Polint

* * *

Crew of the *DeCaal*

Commander Yurrisk	Krai
Petty Officer Sareer	Krai
Lieutenant Beyvek	Krai
Seaman Pyrus di'Himur (pink)	di'Taykan
Mirish di'Yaunah (deep blue)	di'Taykan
Lieutenant Gayun di'Dyon (bright blue)	di'Taykan
Qurn	Druin

* * *

Gunrunners

Mthunz Mackenzie (Mack)	Human
the chief	Human
Shiraz	Human
Harr	Krai
Ferin	Krai
Kid	Krai
Yizaun	di'Taykan

* * *

From *An Ancient Peace* (mentioned)

Joesph Dion	Human
Major Sujunio di'Kail	di'Taykan

From *Valor's Trial* (mentioned)

Staff Sergeant Harnett	Human